NANOFERTILIZERS: Challenges and Prospects

By

J. C. Tarafdar

Ph.D., A.R.S., FDAAD (Germany), FAVH (Germany), FNAAS, FISSS, FISSRS
Former UGC- Emeritus Professor,
Former ICAR-Emeritus Scientist,
National Fellow and Principal Scientist

Published by

SCIENTIFIC PUBLISHERS (INDIA)
5 A, New Pali Road, P.O. Box 91
Jodhpur 342 001 (India)
E-mail: info@scientificpub.com
Website: www.scientificpub.com

ISBN : 978-93-88043-65-6
eISBN : 978-93-88043-88-5

Printed in India

Dedicated to

My beloved Mother.

Preface

A number of excellent books on general nanotechnology are currently available but, to date, none have been dedicated to nanofertilizer. This aspect is very important because nanofertilizer has the potential to replace chemical fertilizer efficiently with less fertilizer requirement, more nutrient use efficiency, more native plant unavailable nutrient mobilization, controlled release throughout the entire crop growth period with minimum loss of nutrients, easy to make bio-nano formulations, plant growth regulation, more stress tolerance, maintenance of soil health with improved yield as compared to the all other generations of fertilizers. Due to its precise nutrient management, it is already having a significant commercial impact, which will certainly increase in the future. It will enable making high quality products at a very low cost and a very fast pace. It may revolutionize the agricultural farming in the entire world. This book has addressed fundamentals and applied aspects of nanofertilizers and their challenges and prospects. The reader should feel invited to learn more about nanofertilizer technology and uses, and to go later on beyond the content of this book. The way I have described each and every aspect related to the nanofertilizer will feel very comfortable to learn on this subject. I hope the readers will find this book inspiring and motivating. The purpose of this book is to provide an up-to-date knowledge to the most important and useful concepts in the study and application of nanofertilizer. My goal is to help you to think through the issue and realize the potential of this unique fertilizer. In writing this book, I have designed the text in such a way that the researchers and industrialists will get a lot of ideas to conceive and work on it for the advancement of the subject. The ultimate purpose of the book is to equip the reader with comprehensive knowledge in nanofertilizer with reference to basic as well as applied aspects and its ethical issues for future implementations. It contains predigested information on nanofertilizer technology, their practical impact on soils and plants for good understanding, assimilation and reproducibility. Critical thinking is so important that I have made it the focus of each chapter. The goal is to provide an instrument of learning useful to educations, research and students today, so that future generations of students will be more informed and able to make judgments concerning the nanofertilizer based upon sound scientific knowledge, so that more industries can come forward. I hope that readers and fertilizer industries will certainly benefited by this book. With the help of this new emerging science, researchers can draw the line that separates them from the rest. This book has been written to provide an overview of many fundamental aspects that underpin all nanofertilizer technology. It includes their uses and examples of how these principles are put into operation. Although I have made every effort to make the book error-free, I am under no illusion; I welcome comments, criticism and creative suggestions from the readers. I am hopeful that this book will find a place on most of the laboratory branches and libraries of different institutes around the world.

November 23, 2020 **J. C. Tarafdar**

Contents

1

Nanofertilizers – Introduction

With the global upsurge in population and rapid urbanization, farmers across the globe are left with the daunting task of feeding more mouths every year from the agricultural fields which are dwindling correspondingly. Satellite images show that the Earth is rapidly running out of fertile land and the food production will soon be unable to keep up with the worlds growing population. The production level of food grains has become a subject of concern as it has been showing a downward trend over the last decade. With natural resources getting exhausted, it is through agriculture that we can envisage a self-sustainable world. With limited availability of land and water resources, a growth in agriculture can be achieved only by increasing productivity through the effective use of modern technology. *Nanotechnology*, focusing on special properties of materials emerging from nanometric size has the potential to revolutionize in the agricultural sector. Nanotechnology applications in agriculture are gradually transforming the theoretical possibilities into the practical applications. The potential is increasing with suitable techniques and sensor being identified for precision agriculture, natural resource management, early detection of pathogens and contaminants in food products, efficient delivery systems for agrochemicals such as fertilizers and pesticides. With increased innovations using nanotechnology in the agricultural sector, it can be envisaged to become the major economic driving force and benefit consumer and farmers with no detrimental effect on the ecosystem.

In India, fertilizers are squarely responsible for enhanced food grain production from 1960s (55 mt) to 2019 (285 mt) that coincides with the exponential increase in fertilizer consumptions from 0.5 mt to 55 mt, respectively. It has been unequivocally demonstrated that fertilizer contributes to the tune of 35–40% of the productivity of any crops. Cosidering its importance, the Government of India is heavily subsidising the cost of fertilizers, particularly nitrogen and phosphorus to encourage farmers to use them to promote productivity of crops. This resulted in imbalanced fertilization and occurance of nutrient pollution in ground waters. In the past few decades, use efficiencies of N, P and K fertilizers remained constant as 30–35%, 18–20% and 35–40%, respectively, leaving a major portion of added fertilizers stay in the soil or enter into aquatic system causing eutrophication. In order to address a number of issue such as low fertilizer use efficiency, imbalanced fertilization, multi-nutrient deficiencies and decline of soil organic matter, it is quite pertinent to evolve a nano-based fertilizer formulation with multiple functions.

Nanofertilizers are made with nanoparticles and can be defined as fertilizers contained within nano-structured formulations which can be delivered to targeted sites to allow the release of active ingredients keeping the plant nutrient demands. Therefore, among different agricultural inputs, nanofertilizers can be quite considered promising in enhancing the growth, nutrition and then productivity of crops as well as regulating the release of nutrients and improving efficiency use of nutrients under controlled environmental conditions. Different crops have been fertilized using nanonutrients or nanofertilizers. Studies include fate and behavior of these nanonutrients in different agroecosystem compartments (soil, water and soil micro-biota), and evaluation of the potential mechanism for toxicity of nanonutrients and their tolerance in both soil–micro-biota and plants. It is found that, the manufactured nanoparticles are not always more toxic than other chemical species containing the same elements. For example, CuO nanoparticles are slightly more toxic comparing with other Cu ions, whereas ZnO nanoparticles have a similar toxicity of Zn ions, but in the case of manganese and iron oxide nanoparticles, the toxicity was less than their ionic counterparts as well as significant enhancement the growth of lettuce seedlings by 12–54%, respectively. That means nanoparticles of Mn and Fe can be considered effective nanofertilizers in increasing the agronomic productivity.

Nano-fertilizer technology is very innovative and scanty reported literature is available in the scientific journals. However, some of the reports and patented products strongly suggest that there is a vast scope for the formulation of nano-fertilizers. An enhanced production has been observed by foliar application of nanoparticles as fertilizer. It has been found that 640 mg ha^{-1} foliar application (40 mg/L concentration) of nanophosphorus gives 80 kg ha^{-1} P equivalent yield under arid environment of clusterbean and pearl millet. Currently, research is underway to develop nano-composite to supply all the required essential nutrients in suitable proportion through smart delivery system. Preliminary result suggests that balanced fertilization may be achieved through nanotechnology. The impact of nano-fertilizer products on physiological, biochemical, nutritional and morphological changes in plants and the fate of nano-products in soil and plant systems needs more study. In addition, the effects of nano-fertilizer products on rhizosphere microorganisms and biogeocycling of nutrients have to be explored under natural field conditions.

Indeed the metabolic assimilation within the plant biomass of the metals, *e.g.*, micronutrients, applied as Nano-formulations through soil-borne and foliar application or otherwise, needs to be ascertained on strong scientific evidence. Further, the Nano-composites being contemplated to supply all the nutrients in right proportions through the "Smart" delivery systems also needs to be examined carefully. It is well known, more often than not the nutrients are required at different growth stages, but all nutrients are not needed at any particular growth stage. One is aware of the simple concept of basal and top-dressing of the nutrients in soil-borne application mode. Even if the application is through the foliar system, the smart delivery system ought to take due cognizance of these finer aspects.

Nanofertilizers have the opportunity to profoundly impact energy, the economy, and the environment by reducing nitrogen loss due to leaching, emissions, and long-term incorporation by soil microorganisms. Currently, the nitrogen use efficiency of plants is low due to the loss of 50–70% of the nitrogen supplied in conventional fertilizers. New nutrient delivery systems that exploit the porous nanoscale parts of plants could reduce nitrogen loss. Fertilizers encapsulated in nanoparticles will increase the uptake of nutrients. In the next generation of nanofertilizers, the release of the fertilizer can be triggered by an environmental condition or simply be time released. Slow, controlled-release fertilizers have the potential to increase the efficiency of nutreint uptake. Nanofertilizers that utilize natural materials for coating and cementing granules of soluble fertilizer have the advantage of being less expensive to produce than those fertilizer that rely upon manufactured coating materials. Slow, controlled - released fertilizers may also improve soil by decreasing toxic effects associated with over application of fertilizer. Zeolites have been used as fertilizer delivery mechanism. One would, however, like to compare the zeolites with the native clay mineral assemblies (present as nanoparticles) of the soils as the effective carrier-cum-delivery material in this regards.

An outburst of world the population in the past decade has forced the agricultural sector to increase crop productivity to satisfy billions of people, especially in developing countries. Fertilizers have an axial role in enhancing the food production, especially in developing countries after the introduction of high yielding and fertilizer responsive crop varieties. In spite of this, it is known that yields of many crops have reduced as a result of imbalance fertilization and a decrease in soil organic matter as well as poor soil health. As land and water resources are depleting day by day, development of agriculture can be achieved exclusively through increasing productivity by effective use of new technologies like nanofertilizers. Use of large scale application of chemical fertilizers to increase the crop productivity is not a suitable action for a long run because the chemical fertilizers are considered as double edged swords, which on the one hand increase the crop production but on the other hand disturb the soil mineral balance, and decrease soil fertility. Large scale application of chemical fertilizers results in an irreparable damage to the soil structure, mineral cycles, soil microbial flora, plants, and even more on food chains across ecosystems leading to heritable mutations in future generation of consumers.

Nanofertilizers are at the leading edge of a rapidly developing field of nanotechnology. The term 'nano' is adopted from the Greek word meaning "dwarf". The word nano means 10^{-9} or one billionth part of a meter. Particles with at least one dimension less than 100 nm are considered as nanomaterials. Nanoparticles have high surface area to volume ratio. They have high penetration and triggering potential as well as other unique properties, which make them highly applicable. Composition of nanoparticles plays a significant role in their application. For example, polymer-coated nanoparticles are used as agrochemical carrier due to its controlled-release ability, whereas metal nanoparticles show size dependent properties, such as magnetism, fluorescence, and photolytic degradation, which have application in sensor development, agrochemical degradation and soil remediation.

Mineral fertilizers are key to food production despite plant low nutrient uptake efficiencies and high losses. However, nanotechnology based fertilizer can both enhance crop productivity and reduce nutrient losses with high efficiency. Due to this, the concept of nanofertilizers has raised interest in nanoscale and nanoenabled bulk fertilizers. The large scale industrial productions of nanofertilizers are very soon to be realized. It is anticipated that the development and validation of nanofertilizers that are non-disruptive to existing bulk fertilizer production system will motivate the industries involvement in fertilizers. Nanofertilizer technology is very innovative and scanty reported literatures are available in the scientific journals. Nutrient use efficiency of conventional fertilizers hardly exceeds 30–35%, 15–20% and 35–40% for N, P and K, respectively. For micronutrients, the use efficiency varies between 3–5%. The data remain constant for the past several decades and research effort did not yield fruitful results. Nanofertilizers have found to be the potential to increase the use efficiency by at least 2 to 20 times of different nutrients as reflected from preliminary results. Nanofertilizers are nutrient carriers that are being developed using substrates with nanodimensions of 1–100 nm. Nanoparticles have extensive surface area and capable of holding abundance of nutrients and release it slowly and steadily, such that it facilitates uptake of nutrients matching the crop requirement without any associated ill-effects of customized fertilizer inputs.

Application of fertilizers can enhance the crop productivity. However, the available nutrients present in the bulk chemical forms are not fully accessible to plants. In addition, the utilization of most of the macro and micronutrients are very low due to their inversion to insoluble forms in soil. Crop plants typically use much less than half of the chemical fertilizers applied. The remaining minerals may leach down and become fixed in soil or contribute to air pollution. So, the use of large scale application of chemical fertilizers to increase the crop productivity is not a suitable option for the long run as it may disturb the soil mineral balance and decrease soil fertility. Considering the deleterious effects of chemical fertilizers, there is an urgent need to develop smart materials that can systematically release chemicals to the specific targeted sites in plants which could be beneficial in controlling nutrition deficiency in agriculture.

Indeed, nanotechnology has provided the feasibility of exploiting nanoscale or nanostructured materials as fertilizer carriers or controlled release vectors for building of so called smart fertilizer as new facilities to enhance nutrient use efficiency, and reduce cost of environmental protection. Encapsulation of fertilizers within nanoparticles is one of these new facilities which are done in three ways 1. The nutrient can be encapsulated inside nanoporous materials, 2. It was coated with thin polymer films, 3. It was delivered as particle or emulsions of nanoscale dimensions. In addition, nanofertilizers will combine nano-devices in order to synchronize the release of fertilizer-N, P , K etc with uptake by crops, so preventing undesirable nutrient losses to soil, water and air via direct internalization by crops, and avoiding the interaction of nutrients with soil, microorganisms, water, and air.

Conventional fertilizers are generally applied on the crops by either spraying or broad-casting. However, one of the major factors that decide the mode of application is the final concentration of the fertilizers reaching to the plant. In a practical scenario, very less concentration (much below to minimum desired concentration) reaches to the targeted site due to leaching of chemicals, drift, run-off, evaporation, hydrolysis by soil moisture, and photolytic and microbial degradation. It has been estimated that around 40–70% of nitrogen, 80–90% of phosphorus, and 50–90% of potassium content of applied fertilizers are lost in the environment under various crops and could not reach the plant which causes sustainable and economic losses. These problems have initiated after repeated use of fertilizer and pesticides, which adversely affects the inherent nutrient balance of the soil. Excess use of fertilizers and pesticides also increases pathogen and pest resistance, reduces soil microflora, diminishes nitrogen fixation, contribute to bioaccumulation of pesticides, and destroys habitat for birds. Hence, it is important to optimize the use of chemical fertilization to fulfill the crop nutrient requirements and to minimize the risk of environmental pollution.

Nanofertilizers are synthesized or modified form of traditional fertilizers in nano size (1–100 nm), fertilizers bulk materials or extracted from different vegetative or reproductive parts of the plant by different chemical, physical, mechanical or biological methods with the help of nanotechnology used to improve soil health and plant production. It has the potential to controlled release of nutrients; site targeted delivery, reduction in toxicity, and enhanced nutrient utilization of delivered fertilizers. These attributes of nanoparticles are due to their high surface area to volume ratio, high solubility, and specific targeting due to small size, high mobility and low toxicity.

Important plant nutrients, such as N, P, K, Ca, Mg, S, B, Cu, Fe, Mn, Mo, Zn can be possible to prepare its nano-form from the respective salts by the help of physical, chemical, aerosol and biological techniques. It is a need of today's nanotechnology to develop reliable, non-toxic, clean and ecofriendly experimental protocols for the synthesis of nanonutrients. These nanonutrients should be of controlled size, shape, stable and monodispersed. Since the nanofertilizers contain nutrients, growth promoters encapsulated in nano-scale polymers, they will also have a slow and a targeted efficient release. When comparing to chemical fertilizers requirements and cost, nano-fertilizers are economically cheap and are required in lesser amount. Nanofertilizers have ability for greater absorbance, high reactivity, small size, high surface to volume ratio and unique optical properties contribute in improving growth and development of crops. Nano-materials have potential contributions in slow release of fertilizers as nano-particles hold the material more strongly from the plant due to higher surface tension of nanoparticles than conventional surfaces. Moreover, nano-coatings provide surface protection for larger particles. Thus, nanotechnology is a leading science, tries to produce less harmful and more effective nano-based fertilizers. The use of nanotechnology in fertilizer sector has begun and will continue to have a significant effect in the areas of development of new fertilizer materials. Nanofertilizer technology is very innovative and some of the reports and patents strongly suggest that there is a vast scope for the formulation of the nanofertilizers. Significant increase in yields has been observed due to foliar application of nanoparticles as fertilizer. Preliminary results suggest that balanced fertilization may be achieved through nanotechnology. Researches show that nanofertilizers have a significant influence on energy, the economy and the environment. Figure 1 depicted their response to global farming.

There is a growing pressure on the agriculture sector to fulfill the continuously increasing demands of the consistently growing human population. As crop usages less than half of the applied amount of chemical fertilizers, which are thought to be indispensable for improving crop productivity, such chemical fertilizer, are extensively applied through different methods. There is generally a need to shift in use of the refind form of nutrients with controlled release. Furthermore, growers tend to use repeated applications of chemical fertilizers in order to achieve desired higher yields, which contrarily can lead to a decrease in soil fertility and increase salt concentrations

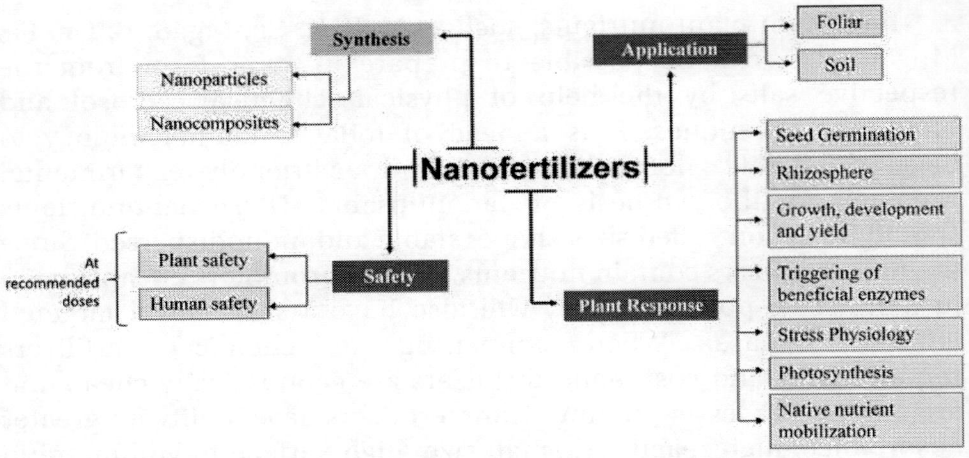

Figure 1: Response of nanofertilizer on global farming

thereby causing future crop losses. Furthermore, uneven use of fertilization without control on nutrient release can deteriorate product quality. Hence, it is crucial to develop slow/control release fertilizers not only to increase crop production and quality, but also to enhance the sustainability in crops and horticultural production. The overall prospects of nanofertilizres are shown as Figure 2. The major benefits are: their low requirement and more use efficiency, encourage enzyme release as well as plant physiological activities, low cost and higher crop yield as compared to the traditional fertilizers. It also helps to maintain soil health.

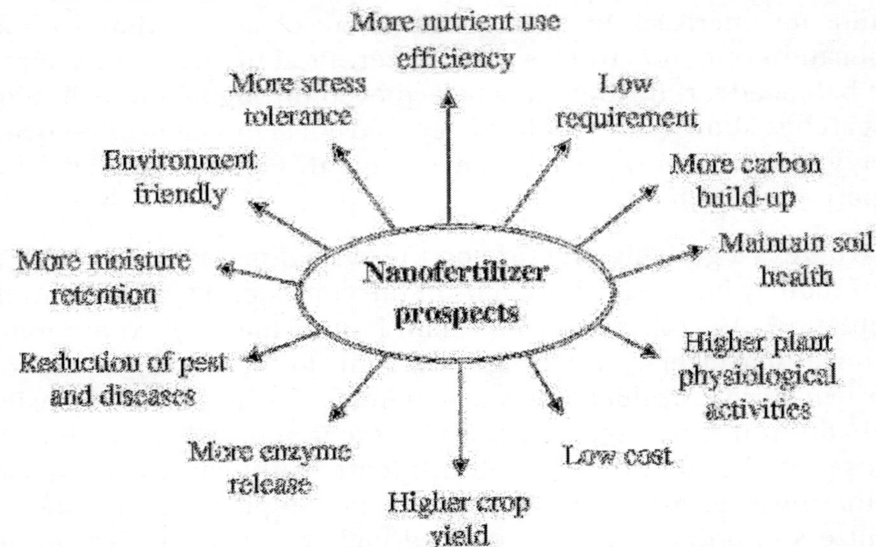

Figure 2: Prospects of nanofertilizer on agricultural production

The horticulture sector today is facing an intense pressure for achieving considerable efficiency in food security using alternatives to chemical fertilizers. New approaches and technologies are required if global horticultural production and demand are to be fulfilled in an economically and environmentally sustainable manner. Materials that are of up to 100 nm particle size in at least one dimension are generally classified as nanomaterial. There are various types of nanomaterials, as single or multiwalled nanotubes, magnetized iron nanoparticles, copper (Cu), aluminum (Al), silver (Ag), gold (Au), zinc (Zn) and zinc oxide (ZnO), silica (Si), cerium oxide (Ce_2O_3), and titanium dioxide (TiO_2), among others . Given the unique properties of nanomaterials, such as high surface-to-volume ratio, controlled-release kinetics to targeted sites and sorption capacity, nano-technology has a high relevance for the design and use of new fertilizers. Nanofertilizers are nutrients encapsulated/ coated with nanomaterial for the control and slow delivery of one or more nutrients in order to satisfy the imperative nutrient requirements of plants. These "smart fertilizers" are currently being regarded as a promising alternative; they are in several cases considered to be the preferred form of fertilizers over the conventional ones. The interaction of nanomaterials and fertilizers, due to the high reactivity of nanomaterials, results in an increased and effective absorption of nutritional elements and essential compounds for plants. Nanofertilizrers are absorbed faster and assimilated more easily into germinated seeds and plants than bulk chemical fertilizers. They allow for controlled nutrient delivery, promote seed germination and plant growth. They help in increasing photosynthetic rate and chlorophyll formation in plants as well as stimulate the rhizospheres and soil microflora. But there are lots of challenges (Figure 3) to establish the nanofertilizer permanently for global farming. The main concerns are their stability, farmer's guidance and ethical aspect.

Uptake, distribution and accumulation of nanofertilizers inefficiencies are typically the result of high release rates of conventional fertilizers overwhelming the actual nutrient absorption rate by plants, and/or the transformation of fertilizers/nutrients to forms that are not bioavailable to crops. As such, there is a great interest towards the development of new innovative fertilizer sources in order to increase the fertilizer use efficiency. Several strategies have been proposed to increase fertilizer use efficiency, such as the use of precision fertilization, split or localized application,

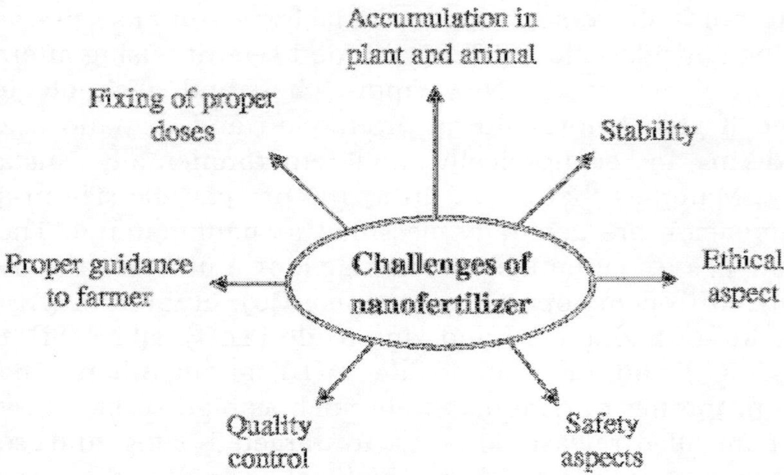

Figure 3: Challenges of nanofertilizer for permanent establishment

fertigation, and the use of nanofertilizers. In the context of sustainable agriculture, the application of nanotechnology for the development of new types of fertilizers is regarded as one of the potentially promising options. It is significantly boosting global production to meet the growing food demands of population with the added benefits of sustainability under the current scenario of climate change. A correct application of nanofertilizers can feed plants gradually in a controlled manner along with the benefits of increasing the fertilizer use efficiency, minimizing volatilization and leaching, and lessening environmental hazards.

The application of nanofertilizers increases efficiency of the elements, minimizes their toxicity in the soil, and reduces the frequency of application. The nanofertilizer application leads to a gradual and controlled release of nutrients in the soil, and prevents eutrophication and pollution of water resources. Several studies, which will be discussed here, have revealed that some nanofertilizers have the potential to increase crop productivity by enhancing seed germination, seedling growth, photosynthesis, nitrogen metabolism, yield, protein and carbohydrate synthesis, aside from improving stress tolerance. Among other advantages, nanofertilizers can be applied in a comparatively smaller amount, ultimately reducing the transport expenditures and increasing ease of application. However, nanofertilizers may also have some disadvantages, which can limit their full implementation in the market.

1.1: Unique Features of Nanofertilizers

1. Due to small size, there are more atoms on the surface compared to the interior of the nanoparticles. This leads to large surface to volume ratio which inturn leads to higher charge density and higher reactivity of nanoparticles.
2. As the surface area increases in comparison to volume, the behaviour of the atoms on the surface of the particles becomes more than the inside the particles. Once the particles become small enough, they exhibit quantum mechanical behaviour.
3. As a result of large surface to volume ratio, there is more interaction between atoms in intermixed with materials in nanoparticles, which leads to increased strength, increased heat resistance, decreased melting point and different magnetic properties of the nano-clusters.
4. Differences in the exposed surfaces of different nanoparticles of various shapes lead to differences in atomic distribution across the nanoparticles. This inturn affects the electron transfer rate kinetics between metal nanoparticles and corresponding adsorbed species.
5. Nanoparticles have higher catalytic activity when they are present in tetrahedral structure followed by cubic and spherical structure, attributed to enhancement of chemical reactivity at the sharp edges and corner of the former.
6. The different nanoparticles except hydroxy apatite have zeta potential in between +30 and -30 mv and show high tendency to agglomerate to higher particles sizes.
7. Nanofertilizers are high surface area, high reactivity; they are the effective catalyst of plant and microbial metabolism. They can better penetrate into the cell and triggers more enzyme release by plants and microorganisms.
8. Nanofertilizers are the effective catalyst of the plant and microbial metabolism.

Further studies

Chhipa, H. (2017). Nanofertilizers and nanopesticides for agriculture. *Environ. Chem. Lett.* **15**: 15-22.

DeRosa, M. C., Monreal, C., Schnitzer, M., Walsh, R. and Sultan, Y. (2010). Nanotechnology in fertilizers. *Nature Nanotech.* **5**: 91.

Liu, R. and Lal, R. (2015). Potentials of engineered nanoparticles for increasing agronomic production. *Sci. Total Environ.* **514**: 131-139.

Tarafdar, J. C. and Adhikari, A. (2015). Nanotechnology in Soil Science. In: Soil Science: An Introduction (ISBN 81-903797-7-1), R, K. Rattan, J. C. Katyal, B. S. Dwivedi, A. K. Sarkar, T. Bhattacharyya, J. C. Tarafdar, S. S. Kukai (eds.). Indian Society of Soil Science, New Delhi, pp.775-807.

Tarafdar, J. C., Sharma, S. and Raliya, R. (2013). Nanotechnology: Interdiciplinary science of application. *Afr. J. Biotech.* **12**: 219-226.

2

Scope of Nanofertilizers

Nanofertilizers play an important role in agriculture to improve crop growth, yield and quality parameters with increase nutrient use efficiency, reduce wastage of fertilizers and cost of cultivation as well as the frequency of application. Nanofertilizers are very effective for precise nutrient management in precision agriculture with matching the crop growth stage for nutrient and may provide nutrient throughout the crop growth period. It increases crop growth up to optimum concentration, a further increase in concentration may inhibit the crop growth due to toxicity of nutrients. It provides more surface area for different metabolic reactions in the plant, which increase rate of photosynthesis and produce more dry matter and yield of the crop. It makes plant more active by triggering different plant and microbial coenzyme system, and prevents plant from different biotic and abiotic stress. Beside tremendous increase in nutrient use efficiency nanofertilizer helps to reduce soil toxicity, minimizes the potential negative effects associated with over dosage and reduces the frequency and amount of application. Nanofertilizer has a great role in nutrition, protection and regulation of agricultural production (Figure 4).

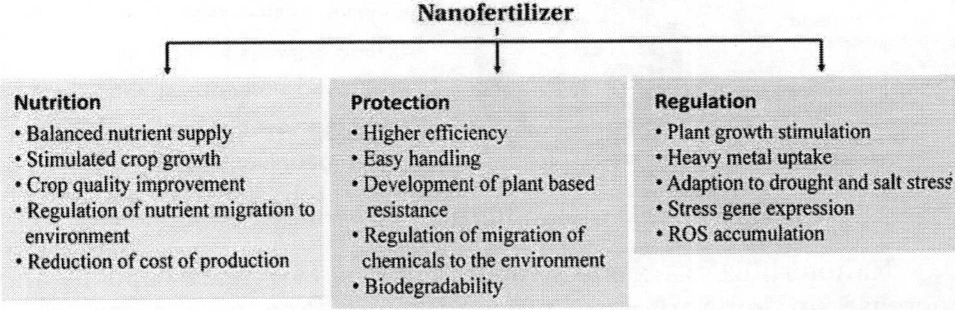

Figure 4: Scope of nanofertilizer under nutrition, protection and regulation of plant growth

They can balance nutrient supply and regulate the nutrient for maximum benefit. They have higher efficiency and can develop plant-based resistance. They are biodegradable and stress tolerant. These nanostructures have shown slow degradation and controlled release of active ingredient for long time. Due to the limitation in arable lands and water resources, the development of agriculture sector is only possible by increasing resources use efficiency with minimum damage to production bed through the effective use of modern technologies. These nano-agro-formulations have high potential for achieving sustainable agriculture, especially in developing countries.

Nanofertilizers are prepared by reducing particle size of nutrients bearing rock, minerals and salts up to nano level. It increases surface area to volume ratio, this means the resultant fertilizer have ability to cover more plant parts for nutrition with same volume. Moreover, nanosize particles can easily pass through cuticle, stomata, membranous structure of plants etc., and it allows delivering of nutrients at their site of action. They may also trigger the specific co-enzyme systems resulted to release more beneficial enzymes by the plant in the rhizosphere, which help in more nutrient mobilization in plant available form as well as more uptake by the plants. Besides soil application and foliar spray, nanofertilizers can also be used as seed treatments. Seed coat acts as good barrier to penetrate an opportunity to increase nano pores on seed surface, which facilitate more moisture and oxygen penetration for better seed germination and establishment (Figure 5).

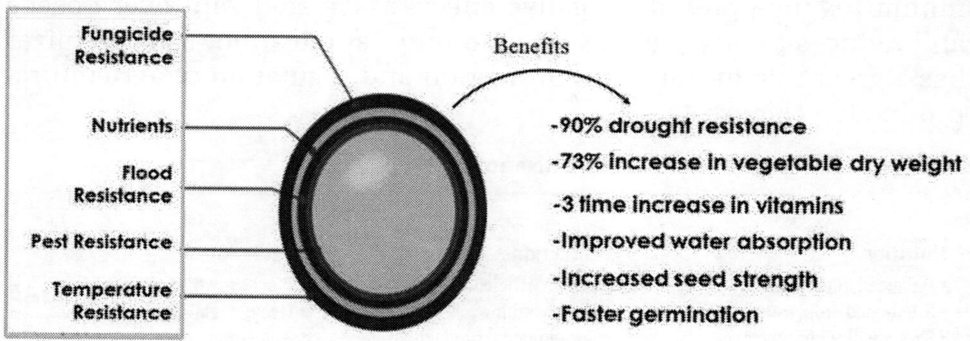

Figure 5: Role of nanofertilizer in seed treatment

Nanofertilizers are effective enough to slow-release capacity and increase in nutrient use efficiency and hence, the doses of application can be reduced considerably than the recommended doses of conventional fertilizers. It proved very effective under foliar application.

Beside this, coating of N-P-K fertilizers with chitosan particles is one of the various formulations of nanofertilizers. Chitosan is a polymer of particular interest in area of fertilizer, because it is biodegradable, bio-absorbable and bactericidal. Due to its polymeric cationic characteristics, chitosan nanoparticles may interact with negatively charged molecules and polymers, showing favourable interaction. Chitosan based nanofertilizers can be applied by foliar sprays of soil fertilization. When chitosan NPK nanoparticles were applied to leaf surfaces, they entered stomata via gas uptake, avoiding direct interaction with soil system. Further, the particles are translocated by the phloem system. The phloem consists of living vascular tissues that translocate photosynthetic product including sucrose, proteins and some mineral ions for plant growth. The nanoparticles are carried in this sugar flow through the phloem sieve tubes to shoots and roots, as a result of pressure differentials between source (leaves) and sink (growing shoot apex etc.) based on mass flow or pressure flow hypothesis, which explains the presence of chitosan-NPK nanoparticles inside the phloem tissue of plants and their absence in the xylem tissue. Phloem tissue is the main and unique pathway for translocation of nanoparticles. Moreover, chitosan is a readily biodegradable product, so they were not destructive to the soils. Nanotechnology can improve performance of fertilizers in other ways, for instance, due to its photocatalytic characteristic; nanosize titanium dioxide has been incorporated into fertilizers as a bactericidal supplement. In addition, nanosilica particles that absorbed by roots have been shown that form films at the cell walls, which can improve the plants resistance to stress and thus increases the crop yield. Some of advantages related to shift formulation of conventional fertilizers using nanotechnology are presented in Table 1

Table 1: A comparison of advantages of nanofertilizers over chemical/ synthetic fertilizer

Desirable Properties	Advantage of Nanofertilizer
Low rate of fertilizer application	The requirement of nanofertilizers are generally 30–100 times less than chemical fertilizer
Nutrient use efficiency	The use efficiency of nanofertilizers are 2–20 times more than chemical fertilizer

Desirable Properties	Advantage of Nanofertilizer
Controlled release formulation	With the application of nanofertilizers, it is possible to control the release speed of nutrients to match the uptake pattern of crops
Solubility and dispersion for micronutrients	Nanosize formulation of mineral micronutrient may improve solubility and dispersion of insoluble nutrients in soil, reduce soil absorption and fixation, and increase the bioavailability
Effective duration of nutrient release	Nanostructured formulation can extend effective duration of nutrient supply of fertilizers into soil

Nanofertilizers have great potential to deliver nutrients to specific target sites in living systems. The loading of nutrients on the nanoparticles is usually done by (a) absorption on nanoparticles, (b) attachment on nanoparticles mediated by ligands, (c) encapsulation in nanoparticles polymeric shell, (d) entrapment of polymeric nanoparticles, and (e) synthesis of nanoparticles composed of the nutrient itself. Nanofertilizers can be used under *in vitro* methods like aeroponics and hydroponics as well as under *in vivo* like soil and foliar application.

Nanofertilizers are more efficacies in terms of nutrients absorption and utilization owing to considerably lesser losses in the form of leaching and volatilization. It records significantly higher uptake owing to free passage from nano sized pores and by molecular transporters as well as root exudates. They also utilize various ion channels which lead to higher nutrient uptake by crop plants. Within plants, nanoparticles may pass through plasmodesmata, resulting in effective delivery on nutrient to sink sites. Due to considerably small losses of nanofertilizers, these can be applied in smaller amounts in comparison to the synthetic chemical fertilizers which are being applied in greater quantities keeping in view their major chunk that gets lost owing to leaching and emission. Nanofertilizers offer the biggest benefit in terms of small losses which lead to lower risk of environmental pollution. It has higher solubility and diffusion impart superiority over conventional synthetic chemical fertilizers.

In future, there is a need to produce nutritiveous agricultural products rich in protein and other essential nutrient required to the human and animal consumption. That is why emphasis should be laid on the production of high quality food with the required level of nutrients and proteins. For solving these problems in crop production nanofertilizers, pesticides and herbicides may be useful tools in agriculture for better nutrient and pest management because these nanomaterials having more penetration capacity, surface area and use efficiency which avoid residues in environment. Size below 100 nm nanoparticles can use as fertilizer for efficient nutrient management which are more ecofriendly and reduce environmental pollution.

Indian agriculture feels the pain of fatigue of green revolution. In the past 50 years, the fertilizer consumption exponentially increased from 0.5 mt (1960) to 55 mt (2019), that commensurate with four fold increase in food grain output (285 mt). Despite the resounding success in grain growth, it has been observed that yields of many crops have begun to stagnate as a consequence of imbalanced fertilization and decline in organic matter content of the soils. The optimum NPK fertilizer ratio of 4:2:1 is ideal for crop productivity, while the current ratio is being maintained at 10:2.7:1 in India. The fertilizer response ratio in the irrigated areas of the country has decreased from 13.4 kg grain/kg nutrient apply in 1970, to just about 5.2 kg in 2019. In other words, more amounts of fertilizers are required to produce the same quantity of grain output. For instance, 27 kg NPK/ha was required to produce one ton of grain in 1970, while the same level of production can be achieved by 109 kg NPK/ha in 2008. The compound annual growth rate (CAGR) of the Indian fertilizer market is expected to increase by 11.9% for the forecasted year 2020–2025 which directly implies incoming revenue to the Indian economy. To transform this forecast into reality, adoption of modern technologies like nanotechnologies, with emphasis to nanoferti-lizers is warranted.

Regarding the potential applications of nanoparticle in agricultural field, it is reported about these impacts including:

- Increasing the crop productivity through using nano-agrochemicals (*e.g.*, nanopesticides and nanofertilizers),
- Improving food security and its productivity,
- Improving soil quality via enhancing the water-holding capacity of soil (*e.g.*, using nanoclays, nano induced polysaccharides, nano-zeolites and hydrogels),

- Stimulating plant growth using nanomaterials by enhancing elemental uptake and use of nutrients (*e.g.*, nano-SiO_2, TiO_2, ZnO and carbon nanotubes),
- Providing smart monitoring using nanosensors by wireless communication devices help farmers in maintaining farm with precise control and report timely needs of plants.

Due to the unique properties of nonomaterials, they are intensively need in agricultureal research both at academic and industrial levels, as well as its suitability candidates in designing and developing such novel nano-tools supporting the sustainability of agriculture.

The extents of multi-nutrient deficiencies are alarmingly increasing year by year which is closely associated with a crop loss of nearly 25–30%. The extent of nutrient deficiencies in the country is in the order of 89, 80, 50, 41, 49 and 33% for N, P, K, S, Zn and B, respectively. In fact, the country is in need of a Second Green Revolution and that can be achieved by nano-fertilizers.

2.1: Potential of Nanofertilizers

Nanofertilizers are envisioned to have the potential to revolutionize agriculture. The following areas can be targeted:

2.1.1: Sensors for quick detection of nutrient deficiency/ nutrient present in plant and soil

There are many nano-sensors now developed to sense nutrient deficiency and nutrient status in plant and soil. They are: carbon nano tube, nano aptamers, smart dust technology, wireless sensor etc. Vo-Dinh, Cillum, and Stocks recently provided an overview of nanosensors and biochips for the detection of biomolecules. There are biosensors that will also provide the very selective, sensitive detection of a broad range of biomelecules. Nanobarcodes can be coated with analyte-specific entities, such as antibodies for selective detection of complex molecules. DNA detection with these nano-scale coded particles has also been possible. Biosensors incorporated with microfluidics and nanomaterials, have diverse application in agricultural production, food processing, and environmental monitoring for in-field, on-line and real time detection of pesticides, pathogens, toxic materials, proteins, antibiotics, odours creating bacteria, microbes etc. in soil, air , water, food, plants and animals.

2.1.2: Nanoformulations to enhance nutrient use efficiency

It is the application of innovative formulation technologies where nano-materials play an important role. It helps to slow release of fertilizer and increases nutrient use efficiency. For example, nano-size Zn has improved the use efficiency by 18–20 times more than bulk size, similarly, nano- phosphorus can increase the use efficiey by 4 times than the bulk fertilizers. Nanoformumations for plant growth regulators (PGR) can be achieved through nanotechnology as nanodelivery systems for controlled release of PGR in the plant cell. Phytosaponin and chitosan based nanoformulations are also possible. In fact, designed nanoparticles can be successfully used to deliver agrochemicals on other substances (elicitors, nucleic acids etc.) into localized areas of plant tissues.

2.1.3: Smart delivery of nutrients

Formulation of nanofertilizer possess all desired properties, such as high solubility, stability, effectiveness, time-controlled release, enhanced targeted activity with effective concentration, and less eco-toxicity with safe, easy mode of delivery and disposal. It has potential to deliver nutrients to specific target sites in living systems. A comparison between nano-technology based formulations and conventional fertilizer application has been shown as Table 2. The comparison clearly depicted the much advantageous nanofertilizers as compared to the conventional bulk particles.

Table 2: A comparison between nano-formulations and conventional fertilizer application

Properties	Nano-fertilizer	Conventional fertilizer
Native nutrient mobilization	In general 30% more than conventional fertilizer	30% less than nano-fertilizer
Nutrient use efficiency	60–85% of different nutrients	3–40% of different nutrients
Requirements	70–100 times less than conventional fertilizers	In general, 80 times more than nano-fertilizer
Release	Release is controlled as per requirement of the crop	Uncontrolled release of fertilizer nutrients
Loss rate	Nutrient loss is minimum	Loss is very high to the applied nutrients

2.1.4: Rhizosphere studies

Nanofertilizer as foliage application, trigger the enzyme system and releasing 18–283 % more beneficial enzymes in different plant rhizosphere. It also helps to release more organic acids in the rhizosphere resulted more solubilization of nutrients and higher uptake especially phosphorus. The nanofertilizer with recommended doses may also help in more microbial build up, especially plant beneficial organisms in the rhizospheres resulted higher nutrient mobilization for plant uptake.

2.1.5: Precision farming

The modern nanotechnology based tools and techniques have the potential to address various problems of conventional agriculture and leads to precision farming. It has been a long desired goal to maximize output (*i.e.,* crop yields) while minimizing inputs (*i.e.,* fertilizers, pesticides, herbicides etc.) through monitoring environmental variable and applying targeted action. Precision farming makes use of computers, global satellite positioning systems, and remote sensing devices to measure highly localized environmental conditions that determine whether crops are growing at maximum efficiency or precisely identifying the nature and location of problems. By using centralized data to determine soil conditions and plant development, seeding, fertilizer, chemical and water use can be fine tuned to lower production costs and potentially increased the production as well as benefiting the farmers. Precision farming can also help to reduce agricultural wastes and thus keep environmental pollution to a minimum. Moreover, tiny sensors and monitoring systems enabled by nanotechnology will have a large impact on future precision farming methodologies. The increased use of autonomous sensor linked into a GPS system for real time monitoring has a major role in precision farming. These nanosensors could be distributed throughout the field where they can monitor soil conditions and crop growth. Wireless sensors are already been used in certain parts of USA and Australia. The union of biotechnology and nanotechnology in sensors will create equipments in increasing sensitivity, allowing on the earlier response to environmental changes. For example;

- Nanosensors utilizing carbon nanotubes or nano-cantilevers are small enough to trap and measure individual proteins or even small molecules.

- Nanoparticles or nanosurfaces can be engineered to trigger an electrical or chemical.
- Signals in the presence of a contaminent such as bacteria.

Other nanosensors work by triggering an enzymatic reaction or by using nanoengineered particles or surfaces. Ultimately precision farming with the help of smart sensors will allow enhanced productivity in agriculture by providing accurate information, thus helping farmers to make better decisions.

2.1.6: Biofortification of nutrients

Biofortification can be done through nanofertilizer to increase the crop nutritional values (Figure 6). It is the better technique than ordinary fortification because it focuses on making plant foods more nutritious as the plants are growing, rather than having nutrients added to the foods when they are being processed. Agrochemical companies also utilizing the advantage of bionanoparticles after development of nano-emulsions and nano-capsules contain active ingredients. Such advanced functionality of bio-nanoparticles may provide the sustainable solution in agricultural sector. It can also solve some limitations of biofertilizer, but this technology needs further refinement. Biofortifications with nano-particles can increase plant phosynthesis rate, promote plant growth, increase plant biomass and plant protein, reducing the plant stress conditions with prolong period of survival under stress environment, synthesis of more plant hormone, accumulation of proline, enhanced siderophore production, nitrogen fixation and P mobilization along with much higher yield.

2.2: Scope of Nanofertilizer in Soil and Water

Nanostructure fertilizer exhibits novel physico-chemical properties, which determine their interaction with biological substances and process. The application of nanotechnological formulation to agricultural crop inputs is one of the proposed tools for sustainable intensifications. It can resolve the issues like low efficiency, imbalance fertilization, multiple nutrient deficiencies and decline in soil organic matter; it is indeed need of the day to evolve the nanobased fertilizer formulations with multiple functions.

The inorganic fertilizers are artificially synthesized and are formulated in appropriate concentrations and the combinations that usually supply three main nutrients: nitrogen, phosphorus, and potassium (N, P, and K) for various crops and growing conditions.

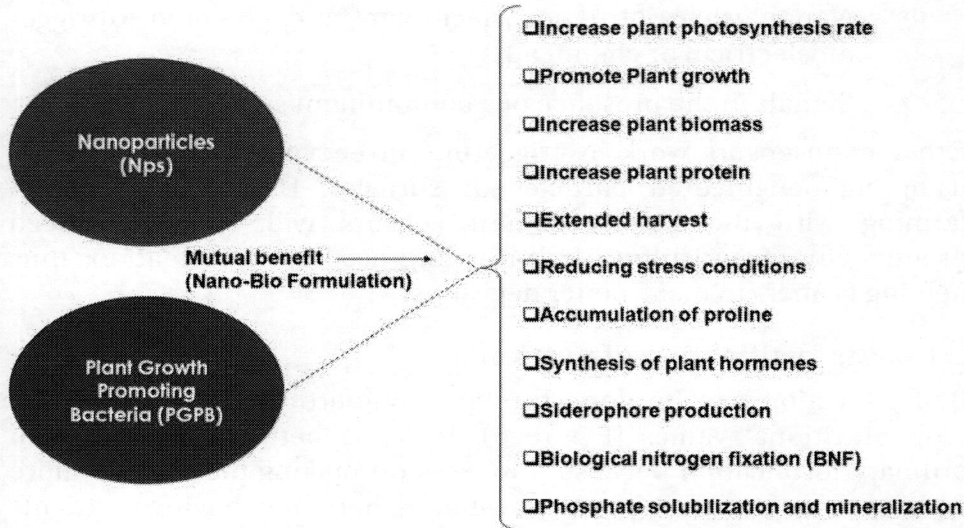

Figure 6: Mutual benefit of bio-nano formulations

Nitrogen promotes leaf growth and forms proteins and chlorophyll. Phosphorus contributes to roots, flower and fruit development. Potassium contributes to stem and root growth and the synthesis of proteins. About 30–60% N, 10–20% P, and 30–50% of K of the applied dose is utilized by plants and the rest is lost to the environment. This causes substantial economic and resource loss as well as serious soil and water contamination. With the application of nanofertilizer, these demerits of conventional fertilizer can be overcome or minimized substantially, so as to utilize the major portion of the applied nutrients. This can be achieved by encapsulating the nanonized nutrients by microbial protein or nanomaterial, coated with a thin protective film, or delivered as emulsions or nanoparticles. The nanobased slow release or controlled release fertilizers have the potential to increase the efficiency of nutrient uptake and to significantly reduce their wastage. The nanotechnology may be applied in the soil nutrition by developing the formulations in two ways, *i.e.,* fertilizer coated, encapsulated or buried in the nanomaterial; and nanoform of fertilizers and other growth promoting materials. Some of the important scope of nanofertilizer under soil and water has been shown as Table 3.

Table 3: Application of nanofertilizer under soil and water

Application area	Probable type of technology
Soil	• Nanozeolites for soil conservation and slow release of fertilizers • Nanomagnets for removal of soil contaminants • Nanoparticles for more release of enzymes by activation of plants and microbes • Nanoparticles for enhanced release of polysaccharides by microbes for soil aggregation, moisture retention and carbon build-up • Nanosensors for automation
Water	• Nanosensors, polymers, clays, zeolites for contaminant detection • Nanomembranes for purification, desalination, detoxification • Nanomolecules for robust water tanks and to prevent leakage • Nanosensors for water flow detection • Nanozeolites for efficient release of water

Biosynthesized nanoparticles can be encapsulated that provide several benefits, such as ease of handling, enhanced stability, protection against oxidation, retention of volatile ingredients, taste making, moisture triggered controlled release, pH triggered controlled release, consecutive delivery of multiple active ingredients, change in flavour character of food and enhanced bioavailability and efficacy. In the biological system, nanocapsules carry the food supplements and help the bioavaibility of the substance. Nanomaterials improve the productivity of crops and efficiently regulate the delivery of nutrients to plants and targeted sites, guaranteering the minimal usage of agrochemicals.

In conventional agriculture, an excess of fertilizer is applied directly into the soil or sprayed on the leaves, which surpass the nutritional need of the plant. This is because a very low percentage of fertilizer reaches its target site, due to leaching of chemicals, evaporation, drift, hydrolysis, run-off, and photolytic or microbial degradation. This excess of chemical fertilizer negatively affects the nutrient equilibrium of the soil, and causes contamination of local water supplies, due to the leaching of toxic materials into water bodies. Nanomaterials can increase crop yield by increasing fertilizer nutrient availability in soil and nutrient uptake by plants. These materials can suppress crop diseases by acting directly on phytopathogens through a variety of mechanisms, including the production of reactive oxygen species. These materials also enhance crop production indirectly by improving crop nutrition and boosting plant defense pathways.

The efficient use of nanomaterials as fertilizer may reduce the negative environmental impact of conventional agricultural practices. In recent laboratory analyses, it has been reported that nano-fertilizers can improve crop productivity by enhancing the rate of seed germination, seedling growth and photosynthetic activity. Nanomaterials incite the plant root and leaf surface, which are the main nutrient gateways of plant systems and highly porous at the nanoscale. The application of nano-fertilizers can enhance the nutrient uptake of the plant through these pores, or the process can facilitate complexation with molecular transporters or root exudates through the creation of new pores, or by the exploitation of endocytosis or ion channels.

Additionally, several researchers have reported that a decrease in the size of nanomaterials facilitates an increase in the surface mass ratio of particles. This enables the absorption of abundant nutrient ions that is later desorbed slowly and steadily for an extended period. Therefore, formulations of nano-fertilizers can provide balanced nutrition for crops throughout the growth cycle, which in turn improves agricultural production. Another important scope is used as a nanobiosensor that react with specific root exudates are also being explored. These techniques are relatively new and have numerous ethical and safety issues that must be carefully studied before implementation.

The global agricultural landscape has radically changed since the revolution of green nanotechnology. Nanofertilizers are now being used in specific concentrations, in accordance with the nutritional requirements of the crops, ensuring minimal differential losses. There are three types of nano-fertilizers: nanoscale fertilizers, nanoscale additive fertilizers, and nanoscale coating fertilizers. Nanoscale fertilizers are made of nanoparticles that contain nutrients. Nanoscale additive fertilizers are traditional fertilizers with nanoscale additives. Nanoscale coating fertilizers are traditional fertilizers coated or loaded with nanoparticles.

The encapsulation of nutrients most commonly produces nanofertilizers with nanomaterials. More recently, the targeted nutrients are encapsulated inside nanoporous materials, coated with a thin polymer film particle, or coated with emulsions of nanoscale dimension. Encapsulation of beneficial microorganisms, such as bacteria or fungi, has shown promise as it can enhance the availability of nitrogen, phosphorus, and potassium in the root zone, thereby improving plant growth. Nanofertilizers can also be classified based on their actions: control or slow-release fertilizers; control active fertilizers; magnetic fertilizers or nanocomposite fertilizers (which use a nanodevice to supply a wide range of macronutrient and micronutrients in desirable concentration).

Porous nanomaterials significantly reduce nitrogen loss by regulating demand-based release, and by enhancing the plant uptake process. Examples of porous nanomaterials include: Ammonium charged zeolites, which can enhance the solubility of phosphate minerals, showing an improvement in phosphorus availability and uptake by crops. Graphene oxide films, a carbon-based nanomaterial, can prolong potassium nitrate release, extending the time of function and minimizing losses by leaching. Nanocalcite ($CaCO_3$-40%) with nano SiO_2 (4%), MgO (1%), and Fe_2O_3 (1%) is another porous nanomaterials which not only improve the uptake of calcium, magnesium and iron, but also notably enhance the intake of phosphorous with micronutrients zinc and manganese.

Concerning the challenges of using nanotechnology in agriculture, some urgent issues should be kept in mind remain to be resolved in the near future including (1) the ecotoxicology of nanomaterials in agroecosystems, (2) the pollution resulted from nanomaterials and its advanced regulations, (3) the sustainability and its biosafety of nanomaterials, (4) the development of carrier of nanomaterials and (5) the investigation of nanomaterial on the applied scale like the industrial level. Therefore, more developed techniques should be saved to monitor the fate of nanofertilizers in different environments under different concentrations. Moreover, more advanced carriers should be used for nanoparticles achieving high delivering the active agents (*e.g.,* pesticides, nutrients, and fertilizers) enhancing the maximization of their efficiency seeking the sustainability. Concerning the applied side, it should be investigated and then evaluated the nanomaterials on the commercial scale by comparing these nanoparticles with the commercial products as well as the interaction between different kinds of nanoparticles.

Regarding the ecotoxicology of nanoparticles, it should be studied the phytotoxic effects of nanoparticles on agroecosystems including plants, soil microbial communities, soil biota, water, air as well as human health. Therefore, it could be concluded that, nanotechnology has a very strong link with the agriculture and penetrates several agricultural fields including fertilization, irrigation and water saving, nanoremediation of soils and water from pollutants as well as the agri-food sector. Many challenges are still needed for more researches and investigations for safe agricultural use.

Further studies

Adhikari, T., Biswas, A. K. and Kundu, S. (2010), Nano- fertilizer- A new dimension in agriculture. *Indian J. Fert.* **6(8)**, 22–24.

Adisa, I. O., Reddy, V. L and Peralta-Videa, J. R. (2019). Recent advances in nano-enabled fertilizers and pesticides: a critical review of mechanisms of action. *Environmental Science: Nano*, **6:** 2002–2030.

Chena, H. and Yada, R. (2011), Nanotechnologies in agriculture: New tools for sustainable development. *Trends in Food Sci. & Tech.* **22:** 585–594.

De Rosa, M. R., Monreal, C., Schnitzer, M., Walsh, R. and Sultan, Y. (2010). Nanotechnology in fertilizers. *Nat. Nanotechnol. J.* **5:** 91–96.

Shang, Y., Hasan, K., Ahammed, G. J., Li, M., Yin H., and Zhou, J. (2019). Applications of Nanotechnology in Plant Growth and Crop Protection: A Review. *Molecules*, **24:** 2558.

3

Synthesis of Nanofertilizers

The nanofertilizers are synthesized by fortifying nutrients singly or in combinations onto the adsorbents with nano-dimension. Both physical (top-down) and chemical (bottom-up) approaches are used to produce nanomaterials, and the targeted nutrients are loaded as it is for cationic nutrients ($NH4^{+}$, K^{+}, Ca^{2+}, Mg^{2+}) and after surface modification for anionic nutrients (NO^{3-}, $PO4^{2-}$, $SO4^{2-}$). Nano-fertilizers are known to release nutrients slowly and steadily for more than 30 days which may assist in improving the nutrient use efficiency without any associated ill-effects.

3.1: Methods of Nanofertilizer Synthesis

Several methods have been applied for nanofertilizer synthesis. The most important feature is the preparation of particles of specified size, shape and stability. To eliminate the inter-particle interactions, nano-fertilizer particles need to be isolated from one another by immobilization on a substrate surface or in the bulk of a stabilizing inert matrix. One should be assured that the distance between the particles in the matrix is controllable and the synthesis process should be relatively simple, inexpensive and reproducible. The method of nanofertilizer synthesis must be cost effective, monodispersed, environmental benign and can be produced in large scale. There are physical, chemical, aerosol (physico-chemical), and biological methods for preparation of nano-fertilizers. The synthesized nanoparticles may be of various shapes (Figure 7) depends on the source and methods used. The particles may be single shaped or in combination of different shapes. The shapes can largely be controlled under chemical synthesis, but most difficult under biological methods of synthesis.

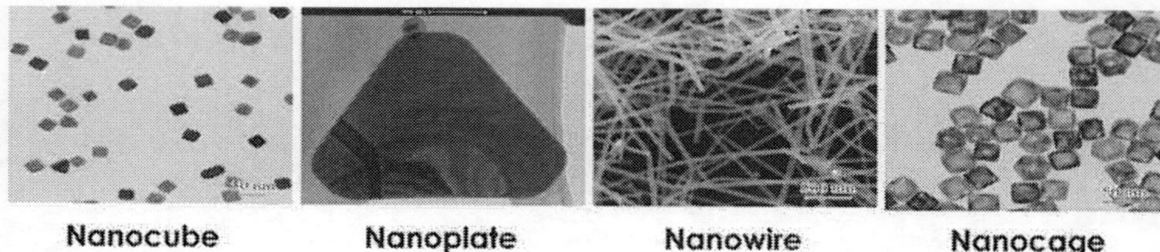

Figure 7: Different types of nanoparticle fabrications obtained by manipulation of matter or salts

3.1.1: Physical method of synthesis

Generally solid precursors are used which are directly converted to nano-size by mechanically or vapourization technique. Many physical methods can be applied for preparation of nanofertilizers; they are grinding, thermolysis or thermal evaporation, sputtering, pulsed laser deposition technique, condensation method, and microwave assisted synthesis

3.1.1.1: Grinding

Natural deposited minerals can be used as nanofertilizer after purification and grinded by ball mill or pot mill. For example, from rock phosphate deposit, phosphorus nanofertilizer can be prepared after purification of rock phosphate and grinded with the help of ball-mill or pot mill (Figure 8).

Figure 8: Ball mill (left) and pot mill (right)

In general, the average size of P nanofertilizer prepared by ball mill was 28 nm and pot mill was 70 nm (Table 4). Similarly, other metal nanofertilizers like Zn, Fe or Cu can also be prepared from their mineral deposits. Two types of grinder may be used *i.e.,* cryo-grinding or hand grinding.

Table 4: Size of nano-P fertilizer produced by grinding

Method used	Average size (nm)
High energy ball mill	28
High energy pot mill	70

3.1.1.2: Thermolysis

It is a process in which nanofertilizers are formed by decomposing the solids into metal cations and molecular anions or metal organic compounds at high temperature. Passivation can be achieved by the introduction of appropriate gas. Particles less than 5 nm can be produced by this method. In general, above 350 degree celsius is required for production. Very few nano-fertilizers can be attempted by this method.

3.1.1.3: Sputtering

A new method is reported to form metal nanofertilizers by sputter deposition inside a reactive ion etching chamber with a very short target-substrate distance. The distribution and morphology of nanoparticles are found to be affected by the distance, the ion concentration, and the sputtering time. Densely distributed nanoparticles of various compositions were fabricated on the substrates that were kept at a distance of 130 micrometer or smaller from the target. When the distance was increased to 510 micrometer, island structures were formed, indicating the tendency to form continuous thin film with longer distance.

The sputtering equipment is less expensive than electron-beam lithography systems. Because the atom vapour is typically generated from targets of pure materials, nanoparticles created by sputtering, usually contain fewer impurities in the composition than those created by chemical methods. Besides, since the different target materials can be sputtered simultaneously, it can produce alloy nanoparticles with easier control on composition than the corresponding chemical reduction methods, in which metal ions are reduced in multiple steps for their different reduction potentials. Therefore, sputtering has been widely studied and applied as a primary physical method for the production of metal nanoparticles.

3.1.1.4: Pulsed laser deposition technique

Pulsed Laser Synthesis (PLS) has been used to fabricate a wide range of nanomaterials, which have shown a variety of chemical, optical, magnetic, and electronic properties. PLS technique for the fabrication of materials can be traced back to the invention of the pulsed ruby laser in the mid-1960s. For decades, researchers have focused on the synthesis of different materials using this technique in liquid and gas phases with solid, liquid and gaseous precursors. There are different experimental set-ups for the synthesis of nanostructures using PLS, which depend on the precursor materials, the laser parameters, and the ambient conditions. The appropriate selection of parameters is essential in minimizing unwanted by-products and increases the yield of the intended nanostructured materials. PLS has demonstrated excellent suitability for the synthesis of a wide range of nanoparticles in terms of the yield and size homogeneity of the produced nanomaterials. Hence, PLS has potential for the industrial manufacturing of such nanostructures. ZnO, TiO2 and Boron nanofertilizer can easily be prepared by using this technique (Figure 9).

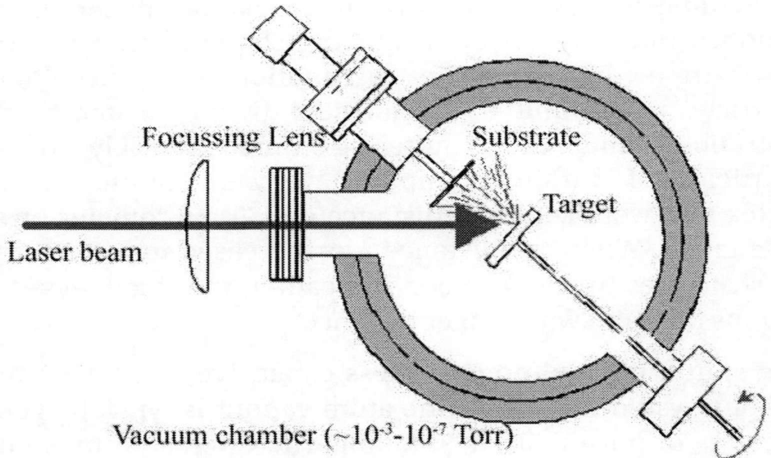

Figure 9: Laser Deposition Technique

The interaction of the laser radiation with matter is the most important issue in the synthesis of nanofertilizers with laser beams. The use of PLS for manufacturing new materials involves the appropriate selection of the laser's processing parameters with respect to the photo-physical properties of the material on which it is applied. When the laser beam interacts with a material in vacuum, gaseous or liquid medium, the incident light can either be reflected or absorbed. The absorbed energy from the laser beam can react with the matter in a form of either a thermal or/and chemical process. Generally, the mechanisms that can possibly occur as a result of the laser- matter interactions can be categorized into two possible mechanisms. These possible mechanisms are known as photochemical, photothermal and photophysical. For interactions between the laser and matter, the following factors should be taken into consideration: Physical state of the material, *i.e.,* solid, liquid or gas; Type of the material, *i.e.,* conductor, insulator, or semi-conductor; Laser beam parameters *i.e.,* excitation wavelength, pulse width, laser influence, and beam diameter; Impurities, defects, and crystal structure of the solid material in two interfaces; Relaxation times of thermalization and initial excitation processes.

Pulsed Laser Synthesis (PLS) is used in the top-down synthesis of nanomaterials, such as nanotubes, nanowires, nanoribbons, quantum dots, and even nano-flakes of the material. The technique is widely used by many researchers in the synthesis of both inorganic and carbon-based materials with varying particle sizes and in high yield.

3.1.1.5: Condensation method

It is a method of nanoparticles synthesis from supersaturated metal vapour. Nanofertilizers are prepared using various ways of metal evaporation, such as laser vaporization, thermal vaporization, plasma vaporization, and solar energy induced vaporization. The size and composition of the deposited nanofertilizers are controlled in the gas phase prior to deposition on the substrate using mass-analyzing system of various types.

3.1.1.6: Microwave-assisted synthesis

Microwaves are a form of electromagnetic energy with the frequencies in the range of 300 MHz to 300 GHz. The commonly used frequency is 2.45GHz. The mechanism involved is dipole interactions and ionic conduction. Heat is generated by molecular collision and friction. Generally, the more polar a molecule, the more effectively it will couple with the microwave field. Iron nanofertilizer can be synthesized by this technique.

Microwave heating is a transfer of electromagnetic energy to thermal energy, and is an energy conversion phenomenon rather than the heat transfer. The electric component of an electromagnetic field causes heating by two main mechanisms, dipolar polarization mechanism and conduction mechanism. In a polarization mechanism; for a substance to generate heat when irradiated with microwaves, it must possess a dipole moment, as in water molecule.

A dipole is sensitive to external electric fields and will attempt to align itself with the field by rotation. The applied field provides the energy for this rotation. As molecules vainly attempt to follow the field, they collide with one another and heating in the sample is observed. High- and low-frequency radiation does not give rise to efficient heating: in the first case, the field oscillates too quickly for the molecules to respond; in the second case, the molecules follow the field so well that there is no random motion generated.

3.1.2: Chemical methods of synthesis

Chemical synthesis of nanofertilizers represents a key approach for the production of materials, which generally involves a number of steps taking place in liquid or gas phase. Here, the liquid and gas precursors are allowed to undergo chemical reaction for achieving the supersaturation that is required to induce homogeneous nucleation of particles. The advantages of using chemical methods are:

- Variety of shapes and sizes of the particle is possible
- Low temperature (<350°C) synthesis
- Techniques are simple
- Large quantities of the material can be obtained with minimum time
- Materials are obtained in the form of liquid, but easily can be converted into dry powder or thin films
- Self-assembly or pattering is possible
- Doping of foreign atoms (ions) is possible during synthesis
- Inexpensive, less instrumentation as compared to the physical methods

A chemical sedimentation technique to prepare nano-zinc fertilizer is shown in Figure 10. To prepare this, the following two solutions of appropriate concentrations were made and mixed at room temperature with continuous stirring:

Solution A – Zinc acetate dehydrate in methanol

Solution B – KOH in methanol

A bulky white precipitate arose immediately, which become translucent after stirring for approximately 70 minutes. The solution was then heated to boiling point, after which the heat source was switched off. The solution was then allowed to stand overnight. The sediment thus obtained was stirred and centrifuged off. The white gel like product obtained was washed with deionized water and then dried to obtain the nano-sized zinc oxide particles.

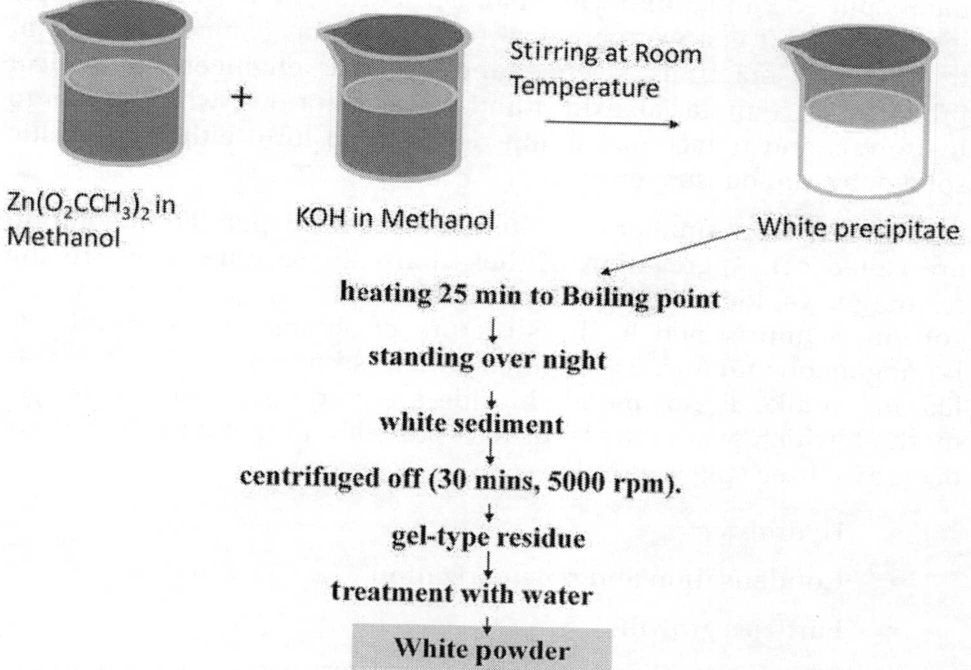

Figure 10: Preperation of zinc oxide nano fertilizer by chemical sedimentation technique

The formation of atoms can be accomplished by using chemical reactions under controlled but mild reaction conditions. Thus, freshly formed atoms can then undergo elementary nucleation followed by growth processes leading to the formation of defined nanofertilizers. In chemical synthesis, role of solvent, capping agent, additive and stabilizer for kinetic controls of tailored nanomaterials are very importatant.

The important chemical processes are: Sol-gel method, Poly Vinyl Pyrolidene (PVP) method, Co-precipitation technique, Micro-encapsulation, Sonochemistry, Colloidal method, Hydrothermal synthesis, and Micro-emulsions method.

3.1.2.1: Sol-gel method

The sol-gel technology is a well established colloidal chemistry technology which offers possibility to produce various materials with novel, predefined properties in a simple process and at a relatively low cost. It consists of a chemical transformation of liquid (sol) into a gel state and with subsequent post-treatment and transition into solid oxide material/fertilizer. The main benefit of sol gel processing is the high purity and uniform nanostructure achievable at low temperature. It is a wet chemical technique and commonly used in the field of materials science and ceramic engineering. Typical precursors are metal alkoxides and metal chlorides, which undergo hydrolysis and polycondensation reactions to form either an elastic solid or a colloidal suspension.

Particle sizes smaller than 50 nm, that are suspended in solution are called sol. Aggregation of these particles together leads to the formation of long chains of particles. With this mechanism, the solvent is imprisoned in the structure of chains and a completely homogeneous mixture is obtained which is known as gel. Reactants like metal alkoxide or metal chloride are commonly used in sol-gel method, which eventually leads to production of metal oxide. Sol-gel occurs in three stages as follows:

- Hydrolysis
- Condensation and polymerization
- Particles growth

A general sol-gel processing of nanofertilizer is shown below (Figure 11).

The entire process may be divided into four parts, *i.e.,* hydrolysis, polymerization, growth of nanostructures and agglomeration. Various factors such as pH, temperature, time, concentration of reagent, concentration of catalysor, H_2O/Si, phase transition sol to gel and drying may affect the quality of the produced nanoparticles, therefore, that may be standardized before synthesis.

3.1.2.2: Poly Vinyl Pyrolidene (PVP) method

This is a solution phase synthesis of many colloidal particles. The hydroxyl end group of PVP is a well-suited reductant for the aqueous synthesis of circular, triangular, and hexagonal nanoparticles of Ag, Au, Pd etc. The morphology can be changed by adjusting the molar ratio of PVP to the salt precursos and by altering the molecular weight of PVP. Figure 12 presents the preparation of Ag nanoparticles by PVP technique of various sizes and shapes.

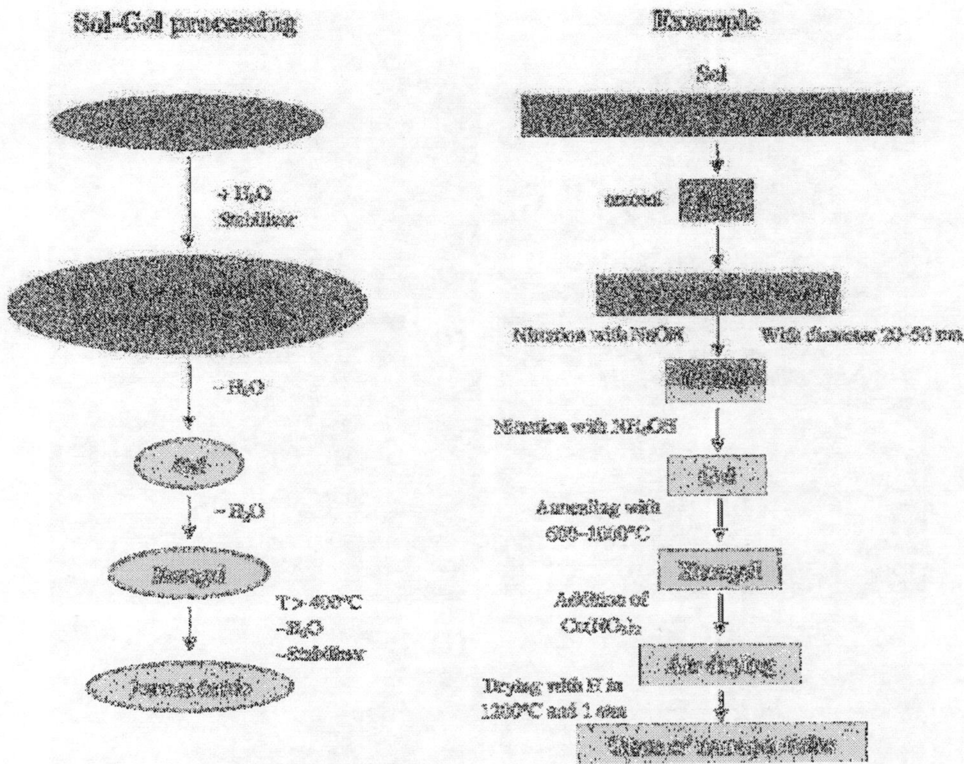

Figure 11: Sol Gel processing of nanomaterials

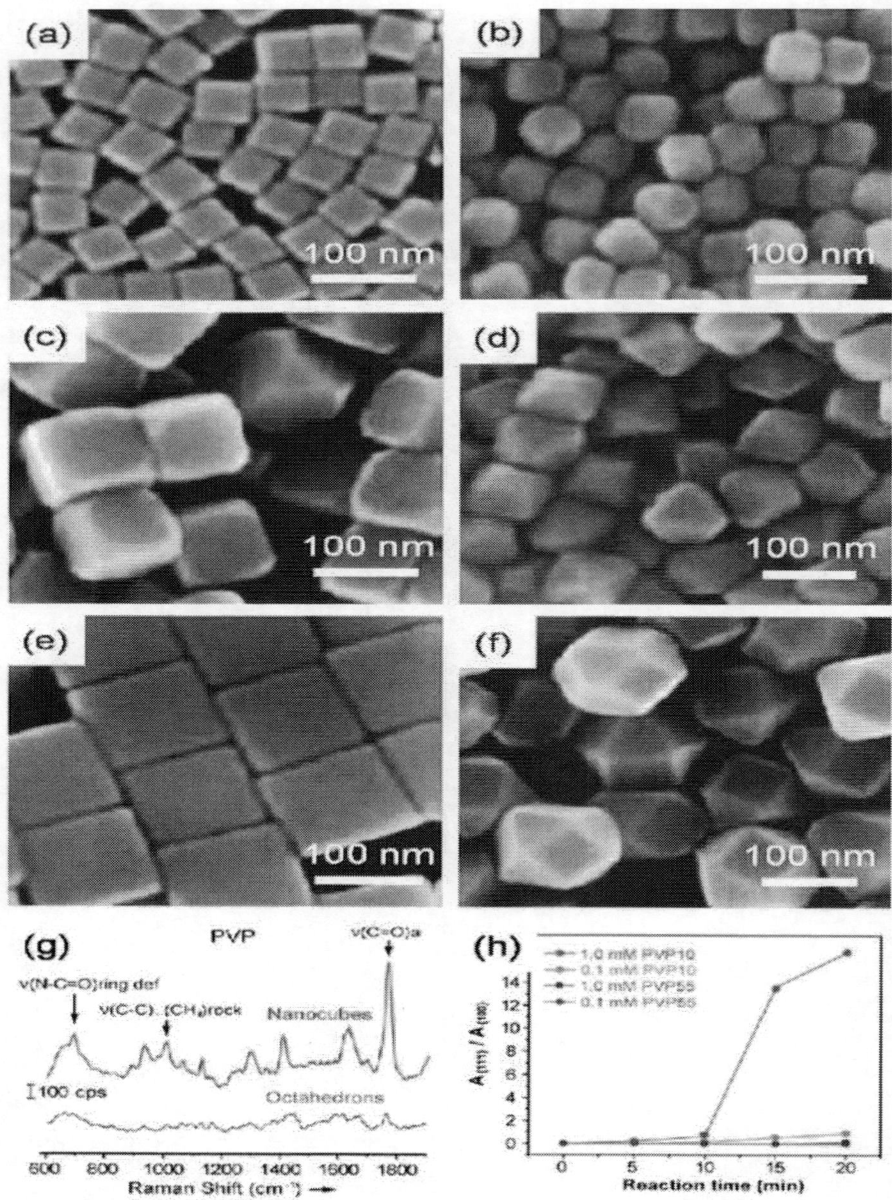

Figure 12: SEM images of Ag nanostructures prepared with varying amounts of PVP and different reaction times including **(a)** cubes (1.0 mM PVP55, 5 min), **(b)** truncated cubes (0.1 mM PVP55, 5 min), **(c)** truncated cubes (1.0 mM PVP55, 20 min), **(d)** octahedra (0.1 mM PVP55, 20 min), **(e)** cubes (1.0 mM PVP10, 20 min), **(f)** truncated octahedra (0.1 mM PVP10, 20 min). SERS spectra of **(g)** cubes and octahedra prepared in the presence of PVP55 and plot of **(h)** A {111} /A {100} versus reaction time of Ag nanocrystals prepared at different concentrations of PVP10 and PVP55.

In another example, the preparation of palladium (Pd) nanoparticles, sodium paladium (II) tetrachloride (Na_2PdCl_4) is used with the molecular weight of PVP ranges between 10,000 and 55,000 (Table 5).

Table 5: Production of different size and shapes of Pd nanoparticles by PVP technique

MW of PVP Na_2PdCl_4 (Metal precursor)	Molar ratio, PVP: metal precursor	Temperature (°C)	Time	Shape, average size, and yield
55,000	1.5	80	5h	Hexagonal, 40nm, 10%
55,000	5	80	5h	Hexagonal, 45nm, 70%
55,000	15	80	5h	Triangular, 25 nm, 30%
29,000	15	80	5h	Triangular, 10 nm, 40%
10,000	15	80	5h	Triangular, 50 nm, 70%
55,000	5	95	5h	Hexagonal, 30 nm, 30%

In general, a specific amount of PVP was dissolved in water and heated to 60°C in air under magnetic stirring and the salt solutions of desired concentration was rapidly added to it. After the viol had been capped, the reaction mixture was continued heating. At different stages of the reaction, the samples were taken from the solution, centrifuged and then washed with water to remove excess PVP before checking for particle size. The morphology of the particle can be changed by adjusting the molar ratio of PVP to the salt precursor and by altering molecular weight of PVP.

3.1.2.3: Co-precipitation technique

Co-precipitation reactions involve the simultaneous occurrence of nucleation, growth, coarsening, and agglomeration processes. Here, the products are generally insoluble species formed under conditions of high super saturation. Nucleation is the key step, and a large number of small particles will be formed. Secondary processes, such as Ostwald ripening and aggregation, dramatically affect the size morphology, and properties of the products. The super saturation conditions necessary to induce precipitation are usually the result of a chemical reaction. There are many co-precipitation techniques:

- Metals formed from aqueous solutions, by reduction from non-aqueous solutions, electrochemical reduction, and decomposition of metallurgic precursors.
- Oxides formed from aqueous and non-aqueous solutions.
- Metal chalconides formed by reactions of molecular precursors.
- Microwave/sonication assisted co-precipitation.

3.1.2.4: Micro-encapsulation

Nanoparticles which can act as an active ingredient can be encapsulated inside a micron-sized matrix or coating using the technique of microencapsulation. This method is useful in controlled release of nanoparticles when and where necessary. For example for TiO_2 nanofertilizer particle, it is a combination of surfactant (*e.g.,* dodecyl dimethylamineoxide and cetyl alcohol), organic solvent (*e.g.,* hydrocarbons or olefins) and water soluble salts (*e.g.,* titanium chloride, titanium tetra-isopropoxide, titanium tetra butoxide). Dissolution of surfactant in organic solvents leads to the formation of spherical particles called micelles which are obtained by dispersion of aqueous phase into organic phase. The resulting mixture is called microemulsion and it is to produce titanium dioxide nanofertilizer.

3.1.2.5: Sonochemistry

Sonochemistry uses high frequency sound waves to power chemical reactions. The ultrasound leads to formation of highly reactive radical species, forming tiny bubbles which collapse quickly, thus generating a chemical reaction. The ultrasonication method is shown in Figure 13.

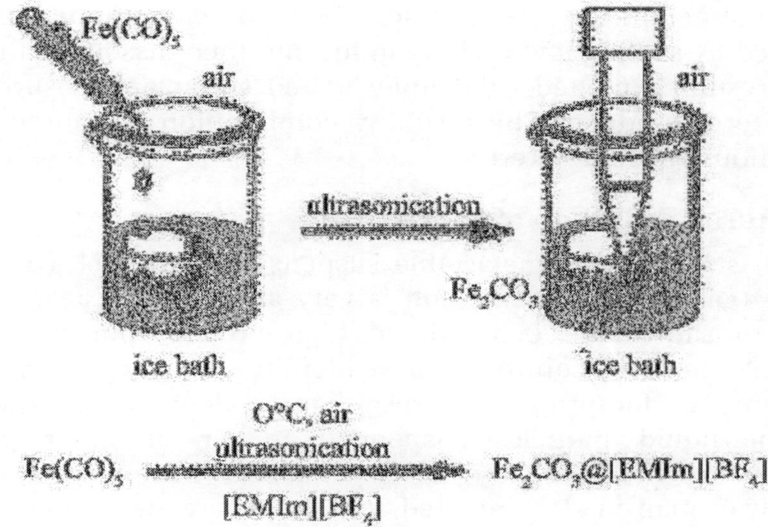

Figure 13: Ultrasonication method for iron nanofertilizer preperation

3.1.2.6: Colloidal method

It is one of the well known methods for the synthesis of nanoparticles. This methodology allows most of the nanoparticles with some preferential orientations/shapes. The main advantage of using this method is the possibility of controlling the shape of the particles. Thus different shapes (cubic, tetrahedral, spherical, octahedral etc.) with different properties can be prepared by this method.

3.1.2.7: Hydrothermal Synthesis

In a sealed vessel (bomb, autoclave etc.), solvents can be brought to temperatures well above their boiling points by the increase in autogenous pressures resulting from heating. Performing a chemical reaction under such conditions is referred to as solvo-thermal processing or, in the case of water as solvent, hydrothermal processing.

3.1.2.8: Micro-emulsions method

By using this method magnetic nanoparticles as small as 1–2 nm and of very uniform size (per cent standard deviation less than 10%) can be prepared. It is generally driven by the diffusion based process. It is an area of considerable current interest due to their unique properties, namely, ultralow interfacial tension, large interfacial area, thermodynamic stability and the ability to solubilize otherwise immiscible liquids.

In microemulsion technique, first a hot microemulsion is produced by stirring low-melting lipids and then dissolving the drug in the resulting melted solid lipids. Then surfactant, cosurfactant, and water are added. The resulting combination is poured into a large amount of cold water.

3.1.3: Aerosol Method of Synthesis

Aerosol is basically a metastable suspension of particles in a gas. Aerosol (gas-to-particle) method is very attractive because of their ability to satisfy the demands of high product purity, scale-up potential, improved control of phase identity, composition, and small environmental footprint. An aerosol can be defined as a system of solid or liquid particles suspended in air or other gaseous environment. Aerosols vary in size and composition, they can be naturally or man-made generated, and thus there are a wide range of them, from flame synthesized nanoparticles and nanomaterials (good aerosols), with fundamentally new properties and functions because of their small size

To capitalize on the unique properties of nanofertilizer particles in applications, will have to be synthesized in large quantities under carefully controlled conditions. For this purpose, aerosol (gas-to-particle) method is attractive because of their ability to satisfy the demands of high product purity, scale-up potential, improved control of phase identity, composition and small environmental footprint. Aerosol synthesis of magnesium nanofertilizer is shown as Figure 14. By changing carrier gas flow, heater and diffusion dryer sizes, different size and shapes of Mg nanofertilizers can be prepared by this technique.

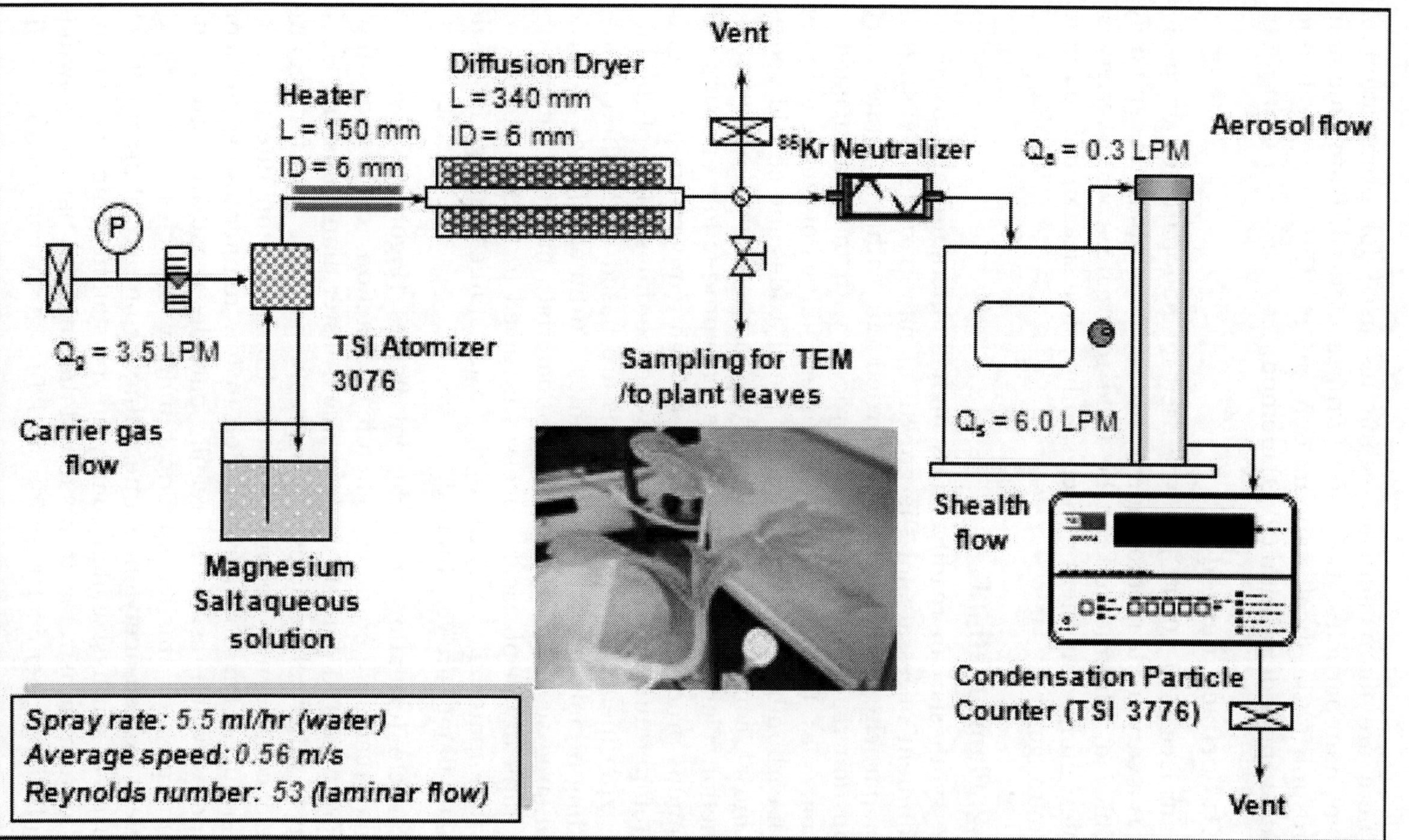

Figure 14: Aerosol synthesis of Mg nanoparticles

There are many aerosol techniques used for preparation of different nanoparticles. They are: Furnace method, Flame method, Electrospray, Chemical vapour deposition (CPD) method and Physical vapour deposition (PVD) method.

3.1.3.1: Furnace method

Tube furnace (500–1000°C) can be used for nanoparticles synthesis. But it is a very crude method and less than 100 nm particle can rarely be achieved. This method can be used to create a wide variety of complex, multicomponent particles with typical sizes in the 50 nm to 500 nm range.

3.1.3.2: Flame method

Flame synthesis can readily offers high temperature and oxidizing environment necessary for the growth of nanoparticles (NPs). In a flame, particle synthesis can be carried out within a flame since combustion reactions provide the process with the heat needed to form particles. Flame synthesis produces millions of metric tons per year of carbon black, TiO_2, and other metal oxides and considered as the most commercially successful method for nanoparticle production. Flame synthesis is a large scale production, one-step and continuous process which provide very high temperature gradient at high temperatures and very short residence times. Most of the lab-scale experiments for flame synthesis are based on laminar flames including premixed and diffusion flames. Turbulent flames are used in real large-scale production of nanomaterials, but they suffer from the complexity of physical processes involved in flame stabilization and nanomaterial's growth and basic studies using lab-scale turbulent flames are very limited.

Particle formation in turbulent flames (Figure 15) within a plasma reactors entrance chemical reaction occurs at high temperatures and particle coalescence is faster than collision, which results in spherical particles. Downstream, particles grow further and particle coalescence becomes slower than collision due to lower temperatures. Far downstream, due to low temperatures, collision dominates which results in irregular particles those are hard or soft agglomerates of smaller spherical unitary particles. A premixed, turbulent oxygen-acetylene flame, using a commercial brazing torch was used to synthesize high quality polycrystalline diamond films on molybdenum substrates. It was found that initially, particle growth is controlled by chemical reaction and Brownian coagulation, while later shear induced coagulation becomes predominant. In general, particle size increases with increasing pressure, while it decreases with increasing temperature, initial mole fraction of the reactants, and turbulence intensity. Away from the burner tip, cooling has a major effect on particle coalescence rate which causes the formation of non-spherical agglomerates.

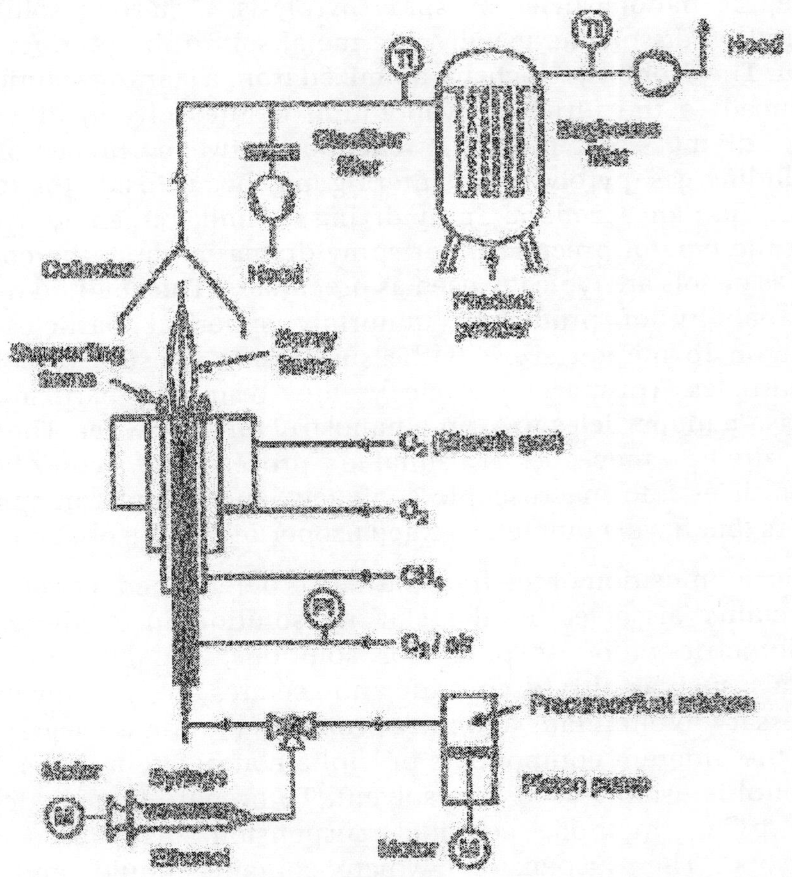

Figure 15: Schematic diagram of synthesis of nanofertilizer using flame synthesis

Due to their unique advantages, flame aerosol reactors can be considered as one of the most important ways to generate nanoparticles. These advantages include; producing a wide range of oxide nanomaterials, self-purifying process, high temperature with high temperature gradient, good control of the nanomaterial properties, and very fast process. In the process of flame spray pyrolysis (FSP), liquid precursor can directly be sprayed into the flame without any need of additional heating to get supersaturated vapour. This method allows the use of viscous precursors that do not have sufficiently high vapour pressure to be delivered as a vapour. The injected precursor enters the flame region as very fine droplets, which later on start to vaporize and react to form the nanomaterials.

3.1.3.3: Electrospray technique

To prepare nanoparticles by spray pyrolysis, a starting solution is prepared by dissolving, usually, the metal salt of the product in the solvent. The droplets, which are atomized from a starting solution, are introduced to the furnace. Evaporation of the solvent, dilution of solute, drying, precipitation, reaction between precursor and surrounding gas, pyrolysis, or sintering may occur inside the furnace to form the nano-product. Spray drying is similar to spray pyrolysis except the type of precursor. For spray drying method, the colloidal particles or sols are typically used as precursors. This method has also the capability of producing uniformly spherical particles from submicron to micron sizes. If the suspension consists of colloidal nanoparticles (primary particles), the resulting particles also comprise nanoparticles to form a nanostructured powder. Therefore, spray drying may be a suitable process for consolidating nanoparticles into macroscopic compacts, and submicron spherical powders that have nanometer-scaled properties can be obtained.

The composition of the final particle is determined by the solutes or reactants dissolved in the starting solution in predetermined stoichiometric ratio. As precursor solutions, usually inexpensive materials, such as nitrate, chloride and acetates are typically used. It is necessary to eliminate solubility problems and phase segregation, where the different components precipitate at deferent times. Water or alcohol is usually used as a solvent. To prepare fine particles by spray drying method, colloidal suspension/sols are used as precursors. The suspension, which contains liquid and solid particles, is sprayed and the liquid phase (the solvent) evaporates from the droplets.

The average size and size distribution of the final particles can be roughly determined from the size of the atomized droplet and the initial concentration of the starting solution. Atomization is the production of droplets and their dispersion into the gas. The size or morphology of the final particles produced can also be determined by the concentration and velocity of the droplet generated by the atomizers. A variety of atomization methods has been used in spray pyrolysis studies, such as air-assist (pneumatic) or a two-fluid nozzle, ultrasonic, vibrating orifice and spinning disk. The main equipment consists of: (i) an atomizer or nebulizer that converts the starting solution to droplets, (ii) carrier gas (es), (iii) tubular furnace/reactor, and (iv) sampler.

The electrospray technique is considered one of the best options for developing micro- and nanospheres above conventional techniques of nanoprecipitation, double-emulsification, and layer by layer, among others. This technique has been applied in the pharmaceutical industry, but today is being applied in the food industry and is seeing great advantages in the encapsulation of bioactive compounds, aromatic enhancers, and manufacturing smart packaging, edible films, and coatings in food. This technique could also be applied in the agrifood sector for the preparation of controlled and/or prolonged release systems of fertilizer or agrochemicals, for which more research is needed.

3.1.3.4: Chemical vapour deposition (CVD) method

This process is mostly used in semiconductor industry for depositing thin films of various materials. This is mainly chemical process. In this process, the substrate is exposed to one or more volatile precursors. These precursors decompose on the substrate and produce the desired deposit. In CVD, the vaporized precursors are introduced into a CVD reactor and adsorb onto a substance held at an elevated temperature. In CVD process, vaporized precursors are made to adsorb onto a substance held at a high temperature. The adsorbed molecules react with other or decompose and produce crystals. There are three main steps in CVD process. 1) Reactants are transported on the growth surface by a boundary layer. 2) Chemical reactions take place on the growth surface. 3) By products formed by the gas-phase reaction has to be removed from the surface. Homogeneous nucleation takes place in gas phase. Heterogeneous nucleation takes place in substrate. The CVD technique is one of the most suitable methods for the synthesis of carbon nanomaterials (CNMs) in terms of product purity and large scale production. The method is simple and economic for synthesizing at low temperature and ambient pressure.

3.1.3.5: Physical vapour deposition (PVD) method

By using this deposition technique, thin films of different metals are deposited on different surfaces. This technique involves condensation from the vapour phase.

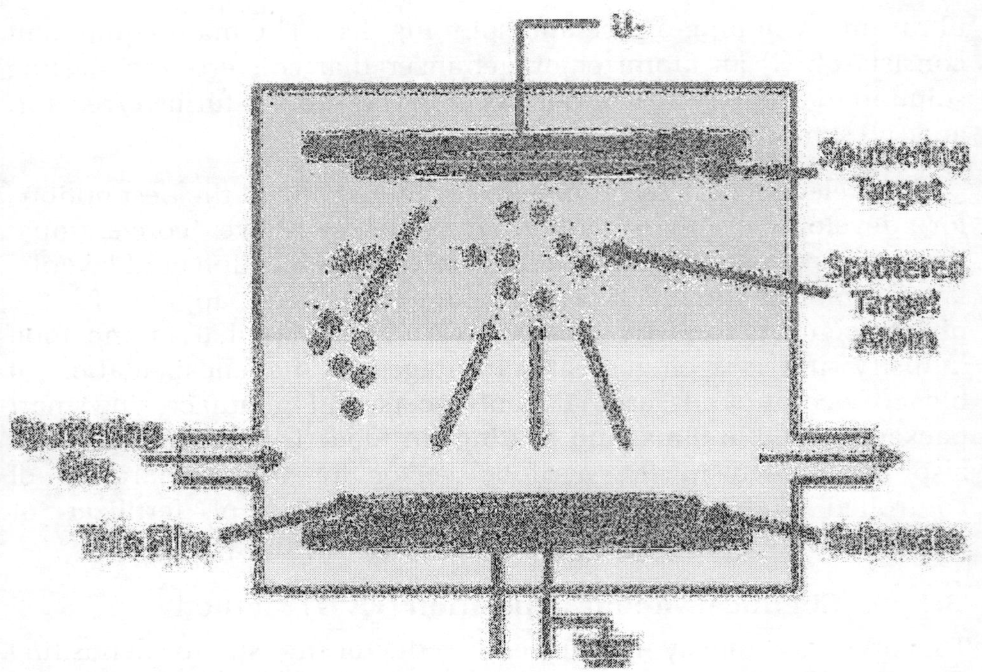

Figure 16: Diagram of physical vapour deposition technique

There are three main steps involved in this process:

- By the sublimation of evaporating of a material corresponding vapour phase
- Transportation of the material to the substrate from source
- Formation of the thin film and particle by nucleation and growth.

The source can evaporate by using electric beams, thermal energy, sputtering technique, cathode arc plasma. Physical Vapour Deposition (PVD) is a collective set of processes used to deposit thin layers of material, typically in the range of few nanometers to several micrometers. PVD processes are environmentally friendly vacuum deposition techniques consisting of three fundamental steps (Figure16):

- Vaporization of the material from a solid source assisted by high temperature vacuum or gaseous plasma.
- Transportation of the vapour in vacuum or partial vacuum to the substrate surface.
- Condensation onto the substrate to generate thin films.

Different PVD technologies utilize the same three fundamental steps, but differ in the methods used to generate and deposit material. The two most common PVD processes are thermal evaporation and sputtering. Thermal evaporation is a deposition technique that relies on vaporization of source material by heating the material using appropriate methods in vacuum. Sputtering is a plasma-assisted technique that creates a vapour from the source target through bombardment with accelerated gaseous ions (typically Argon). In both evaporation and sputtering, the resulting vapour phase is subsequently deposited onto the desired substrate through a condensation mechanism. Deposited films can span a range of chemical compositions based on the source material(s). Further compositions are accessible through reactive deposition processes. Relevant examples include co-deposition from multiple sources, reaction during the transportation stage by introducing a reactive gas (nitrogen, oxygen or simple hydrocarbon containing the desired reactant), and post-deposition modification through thermal or mechanical processing. PVD is used in a variety of applications, including fabrication of microelectronic devices, interconnects, battery and fuel cell electrodes, diffusion barriers, optical and conductive coatings, and surface modifications.

3.1.4: Biological Methods

Living organisms and plants have huge potential for production of nanoparticles of wide applications. By using the specific plants and organisms from simple bacteria or fungi to highly complex eukaryotes in the reaction mixture, the production of nanoparticles with desired shape and size can be obtained. Synthesis of nanoparticles using biological entities has great interest due to their unusual optical, chemical, photo-electrochemical and electronic properties. Nature has devised various processes for the synthesis of nano and micro-length scaled inorganic material which has contributed to the development of relatively new and largely unexplored area of research based on the biosynthesis of nanomaterials. It is well known that biological entities like microorganisms and living cells are the best examples of machines that possess operating parts at the nanoscale level and perform a number of jobs ranging from generation of energy to extraction of targeted material mat with a very high efficiency.

The utilization of such microorganisms like fungi, bacteria, yeast and herbal extracts in the synthesis of nanoparticles is a relatively recent activity. Environment-friendly biological systems now considered as benign nanofactories. Possible mechanisms of nanoparticle biosynthesis by different agents are shown in Table 6.

Table 6: Possible mechanisms biosynthesis of nanofertilizer by different agents

Agents	Possible mechanisms
Plant	Secondary metabolites (alkaloids, flavonoids, saponins, steroids, tannins and other nutritional compounds) act as reducing and stabilizing agents
Algae	Polysaccharides have hydroxyl groups and other functionalities that can play important roles in both the reduction and the stabilization of nanoparticles
Fungi	Fungal proteins or reducing enzymes intracellularly or extracellularly and the procedure of biomimetic mineralization
Yeast	Membrane bound (as well as cytosolic) oxidoreductases and quinine
Bacteria	The microbial cell reduces metal ions of specific reducing enzymes like NADH-dependent reductase or nitrate dependent reductase

3.1.4.1: Nanoparticle synthesis by bacteria

Research has focused heavily on prokaryotes as a means of synthesizing metallic nanoparticles. Due to their abundance in the environment and their ability to adapt to extreme conditions, bacteria are a good choice for study. They are also fast growing, inexpensive to cultivate and easy to manipulate. Growth conditions, such as temperature, oxygenation and incubation time can be easily controlled. It was found that changing the pH of the growth medium during incubation results in the production of nanoparticles of differing size and shape. Controlling such properties is important, as varying sizes of nanoparticles are required for different applications, such as optics, catalysts or anti-microbials. Scientist suggested that metal nanoparticles were synthesised by bacteria via an NADH-Dependant Reductase (Figure 17).

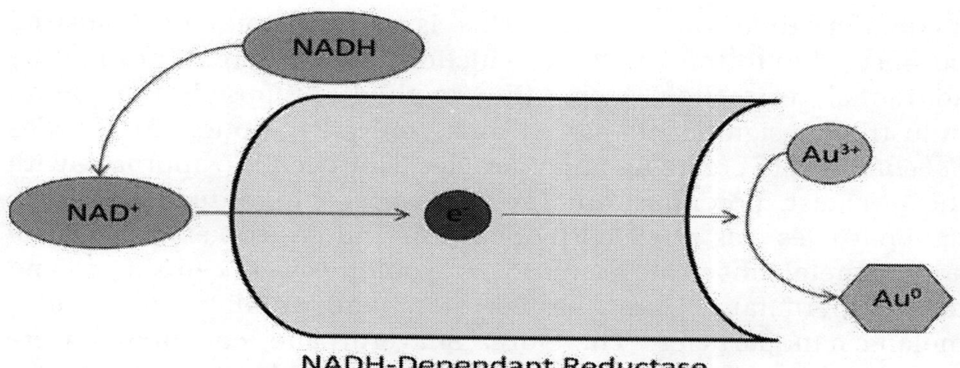

Figure 17: A suggested mechanism for the reduction of gold ions into AuNPs.

The bioreduction is initiated by the electron transfer from NADH by NADH-Dependant Reductase as an electron carrier to form NAD+. The resulting electrons are obtained by metal ions which get reduced to elemental metal nanoparticles.

Pd is a member of the Platinum Group Metals (PGM) which is a collection of highly catalytically active metals and is currently being primarily used as catalysts for dehalogenation and hydrogenation reactions. Recently, it has been shown that zero valent palladium (Pd^0) nanoparticles can be synthesised using bacteria found at Alpine sites heavily contaminated with heavy metals. Of all the variety of heavy metal resistant bacteria that have found in that environment, *Pseudomonas* cells were involved in producing catalytically active nanoparticles, which were successfully used in reductive dehalogenation of tri and tetra-chlorinated dioxin congeners. *Escherichia coli,* can also synthesise Pd^0 nanoparticles with the help of hydrogenases found in the bacterium. In both studies mentioned, the Pd nanoparticles were found to be formed on the cell envelope of the bacteria which makes them attractive as they are easily accessible. Bacteria like *Bacillus megaterium* JCT13, has been reported by Tarafdar and co-workers to produce P nanofertilizer from the calcium phosphate salts of 0.1mM concentration.

The bacterium *Bacillus licheniformis* was observed to produce intracellular Ag nanoparticles. The colour of the culture after addition of silver ions turned a dark brown, indicating the presence of Ag nanoparticles. It was found that intracellular Ag nanoparticles were created when members of the *Bacillus* sp. were sub-cultured into media containing $AgNO_3$. The reaction was slow and the incubation time was 7 days. This is not suitable for industrial purposes due to the lengthy production time. Whereas, despite the additional extraction step, the method proposed was more industrially significant as it took only 24 hours for *Bacillus licheniformis* to create Ag nanoparticles. However, the approach with the greatest potential for the industrial bioproduction of Ag nanoparticles was reported by Shaverdi and co-workers. Instead of adding metal ions to a live culture. The cultures were centrifuged and the supernatant was tested for the ability to create metallic nanoparticles. Therefore, Shaverdi and co-workers were able to create Ag nanoparticles in 5 minutes, without the need for a cell lysis step by the use of culture supernatant. This extracellular type of formation is a more desirable not only because of the simplicity of purification but also due to the observed increased production rate.

An interesting study focused on the production of Ag nanoparticles by lactic acid bacteria. Many bacterial species were tested, but only four were found to synthesise Ag nanoparticles: *Lactobacillus* sp., *Pediococcus pentosaceus, Enterococcus faecium* and *Lactococcus garvieae*. A two-step process of Ag nanoparticle formation was proposed. First, the Ag ions were accumulated at the cell wall via biosorption and then subsequent reduction of those ions formed the metallic nanoparticles. It was also suggested that the cell wall may act as a capping agent for the nanoparticles, which keeps them stable by preventing aggregation and showed that by increasing the pH of the medium, the reduction rate of the nanoparticles increased. The effect of pH on nanoparticle synthesis was also observed. By varying the pH level, the shapes and sizes of nanoparticles may vary. Although Ag and Au nanoparticles are important due to their antimicrobial abilities, many studies have been done on other metals such as uranium (U). A significant amount of research has been done on *Bacillus* species due to their metal bioaccumulation abilities. Pollmann and group studied the ability of *Bacillus sphaericus* JG-A12 to accumulate high concentrations of toxic metals such as Cu, Pb, Al, Cd and U. They found that the S-layer proteins of *B. sphaericus* are responsible for the bioremediation of uranium from aqueous environments. The S-layer is a porous layer consisting of identical proteins which surround the bacterial cell and can contribute up to 15% of the total cell protein content (Figure 18).

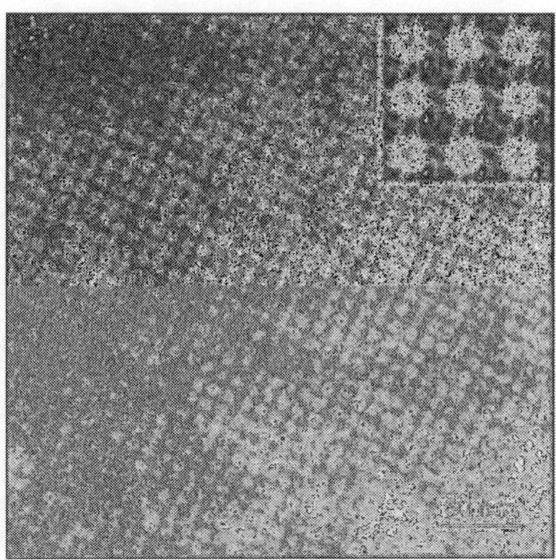

Figure 18: Bacterial S-layer of *Bacillus sphaericus* JG-A12 shown under Transmission Electron Microscopy. The picture illustrates the symmetry of S-layers and allows the visualisation of the pores where metallic nanoparticles are localised.

The synthesis of copper nanoparticles has proven to be a difficult feat in the past, by any available method. Cu is not stable at the nanometre scale and oxidises fairly rapidly to form copper oxide (CuO). Therefore, if Cu nanoparticles are to be used in an application after their synthesis, they need to be stabilised. In 2013, a study suggested the production of pure elemental Cu nanoparticles by biological means, using *Morganella morganii* was possible. Ramanathan and group state that *M. morganii* synthesises the Cu nanoparticles intracellularly by uptake of the Cu ions and subsequent binding of the ions to either a metal ion reductase or similar protein. This results in the reduction of the ion to metallic CuO which then accumulates extracellularly as nanoparticles once effluxed out of the cell; however, this has not been confirmed. Another study based on *Morganella* sp. showed the extracellular synthesis of Ag nanoparticles. It was also suggested the pathway of Ag resistance is also responsible for the formation of the Ag nanoparticles. Bioinformatics analysis of the silE gene from *Morganella* sp. showed 99% similarity with previously reported silver-binding proteins. Further results showed the similarity of the silE homologue to copper binding proteins from other microorganisms. Using this result, Ramanathan and group suggested that *M. morganii* may use the Ag resistance mechanism to produce elemental Cu nanoparticles.

3.1.4.2: Nanoparticle synthesis by fungi

The use of fungi in producing nanoparticles has received significant interest as they offer certain advantages over the use of bacteria for the synthesis of nanoparticles. The ease of scaling up and downstream processing, the economic feasibility and the presence of mycelia offering an increased surface area, are important advantages to consider. It was suggested that because fungi secrete significantly higher amounts of proteins than bacteria, this would amplify the nanoparticle synthesis and may be more resistant and stable than bacteria.

The use of fungi in the synthesis of nanoparticles is a relatively recent addition to the list of microorganisms. However, the genetic manipulation of eukaryotic organisms as a means of over expressing specific enzymes identified in nanomaterial synthesis would be much more difficult than that in prokaryotes. It is potentially exciting since fungi secrete large amounts of enzymes and are simpler to deal with in the laboratory. In addition, to good monodispersity, nanoparticles with well defined dimensions can be obtained by using fungi. This has been shown with an experiment where bioreduction of aqueous $AuCl_4^-$ ions was carried out using the fungus *Verticillium* sp. that led to the formation of gold nanoparticles with fairly well-defined dimensions and good monodispersity. These results have been documented that the trapping of $AuCl_4^-$ ions on the surface of fungal hyphae could occur by electrostatic interaction with positively charged groups *i.e.,* lysine residues in enzymes that are present in the cell wall of the mycelia. In this enzyme mediated reaction, gold ions were reduced by enzymes within the cell wall leading to aggregation of metal atoms and formation of gold nanoparticles. However, they could not find the exact mechanism of formation of gold nanoparticles. It can be concluded from their study that bacteria, fungi could be a source for large scale production of nanoparticles. A shortcoming of the study was that the metallic nanoparticles were synthesized "intracellularly." Indeed, when the site of nanoparticle synthesis is intracellular, downstream processing becomes difficult and often defeats the purpose of developing a simple and cheap process.

In this regard, extracellular biosynthesis of Ag nanoparticles using the filamentous fungus *Aspergillus fumigatus* was reported. Since fungi are known to secrete much higher amounts of proteins, thus might have significantly higher productivity of nanoparticles in biosynthetic approach. Towards elucidating mechanism of nanoparticles formation, an *in vitro* approach was followed, where, species specific NADH dependent reductase, released by the *Fusarium oxysporum*, were successfully used to carry out the reduction of $AuCl_4^-$ ions to gold nanoparticles. This has first time opened up a novel fungal/enzyme-based *in vitro* approach for nanomaterials synthesis. Based on properties of *Fusarium oxysporum*, it was also used in the formation of extremely stable silver hydrosol. The acidophilic fungus *Verticillium* sp. has capability of producing gold as well as silver nanoparticles upon their incubation with Ag^+ and $AuCl_4^-$ ions. However, a novel biological method for the intra and extra cellular synthesis of silver nanoparticles using the fungi, *Verticillium* sp. and *Fusarium oxysporum,* respectively has been documented. This has opened up an exciting possibility, wherein, the nanoparticles may be entrapped in the biomass, in the form of a film or produced in solution, both having interesting commercial potentials. The fungus, *Aspergillus flavus* also resulted in the accumulation of silver nanoparticles on the surface of its cell wall, when incubated with silver nitrate solution.

Extracellularly produced nanoparticles were stabilized by the proteins and reducing agents secreted by the fungus. Four high molecular weight proteins released by the fungal biomass have been found in association with nanoparticles. Further, the reduction of metal ions and surface binding of the proteins to the nanoparticles did not compromise the tertiary structure of the proteins. Biomolecules likes polypeptide and enzymes carry out reduction reaction to synthesize metal nanoparticles by endophytic fungus *Colletotrichum* sp. Instead of fungi culture, isolated proteins from fungi have been successfully used for nanoparticles production.

Growth conditions play an important role during the production of nanoparticles while using the fungi cultures. It was noticed that when gold ions were incubated with the *Trichothecium* sp. biomass under stationary conditions, led to the formation of extracellular nanoparticles. But under shaking conditions, this was resulted in the formation of intracellular gold nanoparticles. These proteins were released into the medium under stationary conditions and did not release under shaking conditions. Research showed the synthesis of magnetic nanoparticles by using *Fusarium oxysporum* and *Verticillium* sp. at room temperature. In an extracellular condition, both fungi secreted proteins which were capable of hydrolyzing iron precursors to form iron oxides predominantly in the magnetite (Fe_3O_4) phase. This was not true with all *Fusarium* sp., *Fusarium moniliforme* produces reductase enzyme, but could not form silver nanoparticles upon their incubation with silver ions. However, by controlling the amount of cofactor NADH, synthesis of quite stable Au–Ag alloy nanoparticles of various compositions have been produced. This approach can be further employed for producing various other composite nanoparticles. Towards elucidating the mechanism of synthesis of nanoparticles, NADPH-dependent nitrate reductase and phytochelatin isolated from *Fusarium oxysporum* has been used for *in vitro* silver nanoparticle was developed, by production. An ecofriendly and low cost protocol for synthesis of silver nanoparticles was developed, by using the cell free filtrate of *Aspergillus flavus* NJP08 when incubated with aqueous silver (Ag^+) ions. The average particle size was 17±5.9 nm. Minimum time, miniaturization and non-hazardous processes are key parameters for any kind of technology acceptance. In this effort, Bhainsa and D'Souza could get fairly monodispersed silver nanoparticles within an hour by using *Aspergillus fumigatus*. The production was even faster compared to the physical and chemical processes of nanoparticles synthesis. Hence, this process could be suitable for developing a biological process for mass production of nanoparticles.

Silver nanoparticle production has been the focus of research by the scientific community, and rightly so. Ag nanoparticles have great potential in a number of industries, such as fertilizers, antimicrobials and electronics. The fungus *Fusarium oxysporum* has been used in a large number of studies attempting to create metallic nanoparticles, especially those made of silver. Pure Ag nanoparticles were synthesised at a size range of 5–15 nm and it was suggested that they were capped in order to stable them by proteins secreted by the fungus. Although, the most important advance in the fungal synthesis of metal nanoparticles was that *F. oxysporum* produced these nanoparticles extracellularly, in contrast to all previous research which reported only the intracellular production of Ag and Au nanoparticles. *Fusarium oxysporum* has also been shown to produce cadmium sulphide (CdS), lead sulphide (PbS), zinc sulphide (ZnS) and molybdenum sulphide (MoS) nanoparticles, when the appropriate salt is added to the growth medium.

A recent study has shown on the production of Ag nanoparticles after using *Aspergillus* fumigatus to synthesise extracellular silver nanoparticles in the size range of 5–25 nm. This range in size was larger than previously reported by *F. oxysporum,* which might cause a disadvantage when application predictability comes into play, as it would be difficult to predict the catalytic activity of the nanoparticles produced if their sizes are different in every batch. However, bio-production by *A. fumigatus* is a very attractive prospect as the organism is able to reduce Ag ions into nanoparticles within 10 minutes of exposure, in contrast of another research which showed Ag nanoparticle production by *F. oxysporum* within an hour.

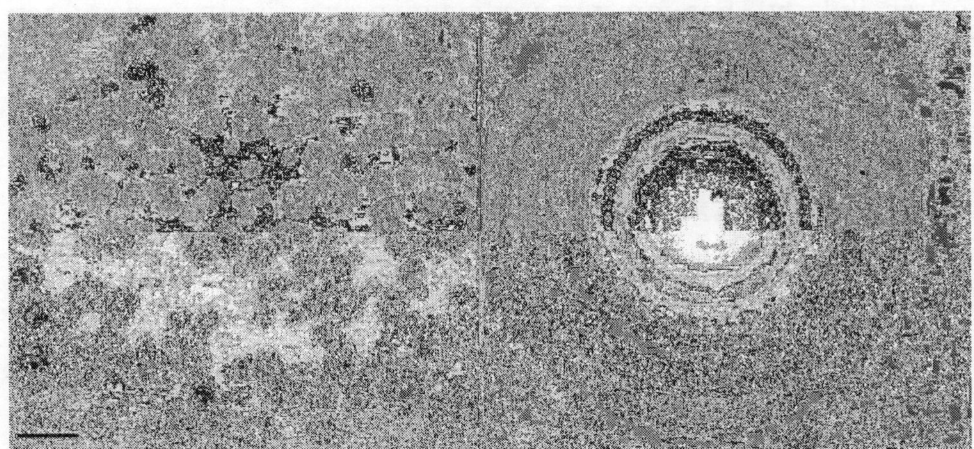

Figure 19: a) TEM micrograph of Ag nanoparticles on a drop-coated carbon grid after 72 hours of synthesis by *Trichoderma reesei*. **b)** SAED pattern from Ag nanoparticle shown in the left picture. The diffraction rings correlate to the silver.

The fungus *Trichoderma reesei* has also been shown to produce extracellular Ag nonoparticles (Figure 19). Bahabi *et al.* were able to produce Ag nanoparticles using this fungus after 72 hours, which was significantly slower than *A. fumigatus* and *Fusarium oxysporum*. However, the use of *T. reesei* does have an advantage over the use of other fungi for the production of metallic nanoparticles. As a comparatively well studied organism, it can be manipulated to produce high quantities of enzymes, up to 100 g/L, which may help to increase the nanoparticle production rate in the future. The size range of the Ag nanoparticles reportedly produced by *T. reesei* was 5–50 nm, which were not as homogenous as those produced by either *A. fumigatus* or *F. oxysporum*.

Although extracellular formation has its advantages, such as lower cost, simpler downstream processing, intracellular formation can also be of great importance. In the case of bioremediation, heavy metals such as Cu and Pt need to be removed from contaminated environments. By using fungi that have the ability to produce intracellular nanoparticles, it would be far easier to remove the fungus and its accumulated metal contaminant from the contaminated sample. A noteworthy study was done on the white rot fungus *Coriolus versicolor* by Sanghi *et al.* (2009) into the production and accumulation of intracellular Ag nanoparticle. The group manipulated the reaction conditions and observed that *C. versicolor* had the ability to produce Ag nanoparticles intracellularly and extracellularly. Therefore, the production process is not fixed and can be adapted according to specific requirements.

Gold nanoparticles have been increasing in their importance over recent years, but despite this, there are fewer examples of their synthesis by fungi than those composed of silver. Their small size makes them reactive, unlike the bulk form of gold, which makes Au nanoparticles ideal for use as catalysts and as precursors for electronics applications. It was reported the synthesis of Au nanoparticles using the fungus *Verticillium* sp. whereby intracellular Au nanoparticles were observed localised on the surface of the mycelia via the biological reduction of $AuCl_4^-$. Tarafdar *et al.* (2013) reported that *Aspergillus oryzae* and *Aspergillus terreus* are able to produce Au nanoparticles within 48 hours. A large number of fungi (73) reported by this group which can produce N, P, K, Mg, S, B, Cu, Fe, Mo and Zn nanoparticles from their respective salt solution, which can be used as nanofertilizers (Table 7).

Table 7: List of some efficient fungi for production of different nanonutrients

Name of the organism	NCBI Accession number	Type of bio-nanofertilizer
Aspergillus flavus CZR2	JF681301	Zn
Aspergillus fumigatus TFR8	JQ675291	Mg
Aspergillus japonicus AJP01	JF770435	Fe
Aspergillus orchraceus TFR23	KC806053	K
Emericella nidulans TFR15	KC175550	P
Emericella quadrilineata TFR25	KC806055	N
Straphylotrichum coccosporum TFR27	KF729586	B
Penicillium limosum TFR26	KF729585	S
Aspergillus terreus TFR2	JN194186	Mo

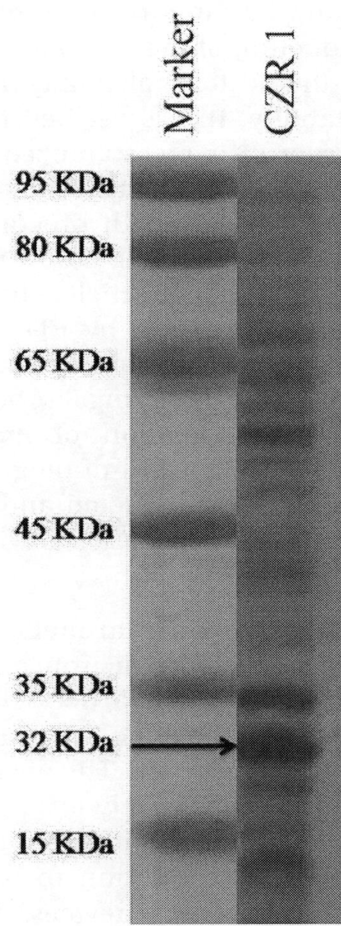

Figure 20: SDS–PAGE profiles of extracellular proteins of *Aspergillus terriues CZR 1* used for

A large number of nanofertilizers can be produced with the help of these isolateed fungi. SDS-PAGE analysis showed (Figure 20), 32kDa protein is responsible for nanofertilizer synthesis. Several studies found that 32 kDa protein showing their intense band in the potential fungal species, upon polyacrylamide gel electrophoresis. This protein play important role in the reduction of precursor metal into nano form, although some other bands were also found but they play role in the encapsulation of metal nanoparticles, which were confirmed by FTIR spectra. However, the actual mechanism of metal biotransformation is yet to elucidated even though many researchers have attempted different ways to support possible mechanism.

The metal ions are reduced in the active sites of enzymes that catalyze their reduction. In nature, biological enzymes operate with very high selectivity. In the active site of an enzyme, chemical transformation such as bond breaking and forming takes place. Thus nanoparticle growth process is thought to occur in the protective environment of the enzyme active site, which could provide control over the nanoparticle size. It was also reported that certain NADH dependent reductase was involved in reduction of silver ions in case of *Fusarium oxysporum*. In this study, the UV–Vis spectrum study provides some clues regarding the mechanism of synthesis of silver nanoparticles. There were two absorbance peaks found in the UV range corresponding to 220 and 280 nm. While the peak at 220 nm may be due to absorption by amide bond, the other peak at 280 nm may be attributed to the tryptophan and tyrosine residues present in the protein. This indicates secretion of some proteinacious components into the medium by the fungal biomass, which might play important role in the reduction of the metal ions in the form of nanoparticles. Consequently, the proteins also may bind to the nanoparticles and enhance the stability. It was verified that silver nanoparticles produced by *F. oxysporum* were stabilized by fungal protein and showed that NADH dependent reductase enzyme secreted extracellularly by fungus make association with precursor metal ion to synthesize metal nanoparticles. A similar study was reported that extracellular synthesis of gold nanoparticles using *Rhizopus oryzae* and characterization the nanoparticle by FTIR, in which the spectra revealed the presence of amide group of peptide bond, supports the present study of the involvement of fungal protein in the metal ion reduction as well as encapsulation of metal nanoparticles. An interesting study of the capping agent of biogenic silver nanoparticles by Raman spectroscopy was conducted and an intense band of Ag–N, and symmetric and asymmetric C=O of CO_2^- ions were verified the protein as capping agent.

Not all metallic nanoparticles are made of elemental metals in their zero-valent form. Magnetite (Fe_3O_4) is a common iron oxide that possesses magnetic properties and magnetite nanoparticles have been shown to be produced by the pathogenic fungus *F. oxysporum* and the endophytic fungus *Verticillium* sp. The use of magnetite nanoparticles has been talked about extensively for use in biomedical applications, such as magnetic resonance imaging and for oscillation damping and position sensing, in addition to non-medical applications, such as in magnetic recording devices. The synthesised magnetite nanoparticles described by Bharde and group were formed intracellularly, which again means an extra step in the purification of magnetite nanoparticles if they are intended to be of commercially use.

To synthesize metal nanoparticles from fungal extracellular enzymes, there was a need to add precursor compound of metal of interest as desired, for the biosynthesis of nanoparticle compounds were selected for the testing of nanoparticles formation on the basis of their ionic potential in water. Sulphates, chlorides, nitrates and oxides of the metal were selected for the synthesis of nanofertilizers on the basis of their ionic strength. It was found that the oxides of the metals having the best potential for reduction of metal into nanoparticles of that metal (Table 8), but nitrate salts are generally used keeping in view of better plant nutrition and solubility.

Table 8: Performance of precursor compounds for biosynthesis of Zn, Mg and Ti nanofertilizers

Metal	Precursor Compound	Avg. nanoparticle size (nm)	PDI*	% nanoparticle yield
Zn	Zinc nitrate	46* ± 1.4	0.122	100.0
	Zinc chloride	74 ± 0.8	0.784	6.5
	Zinc sulphate	88 ± 0.7	1.114	4.1
	Zinc oxide	30 ± 0.6	0.119	100.0
Mg	Magnesium nitrate	15 ± 0.3	0.228	100.0
	Magnesium chloride	69±1.6	0.987	3.8
	Magnesium sulphate	96 ± 2.7	1.418	1.2
	Magnesium oxide	10 ± 0.8	0.236	100.0
Ti	TiO_2 Rutile	17 ± 0.7	0.274	100.0
	TiO_2 Antase	13 ± 0.4	0.261	100.0

*Polydispersity Index; ± = Standard errors of mean

Similar to bacteria, fungi also have an important disadvantage when it comes to safety. Many well studied fungi such, as *F. oxysporum* are pathogenic and therefore might pose a safety risk. For instance, *Trichoderma asperellum* and *Trichoderma reesei* are both fungi that produce Ag nanoparticles when exposed to silver salts and have been proven to be nonpathogenic which makes them ideal for use commercially. In fact, *T. reesei* has already been used widely in sectors, such as food, animal feed, pharmaceuticals, paper and textile industries. But fungi are the favourite choice of microorganisms due to the wide variety of advantages they offer over bacteria, yeast, actinomycetes, plants, and other physico-chemical techniques.

3.1.4.3: Nanoparticle production by actinomycetes

Actinomycetes - mediated biogenic synthesis of metal nanoparticles and their antimicrobial activities are well documented. Actinomycetes facilitate both intracellular and extracellular metal nanoparticles synthesis and are efficient candidates for the production of polydispersed, stable and ultra - small size metal nanoparticles. Secondary metabolites and new chemical entities derived from actinomycetes have not been extensively studied for the synthesis of metal/metal oxide nanoparticles. It has been found that actinomycetes - derived compounds (enzymes, organics acids and bioactive compounds) as metal and metal oxide reducing agents for the synthesis of desired nanoparticles. For example, the formation of silver nanoparticles in the solution of 1 mM silver nitrate and sample of *Streptomyces* sp. was confirmed by change in the colour to yellow after incubation with silver nitrate, while the controls retained the original colour. Nitrate reductase test was confirming the enzyme nitrate reductase present in the extracellular filtrate of the silver nanoparticles synthesized by *Streptomyces* sp.

Actinomycetes are commonly known as filamentous, gram-positive, spore-forming actinobacteria and are considered as both bacteria and fungi. In recent decades, actinomycetes have been used as a resource for the production of bioactive compounds and nanoparticles. Actinomycetes are also rich sources of various enzymes that have industrial importance and of many bioactive compounds with a diverse range of pharmaceutical properties like antibiotics, antimicrobials, antibiotics, etc. *Streptomyces* sp. also found to synthesize silver nanoparticles. Today actinomycetes are also being used in nanotechnology for the synthesis of various nanoparticles used in sustainable agricultural processes as nanofertlizers or biocontrol agents. Using actinomycetes for synthesis of nanoparticles could be advantageous over other environmentally benign biological process of maintaining cell cultures. It can also be suitably scaled up for large scale synthesis of nanoparticles. It has been observed that the extremophilic actinomycete, *Thermomonospora* sp. when exposed to gold ions reduced the metal ions extracellularly, yielding gold nanoparticles with a much improved polydispersity. The antimicrobial activity of silver nitrate and silver nanoparticles synthesized by *Streptomyces* sp were confined against disease causing microorganism. Overall functioning of actinomycetes is shown as Figure 21.

The application of actinomycete (*Streptomyces griseoruber*) mediated extracellular formation of selenium nanoparticles was observed. The biomolecules present in the culture media filtrate produced by actinomycete were responsible for reduction of selenium to generate nanoparticles. The selenium nanoparticles had absorption maxima at 575 nm wavelengths with size ranging from 100–250 nm as revealed through UV-Vis spectrophotometry and TEM investigation. Interestingly, the biologically fabricated selenium nanoparticles had cytotoxic nature against the colon cancer line HT-29. The decreased cellular viability of cell lines with increase in concentration of nanoparticle was documented indicating their exploitation as chemotherapeutic agents.

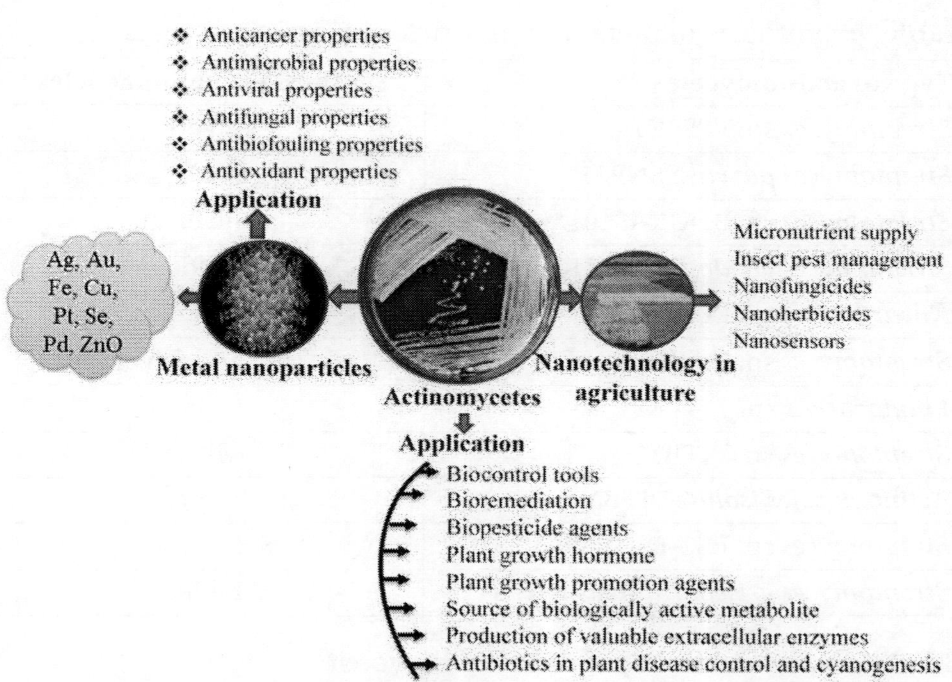

Figure 21: Beneficial role of actinomycetes in different spheres. *Source:* Redrawn from A. Gupta, D. Singh, S. K. Singh, V. K. Singh and A. Kumar (2019). Role of actinomycetes in bioactive and nanoparticles synthesis. In: Role of plant growth promoting microorganisms in sustainable agriculture and nanotechnology, pp. 163–182).

The extracellular synthesis of silver nanoparticles using rhizospheric actinomycete *Streptomyces griseoplanus* SAI25 isolated from rice. The externally produced silver nanoparticles exhibited absorption maxima between 413 and 417 nm. Analytical investigation of cell-free extract through FTIR spectroscopy revealed the presence of biologically important groups such as amine, alcohol, and phenol catalyzing the reduction and thus formation of silver nanoparticles. The actinomycete-induced nanoparticle formation displayed antifungal activity against the plant pathogen *Macrophomina phaseolina* suggesting the application of silver nanoparticles as biopesticides to control the loss of crop productivity. Actinomycetes possess the fascinating potential to generate the nanoparticles of various metals exhibiting antibacterial, antifungal, antioxidant, anticancer, anti-biofouling, and antiparasitic activity. Some actinomycetes as responsible for synthesis of different metal nanoparticles as listed in Table 9.

Table 9: Synthesis of metal nanoparticles by actinomycetes

Type of actinomycetes	Synthesis of nanoparticles
Streptomyces sp. VITPK1	Silver
Streptomyces parvus SSNP11	Silver
Streptomyces sp. BDUKAS10	Silver
Streptomyces viridogens HM10	Gold
Nocardiopsis sp. MBRC-1	Silver
Streptomyces sp.	Zinc and Silver
Rhodococcus sp.	Gold
Streptomyces sp. VITBT7	Silver
Actinomycetes isolate VITBN4	Copper
Streptomyces sp. ES2–5	Selenium
Streptomyces griseobrunneus FSHH12	Selenium

3.1.4.4: Nanoparticle production by yeast

Yeast is easy to control in the laboratory for mass production. They synthesize numerous enzymes rapidly with the use of simple nutrients. In general, the yeast strains possess more benefits over bacteria. Some studies have been conducted to investigate the synthesis of metallic nanoparticles employing the yeast. However, for this aim, through utilizing the eukaryotic systems, namely, *Candida glabrata* and *S. pombe,* one of the primary methods in employing biological material was attained.

The possible applicability of nanoparticles formed by yeast has been shown by a few investigations. For the fabrication of diode cadmium, intracellular synthesized sulfide nanoparticles by *S. pombe* were applied, which had low-voltage operation and high forward current value. It is assumed that these properties can form the artificial structure, a perfect diode. In addition, to synthesize silver and gold nanoparticles, yeast strains have been utilized. It was also reported that with an employment of silver tolerant yeast strain *MKY3*, silver nanoparticles were produced extracellularly, in which the hexagonal silver nanoparticles (2–5 nm) were synthesized in log stage of growth. According to the different warming of the cases, the standardization and documentation of appropriate circumstances for the synthesis of bulky number of silver nanoparticles were also done.

The incubation of *Yarrowia lipolytica* cells was done with changed concentrations of chloroauric acid and formed cell-related gold nanoparticles and nanoplates. Moreover, it is found that the quantity of cells and the utilized salt concentrations can affect the size of nanoparticles. Similarly, to fabricate an air electrode showing outstanding electro catalytic activity for oxygen decrease (ORR), zirconium phosphate with a synthesized mesoporous figuration with the use of yeast as biotemplate was employed. So, as bio-based templates, the synthesis of zinc phosphate nano powders was produced by yeasts. The synthesis of $Zn_3(PO_4)_2$ powders was found in a size ranging from 10–80 nm in width and 80–200 nm in length. Cadmium sulfide nanoparticle synthesis by yeast involves sequestration of Cd^{2+} by glutathione-related peptides followed by reduction within the cell. Since fungi and yeast are very effective secretors of extracellular enzymes and number of species, grow fast and therefore culturing and keeping them in the laboratory are very simple. They are able to produce metal nanoparticles and nanostructure by reducing enzyme intracellularly or extracellularly.

3.1.4.5: Nanoparticle synthesis by plants

Bacteria and fungi have been studied extensively in the past few decades for their ability to synthesise metallic nanoparticles, however there has been less of a focus on plants in this matter. In the past decade, an increasing amount of research is being performed on the green synthesis of metallic nanoparticles using plants or plant extracts. This area is relatively under-explored and offers promising results for the field. A very important aspect of using plants instead of bacteria or fungi for nanoparticles production is the lack of pathogenicity.

An interesting study has illustrated a simple green synthesis route for Ag nanoparticles from $AgNO_3$ salts using the extract from *Jatropha curcas*. The result was the production of fairly homogenous (10–20 nm) Ag nanoparticles in 4 hours. Most bacteria and fungi produce nanoparticles of a much wider size range as described earlier. However, for the synthesis process to occur, the sample needs to be heated to a temperature of 85°C, which would be fairly costly if large volumes are required. Another study, using *Acalypha indica* leaf extracts confirmed that Ag nanoparticles synthesis using plants is possible. The size of the Ag nanoparticles was 20–30 nm, which is again significantly homogenous. It was tested that the synthesised Ag nanoparticles are antimicrobial capacity on *Escherichia coli* and *Vibrio cholera* and observed that the Ag nanoparticles inhibited bacterial growth at a concentration as low as 10 μg/mL. The production of Ag nanoparticles using *Medicago sativa* seed exudates is very popular. Ag^+ reduction occurred almost instantly as Ag nanoparticles were observed within 1 minute of exposure to the silver salt. In less than 50 minutes 90% Ag^+ reduction occurred when the reaction was carried out at 30°C which is significantly lower than the temperature used for *J. curcas*. The nanoparticles produced were spherical, flower-like and/or triangular in shape with a size range of 5–108 nm. In contrast to another study, not only were the purified Ag nanoparticles from *M. sativa* heterogenous in size but also, they did not show significant inhibition of bacterial growth. However, it was suggested that they could act synergistically with the seed exudate in order to eliminate bacteria. Another plant that has the potential to reduce Ag^+ is *Ocimum sanctum*. Its leaf extract was mixed with 1 mM $AgNO_3$ and the production of Ag nanoparticles of 3–20 nm in size was observed. The particles were spherical in shape and it was suggested that they may be stabilised by a component of the leaf broth.

In another study, phyllanthin was extracted from the plant *Phyllanthus amarus* and subsequently used for the production of Ag and Au nanoparticles. This is a unique study as only a single constituent of a plant extract was used for the synthesis of metallic nanoparticles, contrasting other studies in which whole plants or extracts were used. Phyllanthin concentrations played a key role in the size and shape of the nanoparticle produced. Low concentrations resulted in the slow formation of triangular and hexagonal Au nanoparticles and higher concentrations of phyllanthin gave rise to greater levels of spherical nanoparticles which was confirmed by UV-Visible and TEM analysis.

It was talked about the use of plant derived polysaccharides and phytochemicals which can be used for the synthesis of gold and silver nanoparticles. The use of such compounds for nanoparticle synthesis offers advantages, such as decreased use of toxic chemicals and the ability of creating nanocomposites with different metals. Soluble starch, chitosan, cellulose, dextran, alginic acid and hyaluronic acid have been used for the production of silver and gold nanoparticles successfully.

An extensive study was also done by the production of Ag nanoparticles from a number of different plant leaf extracts. It was examined the use of Pine, Persimmon, Ginkgo, Magnolia and Platanus extracts and compared their ability to produce Ag nanoparticles. The Magnolia leaf broth was found to be the best Ag^+ reducer as it took only 11 minutes to reduce 90% of the Ag^+ present in the sample. In this example, once again, reaction conditions such as temperature needed to be controlled closely during the synthesis stage as they affected the rate, size and shape of the nanoparticles. The effect of different reaction temperatures was tested and found that at 95°C Ag^+ conversion was much higher than at 25°C. However, the size of the nanoparticles decreased at higher temperatures. It was hypothesised that this was caused by the increased turnover of the nanoparticles by the reducing agent, as it has less time to build upon the presynthesised nanoparticles before starting to synthesise new ones.

Extracellular synthesis of Au nanoparticles was illustrated using extracts from a lemongrass plant, *Cymbopogon flexuosus*. Liquid-like nanotriangles formed by aggregates of spherical Au nanoparticles were obtained when the extract was incubated with gold tetrachloride ($AuCl_4^-$). This fluidity was attributed to the nanoparticle surface complexation of aldehydes and/or ketones that were present in the plant extract. An additional study showed the synthesis of Au, Ag and bimetallic Au- Ag nanoparticles. The leaf broth of a Neem plant, *Azadirachta indica* was mixed with the salts of Au, Ag and then both metal ions simultaneously. The production was rapid with the synthesis rate starting to plateau at approximately 2 hours. It is believed that the torpedoed and flavanone components of the leaf broth may keep the nanoparticles stable.

It was attempted to create Pt nanoparticles using the leaf extract of *Diospyros kaki*. A greater than 90% reduction of Pt ions into nanoparticles was illustrated in approximately 2.5 hours. It was noted that the reduction of the Pt ions was due to the presence of functional groups within the leaf extract, such as amines, alcohols, ketones and carboxylic acids; as opposed to an enzyme mediated process. This was based on the fact that the reaction temperature was 95°C and no protein peaks were found in their FTIR analysis.

The synthesis of metallic nanoparticles by plants generally involved using part or whole plant extracts, though it is also possible to synthesise metallic nanoparticles inside live growing plants. However there were no reports where plants have been used as the biological producers of Pd nanoparticles until Parker *et al.* (2014) presented a method of creating such nanoparticles using the plant, *Arabidopsis thaliana*. After growing *Arabidopsis*, its medium was then replaced with potassium tetrachloropalladate (K_2PdCl_4) and incubated for 24 hours in the salt solution. Subsequent TEM results visualised the Pd nanoparticles produced were at 2–4 nm size range. The produced Pd nanoparticles were subsequently used in Suzuki-Miyaura coupling reactions where the biogenic Pd nanoparticles were shown to convey a higher catalytic activity that commercially available Pd nanoparticles. Moreover, the synthesis of Au nanoparticles, also by live plants may possible. Alfalfa plant seeds were prepared and grown for two weeks with various concentrations of $K(AuCl_4)$. The nanoparticles were then extracted and analyzed by various methods which confirmed the production of elemental Au nanoparticles. Although an interesting study, the nanoparticle synthesis time was long, over two weeks including the nanoparticle extraction. In order for a biological process for nanoparticle synthesis to be commercially feasible, the production time would need to be shortened significantly. On the other hand, this would be a cheap and green method for synthesising such nanoparticles.

Phytonanotechnology has provided new avenues for the synthesis of nanoparticles and is an ecofriendly, simple, rapid, stable, and cost-effective method. Phytonanotechnology has advantages, including biocompatibility, scalability, and the medical and agricultural applicability of synthesising nanoparticles using the universal solvent, water, as a reducing medium. Thus, plant derived nanoparticles produced by readily available plant materials. The nontoxic nature of plants are suitable for fulfilling the high demand for nanoparticles with applications in the biomedical, fertilizer and environmental areas. Successfully synthesised gold and silver nanoparticles using the leaf and root extract from the medicinal herbal plant *Panax ginseng* suggested the use of medical plants as resources. Additionally, various plant parts, including leaves, fruits, stems, roots, and their extracts, have been used for the synthesis of metal nanoparticles (Table 10) .

Table 10: Synthesis of biological nanoparticles from plants

Plant	Tissues for extraction	Types of nano-particles	Shapes	Size (nm)	Application
Euphorbia prostrata	Leaves	Silver and TiO_2	Spherical	Silver 10–15 TiO_2 82–85	Fertilizer

Plant	Tissues for extraction	Types of nano-particles	Shapes	Size (nm)	Application
Ginkgo biloba	Leaves	Copper	Spherical	15–20	Catalytic, Fertilizer
Panax ginseng	Root	Silver and gold	Spherical	Silver 10–30 Gold 10–40	Antibacterial
Red ginseng	Root	Silver	Spherical	10–30	Fertilizer and antibacterial
Cymbopogon citrates	Leaves	Gold	Spherical, triangular, hexagonal, rod	20–50	Mosquitocidal
Azadirachta indica	Leaves	Silver	Spherical	41- 60	Fertilizer, pesticides
Nigella sativa	Leaves	Silver	Spherical	15	Cytotoxicity
Cocus nucifera	Leaves	Lead	Spherical	47	Antibacterial and photocatalytic
Catharanthu s roseus	Leaves	Palladium	Spherical	40	Catalytic activity
Pistacia atlantica	Seeds	Silver	Spherical	27	Antibacterial
Banana	Peel	Cadmium sulfide	Spherical, round	1- 48	Biological
Nyctanthes arbortristis	Flower	Silver	Spherical	25-35	Antibacterial and cytotoxic
Anogeissus latifolia	Gum powder	Silver	Spherical	5.5-5.9	Antibacterial
Abutilon indicum	Leaves	Silver	Spherical	5-25	Antibacterial
Pinus densiflora	Cones	Silver	Oval, triangular	30-80	Antimicrobial
Artocarpus gomezianus	Fruit	Zinc	Spherical	20-50	Fertilizer, antioxidant

Plant	Tissues for extraction	Types of nano-particles	Shapes	Size (nm)	Application
Citrus medica	Fruit	Copper	Spherical	20	Fertilizer, antimicrobial
Orange and pineapple	Fruits	Silver	Spherical	10-30	Antimicrobial
Lawsonia inermis	Leaves	Iron	Hexagonal	21	Fertilizer
Gardenia jasminoides	Leaves	Iron	Hexagonal	32	Fertilizer, antibacterial

The exact mechanism and the components responsible for plant-mediated synthetic nanoparticles remain to be elucidated. It has been proposed that proteins, amino acids, organic acid, vitamins, as well as secondary metabolites, such as flavonoids, alkaloids, polyphenols, terpenoids, heterocyclic compounds, and polysaccharides, have significant roles in metal salt reduction and, furthermore, act as capping and stabilizing agents for synthesized nanoparticles . For instance, the hydroxyl functional group from polyphenols and the carbonyl group from proteins of *Corallina officinalis* extract could assist in forming and stabilizing gold nanoparticles. The synthesis and stabilization of silver and gold nanoparticles by biomolecule attachment in *Murraya koenigii* leaf extract. Reports also suggest that different mechanisms for synthesising nanoparticles exist in different plant species. For instance, specific components, such as emodin, a purgative resin with quinone compounds that is present in xerophytes plants (plants adapted to survive in deserts or environments with little water) are responsible for silver nanoparticle synthesis; Cyperus quinone, dietched quinone, and remirin in mesophytic plants (terrestrial plants adapted to neither a particularly dry nor particularly wet environment) are useful for metal nanoparticle synthesis. Eugenol, the main terpenoid of *Cinnamomum zeylanicum,* was found to have a principal role in the synthesis of gold and silver nanoparticles. Notably, dicot plants contain many secondary metabolites that may be suitable for nanoparticle synthesis (Table 10).

Further studies

Akhtar, M. S., Panwar, J. and Yun, Y. S. (2013). Biogenic synthesis of metallic nanoparticles by plant extracts. ACS Sustain. *Chem. Eng.* **1:** 591–602.

Ealias, A. M. and Saravansakumar, M. P. (2017). A review on the classification, characterization, synthesis of nanoparticles and their application. IOP Conference Series: Mat. Sci. Eng. 263, 032019, doi: 10.10sc/1757–899X/263 /3/032019.

Li, X., Xu, H., Chen, Z. S. and Chen, G. (2011). Biosynthesis of nanoparticles by microorganisms and their applications. *J. Nanomater.* **11:** 1–16.

Satyanarayan, T. and Reddy, S. S. (2018). A review on chemical and physical synthesis methods of nanoparticles. *Int. J. Res. Appl. Sci. Eng. Tech.* **6:** 2885–2889.

Winterer, M, Hahn, H. and Metallked, Z. (2003). Chemical vapour synthesis of nanocrystalline powder. *Chem. Vapour Syn.* **94:** 1084–1090.

4

Biosynthesis of Liquid Nanonutrients

Liquid nanonutrients can be prepared either by microbial technique or by plasma techniue. An essential feature of their synthesis is the preparation of the particles of specified size and shape. The synthetic procedure should be relatively simple, inexpensive and reproducible.

4.1: Microbial Technique

Living organisms have huge potential for production of different nanofertilizers of wide applications. By using the organisms from simple bacteria to highly complex eukaryotes in the reaction mixture, the production of nanoparticles/nanodevices with desired shape and sizes can be obtained. Synthesis of nanoparticles using biological entities has great interest due to their unusal optical, chemical, photoeletrochemical and electronic properties. Liquid nanonutrients were biosynthesised after isolating suitable microorganisms from the native soil and using their spore for preparation of microbial balls (fungi as an example). The fungal balls were allowed to release the enzymes/proteins in the water bodies which were then used to breakdown the desired salts into nano (Figure 22).

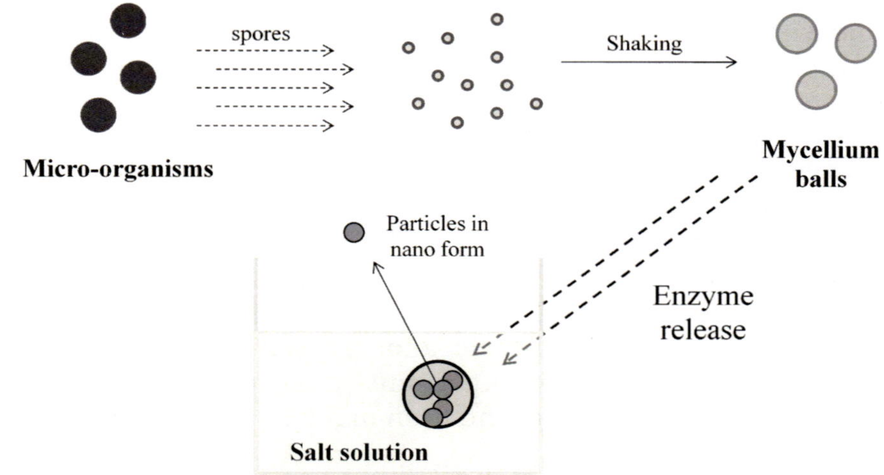

Figure 22: Biosynthesis process for preparation of liquid nanofertilizers

The environmentally-friendly microorganisms could minimize the toxicity in the process of nanofertilizer production (Figure 23) by reduction of the metal ions or by formation of insoluble complexes with metal ions. Liquid culture nanoparticles can be prepared by selecting microorganisms to grow in their preferred salt solution. The nanoparticles prepared in solution can easily be make solid or may be kept as liquid as shown below.

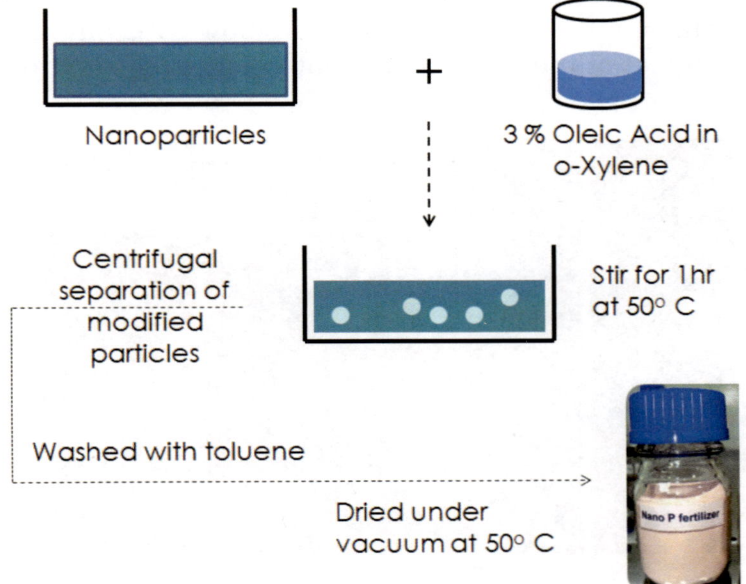

Figure 23: Preperation of solid phosphorus nanofertilizer from liquid phosphorus nanosolutions

Biosynthesised nanofertilizers are relatively stable and ecofriendly as they are naturally encapsulated by mother protein. The protein layers formed around the nanoparticle surfaces create a protective shell for protection and stabilization of the particle. Encapsulation of the nanoparticles makes them stable against oxidation, corrosion and spontaneous aggregation, which allow them to retain a single domain structure. These environmentally-friendly biological systems may be considered as benign nanofactories. The synthesis and assembly of nanoparticles would benefit from the development of clean, non-toxic and environmentally acceptable green procedure; probably organisms involved in this process ranging from bacteria to fungi and even plants.

In general, 15 days old fungal culture was used for the process. The salt concentration was found best between 0.1 and 0.5mM. The reaction period varies between 48–72 hours with a temperature 28°C in a shaking speed of 60–90 rpm. The best pH for production of liquid nanonutrients is between 5.5 and 7.0.

The advantage of the process is the nanoparticles after production are naturally encapsulated by mother protein (Figure 24) which helps them to keep the particle under stable condition with the nano-size. The fungal protein to salt solution used for nanoparticles production is generally optimum for 1:1 ratio. The particles are generally stable for 1–2 years if it stored below 35°C but the stability varies in respect of the pH and zeta potential of the solution (Figure 25).

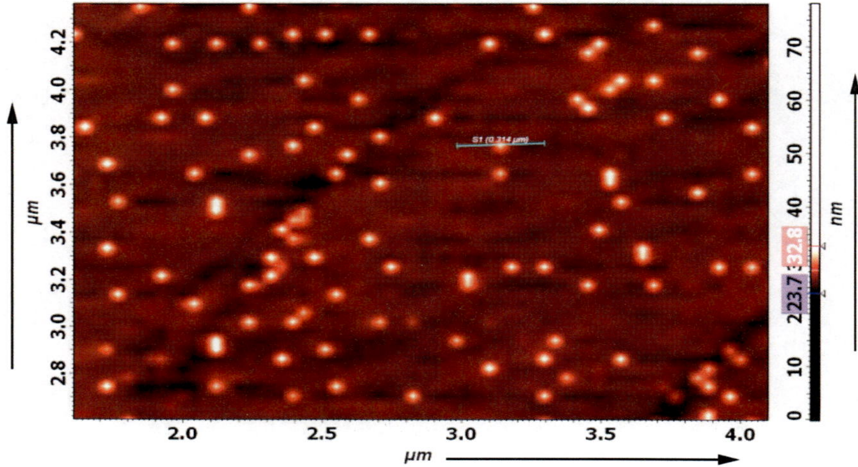

Figure 24: TEM image of nanonutrients surrounded by naturally encapsulated mother protein.

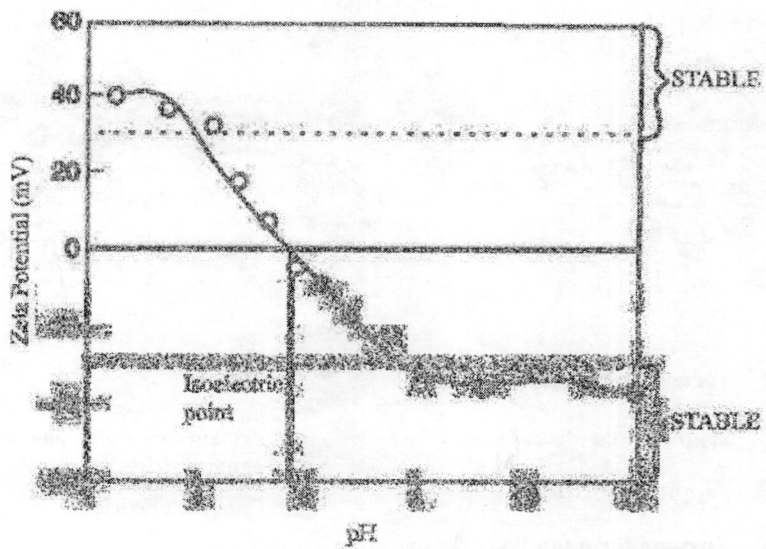

Figure 25: Stability of nanoparicles in suspension varies with respect to pH and zeta potential

Types of bio-fertilizers such as *Rhizobium*, *Azospirillum*, Cyanobacteria, Azolla, *Azotobacter* and *Acetobaceter* as well act as P, K and Zn solubilizer microbes. They are the most important microbial traits in soil biogeochemical cycles and plant growth promotion. The biosynthesis of nanomaterials using bacteria, algae, yeast, fungi, actinomycetes and plants has led to a new area of research for the formation of inorganic nanoparticles as eco-friendly fertilizers. For example, a proposed microbial synthesis through biological route for P nanoparticles is shown as figure 26.

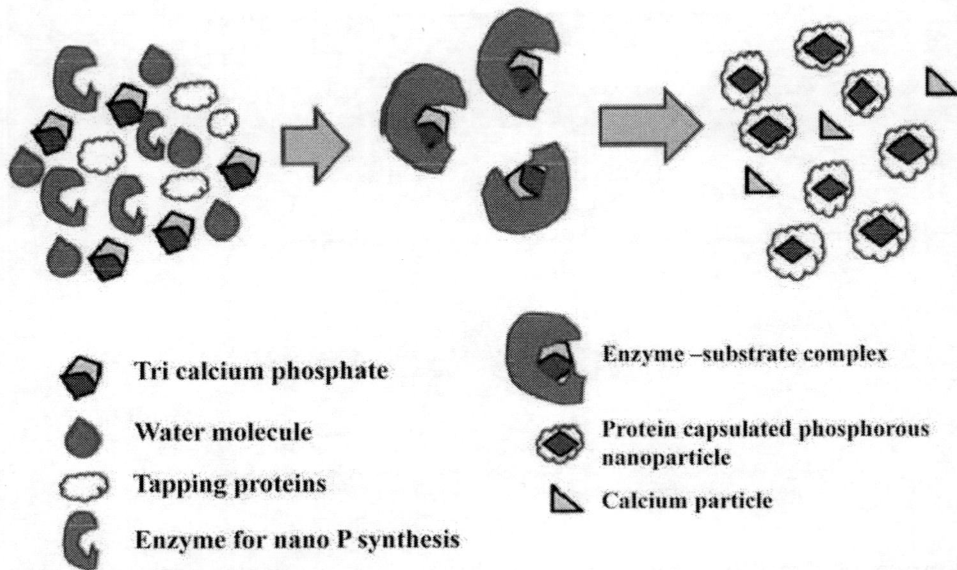

Figure 26: Microbial synthesis of phosphorous nanoparticles using *Aspergillus tubingensis* TFR-5 from tri calcium phosphate.

Now, use of nanofertilizers in plant nutrition is one of the major roles of nanotechnology in agriculture and soil sciences. Thus, the application of biosynthesized nanoparticles in agricultural sector may lead to sustainable development. Hence, this leads to sustainable agriculture through putting less inputs and generating less wastes, minimizing nutrient losses, and release nutrients at a proper rate for plant demand comparing with conventional farming. There is very minor difference between bio-and nano-fertilizers depending on their mechanisms in the soil and plant, application methods, effective rates as well as their impact on the environment. However, both nano-and bio-fertilizers and their interaction are needed further studies. It was found that both unicellular and multicellular organisms produce inorganic nanoparticles, either intra- or extracellularly.

Although a lot of reports to the synthesis of nanoparticles using the biological route is available, but there was not enough data for understanding the exact mechanism for the same. In general, the interpretation is the involvement of proteins like enzymes and cofactors that has redox potential as well as act as electron shuttles play a key role in metal reduction. In recent study, it was proposed that di-picolcomic acid moiety as main mechanism for production of silver nanoparticles by *Bacillus stratophericus* spores. The SDS-PAGE profile study revealed presence of two extracellular proteins of 32 and 35 kDa group found to be responsible for synthesis and stability of silver nanoparticles. In order to investigate the mechanism behind the formation of silver nanoparticles using *Syzygium cumini* fruit extract, different fractions employing water, hexane, ethyl acetate, chloroform and dichloromethane were extracted and evaluated. It was found that the members of flavonoids family (molecules F_1 and F_2) as major components mainly responsible for the synthesis of nanoparticles. It was hypothesized the mechanism showing the reduction of cationic silver to metallic silver by the flavonoid compounds through redox reaction.

4.2: Plasma Technique

Among the various methods available today for nanoparticle synthesis, techniques for plasma generation in liquid are relatively new. Various types of plasma, such as arc discharge and glow discharge can be applied to produce metal, alloy, oxide, inorganic, carbonaceous, and composite nanomaterials. Many experimental set-ups have been reported, in which various parameters, such as the liquid, electrode material, electrode configuration, and electric power source are varied. All available plasma in liquid set-ups into four main groups: (i) gas discharge between an electrode and the electrolyte surface, (ii) direct discharge between two electrodes, (iii) contact discharge between an electrode and the surface of surrounding electrolyte, and (iv) radio frequency and microwave plasma in liquid. Nanomaterials of metal, alloy, oxide, silicon, carbon, and composite produced by techniques for plasma generation in liquid are available, where the source materials, reaction media, and electrode configurations are discussed in detail. By examining electrode configurations and power sources, plasma in liquid generation can be sub-divided into four main groups as stated above.

Metal nanoparticles containing Au, Ag, Pd, and Pt were synthesised using various types of plasma techniques. Size control of the synthesised nanoparticles is the main subject of this category. In most cases, spherical nanoparticles with diameter less than 10 nm are produced. In other instances, nanorods and polygonal nanoparticles are generated instead. The ultraviolet-visible (UV-Vis) spectra of nanoparticles dispersed in solution were measured due to their observable surface plasmon resonances; it is well known that the resonance wavelength varies with size and shape of the particle and hence this method of analysis can be used as another way to confirm a desired synthesis. To supply the source materials, two methods are commonly used: metallic electrodes and ions of chlorides or nitrates. Nanoparticle synthesis using metallic electrodes is a surfactant-free and high-purity method; however, size control of the synthesised particles is difficult in this technique. Conversely, the plasma reduction of metal ions with a surfactant allows for synthesis with a high degree of size control. Recently, this technology has extended to the field of noble metal alloy and composite nanoparticles such as Pt supported on carbon.

In spite of their instability at high temperatures, nanoparticles of Ni, Cu, Zn, and Sn have been synthesised by plasma in a solution. Generally, these materials can react with water vapour, which causes them to undergo oxidation. However, in the plasma solution the short reaction time and cooling effect of surrounding water might prevent the produced nanoparticles from undergoing oxidation. Additionally, experimental conditions, such as electrode temperature, electrolyte additives, and solution pH were carefully optimized to fabricate metal nanoparticles. For producing Cu nanoparticles, surfactant of cetyl trimethyl ammonium bromide (CTAB) and ascorbic acid as the reducing agent were added to the solution. Gelatin and ascorbic acid were selected as the capping agents to protect the particles against coalescence and oxidation side reactions. In addition, Cu nanoparticles were formed in a citrate buffer, where Cu_2O was stable at low concentrations in a K_2CO_3 electrolyte (0.001 M); the clear formation of CuO was observed with increasing K_2CO_3 electrolyte concentration (0.01–0.5 M). Compared to the afore mentioned materials, Al, Ti, Fe, and Zr are more active and undergo oxidation more readily. Therefore, a solution-based synthesis has not been reported for these metals, and molten salts are most frequently used to synthesise them as these salts do not contain oxygen.

Techniques for plasma generation in liquid have been used to synthesise alloy nanoparticles with unique properties suitable for many applications. Noble bimetallic alloy nanoparticles have been investigated for plasmonics related applications, catalysis, and biosensing, utilizing properties that can be tuned by changing the composition. Similar to noble-metal nanoparticle synthesis, metal ions in solution or solid electrodes were supplied as raw materials. In the case of the solid electrodes, an alloy electrode and a pair of two different pure-metallic electrodes were used. Magnetic nanoparticles of Co-Pt, Fe-Pt, and Sn-Co have also been synthesised using plasma in molten salt for use in ultra-high density hard drives, bioseparations, and sensors applications. When oxide nanomaterials were synthesised via techniques for plasma generation in liquid, the solid electrode was consumed under high-temperature plasma conditions (such as arc discharge), followed by the subsequent reaction of produced nanoparticles or generated metal vapour with the surrounding electrolyte to synthesise oxide nanoparticles. During the oxide precipitation, the surfactant and liquid temperature also affected the final morphology and composition of the synthesised nanomaterials. Since the products are quickly synthesised and immediately cooled down in the plasma process in liquid, the synthesised oxide crystals tend to have lower crystallinity. Sometimes, metastable phases and defect structures were observed after synthesis.

A wide variety of carbon nanomaterials have been synthesised using plasma in liquid, in which the graphite electrode was consumed during arc discharge. The organic solvent was also used as a carbon source. Under high temperature plasma conditions, carbon atom sublimation and subsequent reaggregation into solid carbon have been shown to occur. When catalytic particles such as Ni, Fe, and Co are present reaction field, carbon nanotubes (CNT) generation can be enhanced as well. The decomposition of organic solvents such as toluene, ethanol, and butanol are also used for synthesizing carbon nanomaterials. For toluene in particular, the decomposition is expected to occur uniformly because of its symmetric structure. During plasma generation in alcohol, 2,3-butanediol, phenylethylene, indene, naphthalene, and biphenylene can be formed as by-products.

4.3: Stability of Nanoparticles

The stabilization of nanofertilizer particle in suspension is very important for both controlling the particle size and for developing process based application of these suspension to achieve a desired result. The formation, stabilization and sedimentation of nanoparticles depend upon the discreet steps of nucleation, condensation and coagulation into larger particles. Therefore, stabilization requires the optimization of these competing factors. If the size of the particles is reduced to the nano level, the ratio of surface to the bulk atoms increases, thereby increasing energy of the system as a whole, leading to a decrease in the system stability. A number of laboratory studies indicate that nanoparticles aggregate to some extent following their release to water. Therefore, it is necessary to get stable suspension of nanoparticles of desired size for harnessing the benefit of nanoparticles as nanofertilizer. It has been found that the negative charged surface of the particles had strong affinity towards water and the hydrated radii of the particle is much higher than their unhydrated counterparts.

The zeta potential of the nanoparticles is another important parameter that has been extensively investigated for stability of nanoparticles (Table 11). High zeta potential (negative or positive), preferably above 30 mV, will impart stability of nanofertilizer in suspension, whereas nanoparticles with low zeta potentials tend to coagulate, flocculate or aggregate. pH is a measure factor in determining the zeta potential of nanoparticles. For example, if the pH is at point of zero charge (pzc) or isoelectric point, the nanoparticle exhibits minimum stability. When the pH is lower than the pzc value, the nanoparticles surface is positively charged and the zeta potential will increase with decreasing pH value below the pzc. Conversely, at pH above pzc, the surface is negatively charged and the zeta potential will be more negative with increasing pH.

Table 11: Effect of zeta potential on stability of nanoparticle

Zeta potential (mV) either positive or negative	Stability behaviour
< 5	Rapid aggregation or coagulation
5–10	Unstable
10–30	Incipient instability
30–40	Moderate stability
40–60	Good stability
>60	Excellent stability

The ionic strength of the surrounding aqueous medium is also important factor affecting the stability of the nanoparticles. For example, if ionic strength of the surrounding aqueous medium is increased and/or the zeta potential is reduced (both effects usually result from an increased salt concentration), attractive force between colloids will outweigh the repulsion, and the particles can then adhere each time they collide. The surface charge is also another important property that can also influence the stability. Nanofertilizers with higher negative charges are believed more mobile in the soil matrix because of the longer electrostatic repulsion between the nanofertilizer particles and soil particles and between nanoparticles themselves as well.

Further studies

Ashkarran, A. A. (2011). Metal and metal oxide nanostructures prepared by electrical arc discharge method in liquids. *J. Cluster. Sci.* **22:** 233-266.

Baba, K. Kaneko, T., Hatakeyama, R., Motomiya, K. and Tohji, K. (2010). Synthesis of monodispersed nanoparticles functionalized carbon nanotubes in plasma-ionic liquid interfacial fields. *Chem. Comm.* **46:** 255–257.

Hu, X., Zhang, X., Shen, X., Li, H., Takai, O. and Saito, N. (2014). Plasma-induced synthesis of CuO nanofibres and ZnO nanoflowers in water. *Plasma Chem. Plasma Process.* **34:** 1129–1139.

Kulkarni, N. and Muddapur, U. (2014). Biosynthesis of metal nanoparticles: A review. *J. Nanotech.* 2014, Article ID 510246, 8 pages.

Tarafdar, J. C. and Rathore, I. (2016). Microbial synthesis of nanoparticles for use in agriculture ecosystem. In: Microbes for plant stress management, Bagyaraj, D. J. and Jamaluddin (eds.)., New India Publishing Agency, New Delhi, India, pp. 105–118.

5

Factors Affecting Nanofertilizer Production

There are many factors which affect the nanofertilizer production; the production of nanofertilizer depends on all the following factors stated below.

5.1: pH

If it is chemical synthesis then during the preparation of nanoparticles the pH may alter the shape and size of chemical synthesised nanoparticles. However, because certain chemical reactions are pH dependent and growth of nanoparticles vary according to pH. Generally, in basic medium rate of reaction is higher and it can affect the size. The pH plays an important role in nanoparticle synthesis by inducing the reactivity of extract with respective salt ions. It has been found that alkaline pH 8.2 showed a sharp peak at 460 nm with maximum production of silver nanoparticles. The sharp peak indicated the formation of spherical shape of silver nanoparticles. Thus indicating alkaline pH is more suitable for synthesis of silver nanoparticles. In general, the pH 5.5 was found most suitable for most of the nanoparticle production biologically. The stability of nanofertilizer particles in the solution depends upon the pH.

To standardize the pH of the reaction mixture of precursor compounds with extracellular enzyme secretes, only the best precursor compound of each metal (ZnO for Zn; MgO for Mg; TiO_2 antase for Ti) was tested at various pH levels ranging from 4 to 8 with an increment of 0.5. The average nanoparticle size was reduced with increase in pH from 4.0 to 5.5, which was further enhanced up to pH 8.0 (Table 12). In general, the pH 5.5 was found the most suitable not only for accelerating the rate of reaction but also to substantially reduce nanoparticle size as compared to other pH levels. The least nanoparticle size of 8 nm was recorded using MgO for magnesium nanoparticles followed by 11 nm by TiO_2 (antase) for Titanium and 26 nm by ZnO for zinc metal nanoparticles at 5.5 pH.

Table 12: Average distribution of Zn, Mg and Ti nanoparticle at different pH of the reaction medium

Precursor compound	pH								
	4.0	4.5	5.0	5.5	6.0	6.5	7.0	7.5	8.0
ZnO	45 ± 0.3*	30 ± 0.4	28 ± 0.2	26 ± 0.1	30 ± 0.2	38 ± 0.2	59 ± 0.3	63 ± 0.1	98 ± 0.8
MgO	35 ± 0.3	30 ± 0.2	12 ± 0.1	8 ± 0.1	14 ± 0.2	25 ± 0.3	35 ± 0.4	54 ± 0.7	72 ± 0.7
TiO_2 (Antase)	39 ± 0.4	21 ± 0.2	12 ± 0.3	11 ± 0.1	24 ± 0.1	37 ± 0.2	62 ± 0.2	83 ± 0.6	98 ± 0.9

*Particle size in nm; ± =Standard errors of mean

This pH not only for accelerating the rate of reaction but also substantially reduces nanoparticle size as compare to other pH level. The least nanoparticle size was recorded at pH 5.5, where average nanoparticle size of 8.0 nm of MgO and 11.0 nm of TiO2 and 26.0 nm of ZnO were noticed (Figure 27).

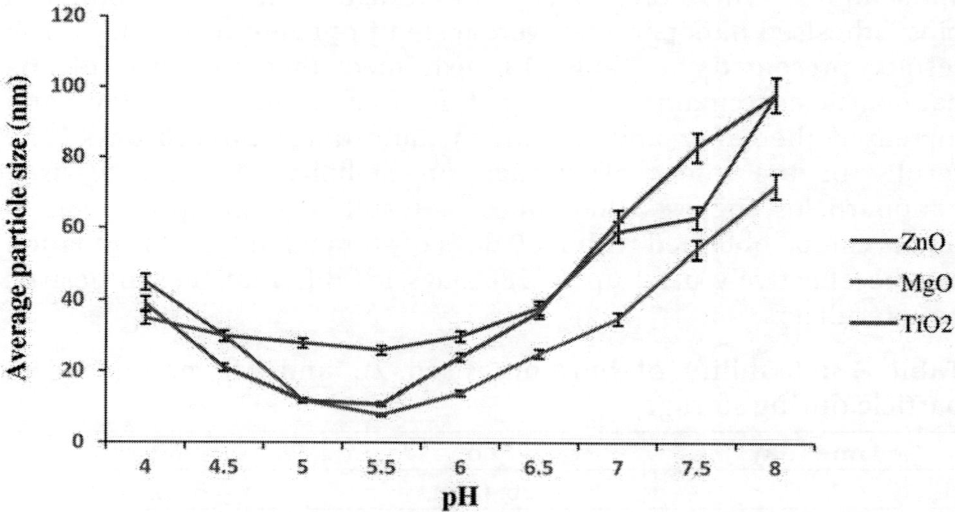

Figure 27: Effect of pH on nanoparticle production

5.2: Shaking Speed

Low speed is always better. The best speed for production of nanoparticles was observed between 70 and 125 rpm of different nanonutrients. In general, 80 rpm is the most preferable shaking speed for best yield of nanoparticles.

5.3: Reaction Time

It depends on the type of nanoparticles synthesising. In general, 12–72 h time is needed for the production of different nano-nutrients. On an average 24 h is the optimum reaction time for nanoparticles production biologically. The optimization of reaction period of precursor compound and extracellular enzyme secrets for biosynthesised of nanoparticles revealed that very few nanoparticles could produce up to incubation period of the reaction. In general, most of the biosynthesis of nanoparticle detected after 36 hours of incubation and the size stabilizes at 48 hours in most of cases. It was also observed that further incubation beyond 48 h, up to 120 h neither reduces the size of nanoparticles nor was economical to harvest appreciable yield. Over all 48 h was found to the optimum time for biosynthesis of reaction period. The shaking speed for nanofertilizer production during biosynthesis process over optimum time was found between 70–80 rpm.

Once the nanoparticle synthesised in solution, agglomeration is a recurrent problem with storage and adversely affect field applications. Therefore, effect of storage time on stability of biosynthesised nanoparticles were tested from zero to 180 days. The results presented in Table 13, exhibited that the size of the nanoparticles remains unchanged from zero to seven days and increased thereafter and remain in nano-range till 180 days. The results of the effect of storage on stability of biosynthesised nanoparticles suggests that the best result of their application to plants can be obtained within 30 days of storage, but the formulation can be effectively used up to 180 days for different biosynthesised nanofertilizers.

Table 13: Stability of biosynthesised Zn and Mg nanofertilizer particle during storage

Time (days)	Zn	Mg
0	26.4 nm	8.1 nm
7	26.8 nm	8.3 nm
15	27.1 nm	8.8 nm
30	28.9 nm	14.6 nm
60	36.8 nm	22.6 nm
90	56.4 nm	56.3 nm

Time (days)	Zn	Mg
120	76.3 nm	68.4 nm
150	88.4 nm	79.1 nm
180	100.2 nm	94.5 nm

5.4: Protein Quality

It depends on the source of protein and the microorganisms involved. 32 kDa group of protein was found to be most efficient for synthesis of most of the nano-nutrients. However, 33–35 kDa protein groups also reported to be useful for some of the nanonutrients production.

5.5: Ratio of Extract: Salt Solution

Generally 1:1 ratio between extract and salt solution was found to be the best for smallest nanoparticles production and encapsulation by microbial protein to each particle. In general, nanoparticle yield was found to be more at 1: 1 protein: salt ratio as compared to 1:2, 2:1 and 1:1.5 protein salt concentrations tried for all the biosynthesised nanofertilizers (Figure 28). The reasons may be free salt particle at 1:2 ratio, which helps in agglomerating themselves due to lesser availability of protein enabling less nanoparticles production. Similarly, at 2:1 and 1:1.5 levels, there may be more free proteins in the solution resulting in fewer yields of particles in nano form. From the present results, it can be inferred that 1:1 protein: salt ratio is the ideal for biosynthesis of different nanoparticles from the respective salts.

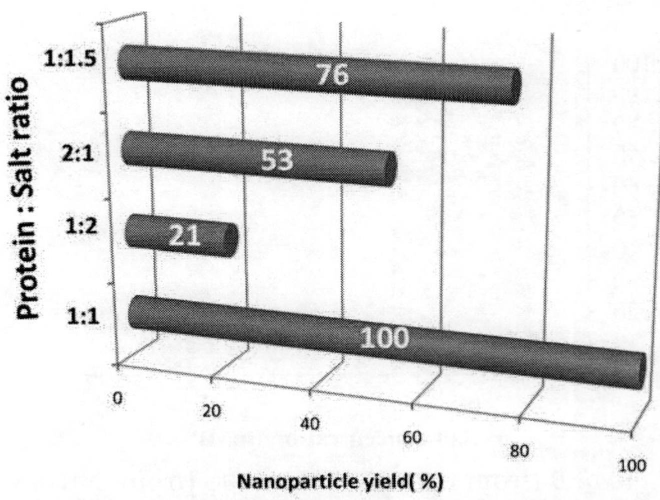

Figure 28: Effect of fungal protein on salt concentration for nanofertilizer production

5.6: Salt Concentration

Increasing the concentration of salts from 0.01–5.0 mM in reaction mixture, revealed that in 0.1 to 0.5 mM concentration of salts yield more nanoparticles and size reduction starts quickly due to more availability of functional groups in the extract. An increasing the substrate concentration further, the large size and aggregation of nanoparticles occurred due to the occurrence of competition between respective salt ions and functional group. Figure 29 and 30 presented that yield of different nanofertilizers vary with the variation of the salt concentration.

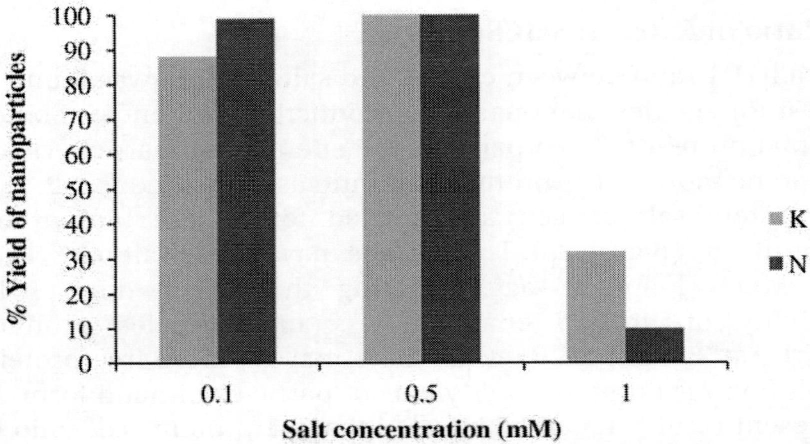

Figure 29: Yield of N (from NH_4NO_3 salt) and K (from KNO_3 salt) nano-particles with different concentration of salt at similar incubation time

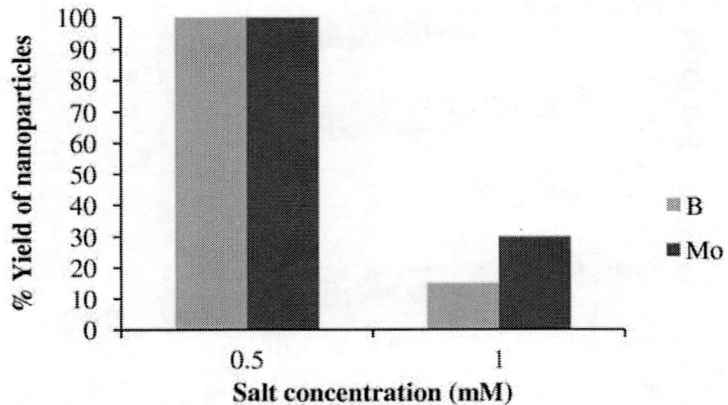

Figure 30: Yield of B (from H_3BO_3 salt) and Mo (from $(NH_4)_6Mo_7O_{24}$ salt) nanoparticles with different concentration of salt at similar incubation time

Different salts were tested for preparation of desired nanofertilizers. The concentration for production of different nanoparticles and their optimum pH, temperature were standardized. The proteins responsible for the production of nanoparticles were identified for important particles used for plant nutrition. A list of compounds used for different nanofertilizer production by the action of microbial protein is listed in Table 14.

Table 14: Effect of different compounds on yield of nano-particles from microbial protein

Metal	Concentration	Compound
Mg	0.1 mM	MgO, $MgNO_3$, $MgSO_4$
Zn	0.1 mM	ZnO, $ZnCl_2$, $ZnSO_4$, $ZnNO_3$
Fe	0.1 mM	Fe_2O_3,$FeCl_3$, $K_3Fe(CN_6)_6$
P	0.1 mM	$Ca_3(PO_4)_2$, phytin
Ti	0.1 mM	TiO_2
Ag	0.1 mM	$AgNO_3$
B	0.1 mM	H_3BO_3
K	0.1 mM	KNO_3
N	0.1 mM	$NH_4(NO_3)_2$
Pt	0.1 mM	$Pt(OH)_2$

5.7: Temperature

A range of teperature was tested for production of nanofertilizer during biosynthesis. In general, 28°C was recorded for smallest size of nanoparticle production. The average distribution with regard to size of Zn, Mg, Ti nanoparticles increased when temperature deviated from 28°C (Figure 31). The optimum temperature between 28 and 30°C for biosynthesis of metal nanoparticles was also reported.

In general, high temperature is conducive to nucleation while low temperature is conducive to growth. It is observed that the total reaction rate is increased with the increasing reactive temperature. To standardize the temperature for optimum production of different nanoparticles, various temperatures between 20 and 40°C were tested. The maximum production of Zn, Mg and Ti nanoparticles were recorded at 28°C temperature (Table 15). The average distribution with regards to size of Zn, Mg and Ti nanoparticles increased when temperature deviated from 28°C.

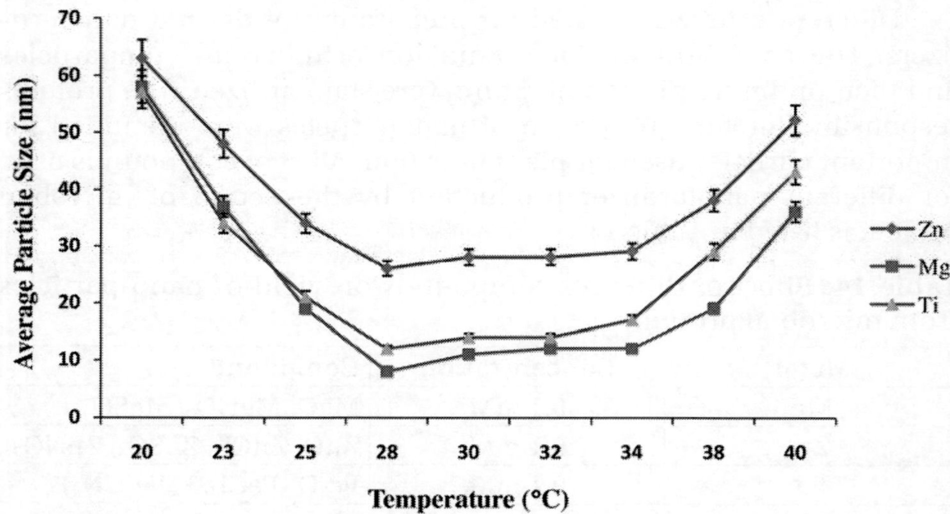

Figure 31: Effect of Temperature on biosynthesise of nanofertilizer production

Table 15: Effect of temperature for biological production of metal nanoparticles

Precursor Compound	Temperature (°C)									
	20	**23**	**25**	**28**	**30**	**32**	**34**	**38**	**40**	**20**
ZnO	63 nm	48 nm	34 nm	26 nm	28 nm	28 nm	29 nm	38 nm	52 nm	63 nm
MgO	58 nm	37 nm	19 nm	8 nm	11 nm	12 nm	12 nm	19 nm	36 nm	58 nm
TiO_2	57 nm	34 nm	21 nm	12 nm	14 nm	14 nm	17 nm	29 nm	43 nm	57 nm

5.8: Effect of Sonication on Nanoparticle Size

In order to overcome recurrent problem of agglomeration, the effect of sonication time on the size of biosynthesised nanoparticles were evaluated on the final product. It was clear from the result that 15 minutes of continuous sonication produces least average size of nanoparticles, but with further increase in sonication time no significant enhancement in reduction of average particle size was observed. In general, 50–80% of size reduction of different nanofertilizer particles is possible through sonication.

Further studies

Li, H.,Yao, Q. H., Dong, Z. M., Zhao, T. L, Zhou, G. T. and Fu, S. Q. (2019). Controlled synthesis of struvite nanowires in synthetic wastewater. ACS Sustain. *Chem. Eng.* **7:** 2035–2043.

Shakiba, S., Carlos, E. A., Paudel, S., Sabliov, C. M., Debora, F. and Rodrigues, S. M. (2020). Emerging investigator series: polymeric nanocarriers for agricultural applications: synthesis, characterization, and environmental and biological interactions. *Environ. Sci.: Nano* **7:** 37–67.

Sun, H., Du, W., Peng, Q., Zhiyuan L. V., Mao, H. and Peter, M. K. (2020). Development of ZnO Nanoparticles as an efficient Zn fertilizer: using synchrotron-based techniques and laser ablation to examine elemental distribution in wheat grain. *J. Agril. Food Chem.* **68:** 5068–5075.

Zhao, L., Li, Lu., Wang, A., Zhang, H. Huang, M.,Wu, H., Xing, B., Z. and Wang, R. J. (2020). Nano-Biotechnology in Agriculture: Use of nanomaterials to promote plant growth and stress tolerance. *J. Agril. Food Chem.* **68:** 1935–1947.

Zulfiqar, F., Navarro, M., Ashraf, M., Akram, N. A. and Munné-Bosch, S. (2019). Nanofertilizer use for sustainable agriculture: Advantages and limitations. *Pl. Sci.* **295:** 289–299.

6

Characterization of Nanofertilizers

The synthesised nanoparticles were characterized for their size, shape, morphology, structure and concentration after using the following instruments: Particle size analyzer (PSA) / Dynamic light scattering (DLS), Scanning electron microscopy (SEM), Transmission electron microscopy (TEM), Atomic force microscopy (AFM), UV-VIS absorption spectroscopy, X-ray diffraction (XRD), Fourier transform infrared spectroscopy (FTIR), Energy dispersive X-ray spectroscopy (EDS), Lithography, Inductively coupled plasma mass spectrophotometer (ICPMS) and Inductively coupled plasma optical emission spectrophotometer (ICPOES).

The properties of nanofertilizers depend upon a variety of parameters such as particle size, dispersity index, surface area, porosity, solubility, aggregation, zeta potential etc. The detail of instrument used for characterization is as follow:

6.1: Particle Size Analyzer/ Dynamic Light Scattering (DLS)

Particle size can be determined by measuring the random changes in the intensity of light scattered from a suspension or solution. This technique is commonly known as dynamic light scattering (DLS), but is also called photon correlation spectroscopy (PCS) and quasi-elastic light scattering (QELS). The later terms are more common in older literature.

DLS is most commonly used to analyze nanoparticles. Examples include determining nanonutrient size, protein size, latex size, and colloid size. In general, the technique is best used for submicron particles and can be used to measure particle with sizes less than a nanometer. In this size regime (microns to nanometers) and for the purposes of size measurement, (but not thermodynamics) the distinction between a molecule (such as a protein or macromolecule) and a particle (such as nanonutrient) and even a second liquid phase (such as in an emulsion) becomes blurred. Dynamic light scattering can also be used as a probe of complex fluids such as concentrated solutions. However, this application is much less common than particle sizing.

Small particles in suspension undergo random thermal motion known as Brownian motion. This random motion is modeled by the Stokes-Einstein equation. Below the equation is given in the form most often used for particle size analysis (Figure 32).

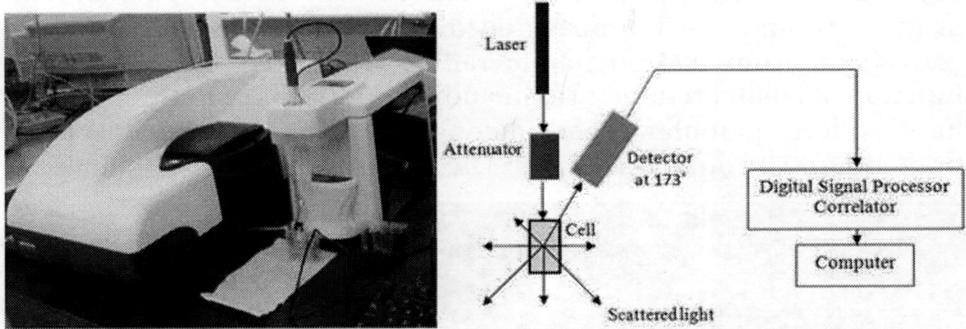

Figure 32: Particle Size Analyzer and the principle involved

$$D_h = \frac{k_B T}{3\pi\eta D_t}$$

The Stokes-Einstein relation that has connects diffusion coefficient measured by dynamic light scattering to particle size.

where

D_h is the hydrodynamic diameter (this is the goal: particle size)

D_t is the translational diffusion coefficient (we find this by dynamic light scattering)

k_B is Boltzmann's constant (we know this)

T is thermodynamic temperature (we control this)

η is dynamic viscosity (we know this)

The calculations are handled by instrument software. However, the equation does serve as important reminder about a few points. The first is that sample temperature is important, at it appears directly in the equation. Temperature is even more important due to the viscosity term since viscosity is a stiff function of temperature. Finally, and most importantly, it reminds the analyst that the particle size determined by dynamic light scattering is the hydrodynamic size. That is, the determined particle size is the size of a sphere that diffuses the way as your particle.

For those who work with protein sizing and other areas where hydrodynamic radius is more commonly used, note that the development here is around diameter. Radius calculations are the same except for a factor of two. Also, a note to those interested in polymer size. The hydrodynamic radius is not the same as the radius of gyration. Hydrodynamic sizes are more easily measured than radii of gyration and can be measured over a wider range of sizes. The conversion from hydrodynamic radius to radius of gyration is a function of chain architecture (including questions of random coil vs. hard sphere, globular, dendrimer, chain stiffness, and degree of branching). The optical set up for DLS (Figure 33) is shown below.

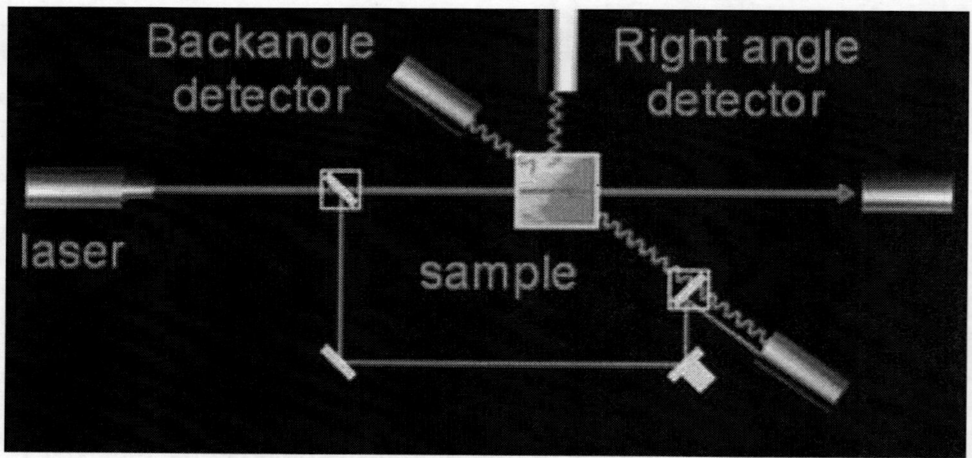

Figure 33: Optical set up of DLS

Light from the laser light source illuminates the sample in the cell. The scattered light signal is collected with one of two detectors, either at a 90 degree (right angle) or 173 degree (back angle) scattering angle. The provision of both detectors allows more flexibility in choosing measurement conditions. Particles can be dispersed in a variety of liquids. Only refractive index and viscosity the liquid need to be known for interpreting the measurement results.

The obtained optical signal shows random changes due to the randomly changing relative position of the particles. This is shown schematically in the graph (Figure 34) below.

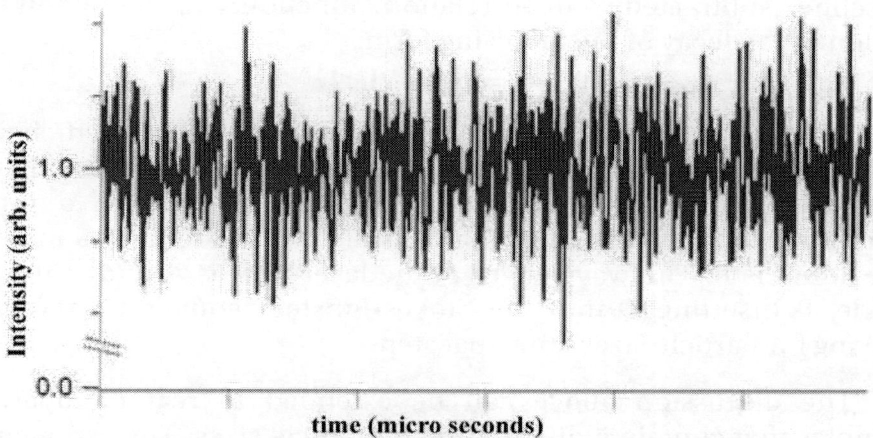

Figure 34: Optical signal from a nanoparticle sample on a microsecond timescale

The "noise" is actually due to particle motion and will be used to extract the particle size. In contrast to laser diffraction, DLS measurements are typically made at a single angle, although data obtained at several angles can be useful. In addition, the technique is completely noninvasive; the particle motion continues whether or not it is being probed by DLS. The variations in the signal arise due to the random Brownian motion of the particles. The signal can be interpreted in terms of an autocorrelation function (Figure 35). Incoming data is processed in real time with a digital signal processing device known as a correlator and the autocorrelation function as a function of delay time, τ, is extracted.

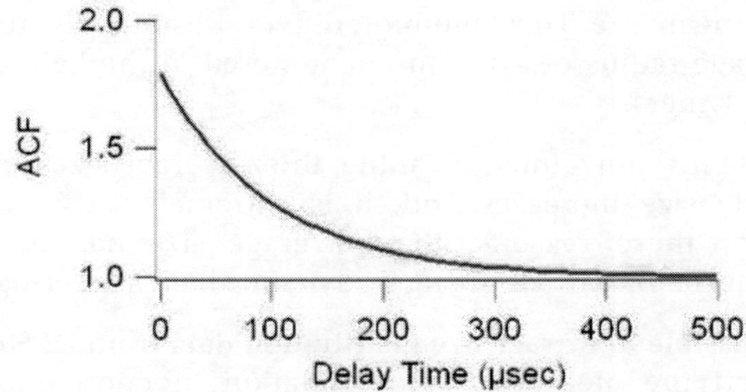

Figure 35: Autocorrelation function from dynamic light scattering. The decay of this function is used to extract particle size. Faster decays correspond to smaller particles.

For a sample where all of the particles are the same size, the baseline subtracted autocorrelation function, C, is simply an exponential decay of the following form:

$$C = \exp(-2\Gamma\tau)$$

The decay constant is proportional to the diffusion coefficient. Γ is readily derived from experimental data by a curve fit. The diffusion coefficient is obtained from the relation $\Gamma = D_t q^2$ where q is the scattering vector, given by $q=(4\pi n/\lambda)\sin(\theta/2)$. The refractive index of the liquid is n. The wavelength of the laser light is λ, and scattering angle, θ. Inserting D_t into the Stokes-Einstein equation above and solving for particle size is the final step.

The discussion above can be extended to real nanoparticle samples that contain a distribution of particle sizes. The exponential decay is rewritten as a power series:

$$C = \exp(-2\overline{\Gamma}\tau + \mu_2\tau^2 - ...)$$

The linear decay constant is proportional to the average diffusion coefficient and is used to extract average particle size.

Once again, decay constant is extracted and interpreted to obtain particle size. However, in this case, the obtained particle size, known as the z-average size, is a weighted mean size. Unfortunately, the weighting is somewhat convoluted. Recall that the decay constant is proportional to the diffusion coefficient. So, by dynamic light scattering one has determined the intensity weighted diffusion coefficient. The diffusion coefficient is inversely proportional to size. So, in truth, the "z-average size" is the intensity weighted harmonic mean size. This definition differs substantially from that of the z-average radius of gyration encountered in the light scattering study of polymers.

Despite the convoluted meaning, the z-average size increases as the particle size increases. And, it is extremely easy to measure reliably. For these reasons, the z-average size has become the accepted norm for particle sizing by dynamic light scattering.

It is possible to extract size distribution data from DLS data. One can convert the measured autocorrelation function into what is known as an electric field autocorrelation function, $g_1(\tau)$. Then use the following relationship between $g_1(\tau)$ and the scattered intensity, S, for each possible decay constant, Γ. The overall electric field autocorrelation function is the intensity weighted sum of the decays due to every particle in the system.

$$g_1(\tau) = \int S(\Gamma)\exp(-\Gamma\tau)\,d\Gamma$$

Electric field autocorrelation function as a sum of exponential decays. The decay constants are inversely proportional to the particle size.

Inversion of this equation, that is using experimentally determined values of $g_1(\tau)$ to find values of $S(\Gamma)$, will lead to information about the size distribution. Unlike the cumulants analysis discussed above, this is an ill-posed mathematical problem. Even so, the technique remains useful for interpreting DLS data. All of these equations and the analysis are handled automatically in the HORIBA software. As such, dynamic light scattering has found application for determining protein size, nanoparticle size, and colloid size.

Dynamic light scattering (DLS) sometimes also referred as photon correlation spectroscopy (PCS) or quasi elastic light scattering (QELS) is a non-invasive, well-established technique for measuring the size of molecules and particles typically in the nanometer size. DLS is an important tool for characterization of nanoparticle in solution, measures light scattered from a laser that passes through a solution by analyzing the modulation of the scattered light intensity as a function of time, information can be obtained on the size of the particle in solution. The analysis is based on the diffusion motion of particle in solution. The hydrodynamic diameter measured by DLS is an important complement to other sizing measurement such as TEM because it provides information on aggregation state of nanoparticle solutions. DLS measures light scattered from a laser that passes through a colloidal solution. The primary results obtained from the DLS study for biologically synthesised boron (B) nanoparticles, for example, figure 36, gives a clear indication regarding the size distribution on the basis of intensity, number and volume, and yield. As shown in DLS histogram, majority of the particles were below 30 nm in size, which appeared to be highly mono-disperse. Monodispersity of nanoparticle could be further confirmed by polydispersity index obtained by DLS measurements.

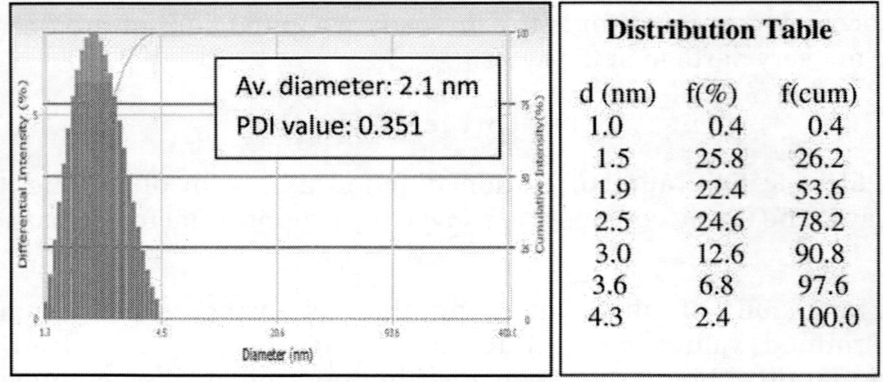

Distribution Table

d (nm)	f(%)	f(cum)
1.0	0.4	0.4
1.5	25.8	26.2
1.9	22.4	53.6
2.5	24.6	78.2
3.0	12.6	90.8
3.6	6.8	97.6
4.3	2.4	100.0

Figure 36: Particle size distribution of nano-boron fertilizer from 0.5mM H_3BO_3 salt by *Aspergillus tubingenesis* TFR-29

The results showed the particles are monodispersed (PDI value 0.351) and diameter of 100% particles is within 4.3, with average nanoparticles size of boron is only 2.1 nm.

The particle size analysis data of nano-nitrogen (Figure 37) distribution table clearly indicates that 10% nanoparticles were 1.1 nm sizes, almost 50% nanoparticles were 1.3 nm size and almost 90% nanoparticles were 1.6 nm size; with the average diameter of the entire particles were 1.4 nm. The polydispersity index (PDI) value was 0.131, indicates the entire particles were monodispersed. The results clearly showed that developed organism *Aspergillus tubingenesis* TFR 29 is a very good organism for the production of N nanoparticles, which can be exploit commercially.

The Zeta (ξ) potential is the electrostatic potential that exists at the shear plane of a particle, which is related to both surface charge and the local environment of the particle. Zeta potential measurements provide an important criterion for the stability of a colloid system. In cell biology, the concept of Zeta potential has been used to study cell biological activation, cell agglutination and cell adhesion, which are related to cell surface charge properties. Although Zeta potential measurements have been utilized in probing the interaction between plant cells and biomolecules, it is important for studying the interaction between nanoparticles and biological cells. For more stable of the nanofertilizer particle, the zeta potential value should be apart + 30 and – 30 mV.

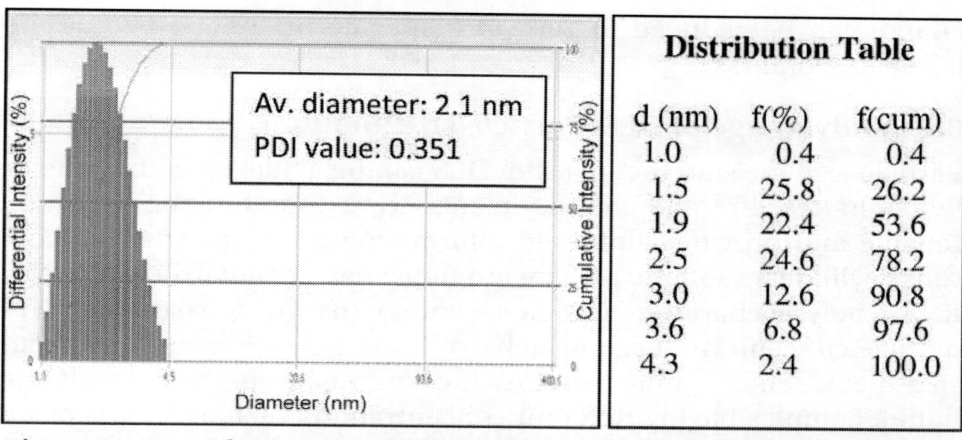

Distribution Table

d (nm)	f(%)	f(cum)
1.0	0.4	0.4
1.5	25.8	26.2
1.9	22.4	53.6
2.5	24.6	78.2
3.0	12.6	90.8
3.6	6.8	97.6
4.3	2.4	100.0

Figure 37: Particle size distribution of Nitrogen nanoparticles produced by the action of fungal protein of *Aspergillus tubingenesis* TFR29 from 0.5mM NH_4NO_3 salt

6.1.1. Zeta potential

The zeta potential of a system is measure of charge stability and controls all particle-particle interactions within a suspension. Understan-ding zeta potential is of crucial importance in controlling dispersion and determining the stability of a nanoparticles in suspension *i.e.,* to what degree the aggregation will occur over time. The zeta potential is the measure of the elecric potential at the slip plane between the bound layer of dilute molecules surrounding the particle, and the bulk solution. This can be closely linked to the surface charge of the particle in simple systems, but is also heavily dependent on the properties of the diluents solution. A higher level of zeta potential results in greater electro-static repulsion between the particles, minimizing aggregation/flocculation (Table 16).

Table 16: Zeta potential value and nanofertilizer nature

Zeta potential	Nanofertilizer nature
0–5 mV	Tend to agglomerate or aggregate
5–20 mV	Minimally stable
20–40 mV	Moderately stable
> 40 mV	Highly stable

The magnitude of the zeta potential provides information about particle stability, with higher magnitude potential exhibiting increased electrostatic repulsion and therefore increased stability. It is important to consider that the magnitude of the charge on the nanoparticles surface depends on the solution pH. In fact the surface charge can be reduced to zero of a specific pH referred to as the isoelectric point.

6.1.2: Advantage of DLS/Particle size analyzer

Particle size measurements using DLS can be achieved within one or two minutes and the measurements were found to be accurate, reliable and provided unbiased information on shape (flexible coil, rods, cylinders, spheres) of macromolecules (nanofertilizers, DNA, RNA, polysaccharides, proteins, and viruses). Measurement of particle concentration can be achieved calibration free and the mean size measurement, only requires the knowledge of viscosity of the liquid. Samples having different concentrations such as simple or no sample preparation, high concentration and turbid samples can be analysed directly using DLS. This technology is also suitable for molecular weight determination and size measurements of molecules in the range of 10μm to less than 1 nm, and having molecular weight less than 1000 Da can be determined. DLS's major advantage is that very less amount of sample (< 3μL) is essential for analysis. DLS is found to be suitable for protein–RNA interaction studies and modified DLS in-line process systems are found to be suitable. It is possible to conduct experiments on wide range of sample buffer and wide temperature range using DLS technique.

6.1.3: Limitations of DLS

Measurements obtained by DLS are altered by variations in temperature and solvent viscosity as it is highly sensitive to it. Therefore, both temperature and solvent viscosity must be kept constant for reliable results. DLS cannot differentiate molecules that are closely related (*e.g.,* monomer and dimer) since it is a low resolution method. DLS must be used on highly dilute solutions of food and fertilizer particles. It is difficult to dilute many food particles as it may result in variation of structural properties, for instance complex particles may dissociate. So, it is essential to ensure that the particles under study in a DLS experiment must retain their structures which are native to the food. Presence of large aggregates even in a small amount will adversely affect the measurements since scattering intensity is proportional to 6th power of the size of macromolecules (I α d6, where d is a diameter of macromolecule). Therefore, proper cleaning of the sample-holding cuvette prior to measurement is essential. The signals from DLS vary with changes in size and concentration of macromolecules and so the concentration range is to be optimized prior to measurement for attaining reliable results.

6.2: Scanning Electron Microscopy

A scanning electron microscope (SEM) is a type of electron microscope that produces images of a sample by scanning the surface with a focused beam of electrons. The electrons interact with atoms in the sample, producing various signals that contain information about the surface topography and composition of the sample. SEM carry significant amounts of kinetic energy, and this energy is dissipated as a variety of signals produced by electron-sample interactions when the incident electrons are decelerated in the solid sample. A scanning electron microscope (SEM) scans a focused electron beam over a surface to create an image. A scanning electron microscope scans a beam of electrons over a specimen to produce a magnified image of an object. Electrons from the beam hit the surface of the object and bounce off it. A detector registers these scattered electrons and turns them into a picture. SEMs are expensive, large and must be housed in an area free of any possible electric, magnetic or vibration interference. Maintenance involves keeping a steady voltage, currents to electromagnetic coils and circulation of cool water.

Scanning electron microscopy (SEM) is extremely useful for the determination of topology and observations of surfaces as they offer better resolution and depth of field than optical microscope. SEM is essentially a high magnification microscope, which uses a focused scanned electron beam to produce images of the sample, both top-down and, with the necessary sample preparation, cross-sections. The primary electron beam interacts with the sample in a number of key ways:

- Primary electrons generate low energy secondary electrons, which tend to emphasis the topographic nature of the specimen.
- Primary electrons can be backscattered which produces images with a high degree of atomic number (Z) contrast.
- Ionized atoms can relax by electron shell-to-shell transitions, which lead to either X-ray emission or Auger electron ejection. The X-rays emitted are characteristic of the elements in the top few μm of the sample and are measured by the EDX detector.

6.2.1: Instrument

The two major components of an SEM (Figure 38) are the electron column and control console.

- The electron column consists of an electron gun and two or more electron lenses, which influence the path of electrons travelling down an evacuated tube. Generally, tungsten or Lanthanum hexaboride thermionic emitters are used as electron gun. The purpose of electron gun is to provide a stable beam of electrons.
- The control console consists of a cathode ray tube (CRT) viewing screen and computer to control the electron beam. The base of the column is usually taken up with vacuum pumps.

The Scanning Electron Microscope developed by Professor Dr. Charles Oatlev with the assistance of graduate students in the 1950s, are one of the three types of electron microscopes (EM). It utilizes the same basic principles as light microscopes, but focus beams of energetic electrons rather than photons, to magnify an object. The scanning electron microscope images the sample surface by scanning with a high energy beam of electrons. When the beam of electrons strikes the surface of the specimen and interacts with the atoms of the sample, signals in the form of secondary electrons, back scattered electrons and characteristic X-rays are generated that contain information about the sample's surface topography, compositions etc. The results obtain from the SEM study gives the characteristic three dimensional appearance useful for the understanding of the surface structure of the synthesised nanoparticles/ nanofertilizers. Zinc and silver nanoparticle, as showed in the figure 39 further confirmed the surface topography of the respective nanofertilizers.

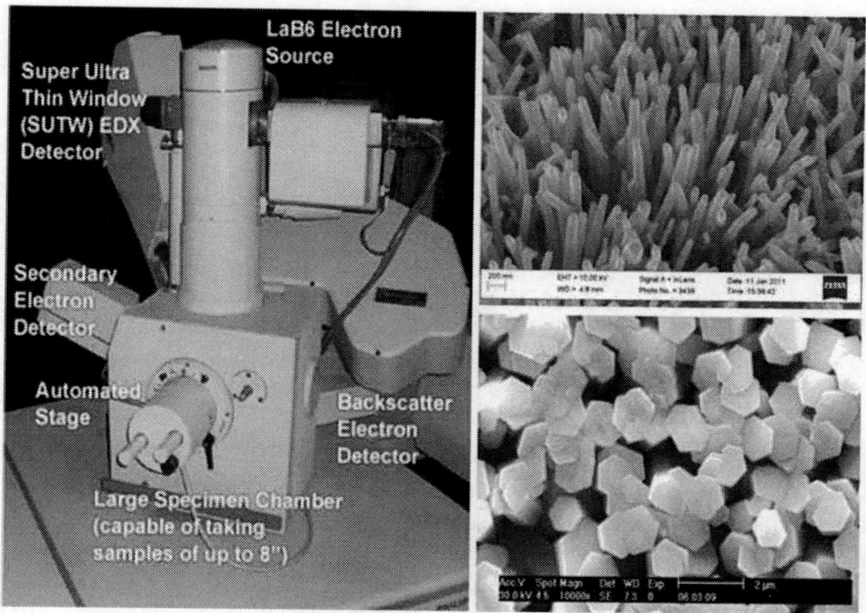

Figure 38: SEM instrument with image

SEM samples can be prepared by placing the sample on a double-side adhesive tape which has to be further placed on an aluminum stub. Sputter coating can also be carried out to enhance surface charge distribution, if necessary.

SEMs consist of the following components:

- Electron Source
- Thermionic Gun
- Field Emission Gun
- Electromagnetic and/or Electrostatic Lenses
- Vacuum chamber
- Sample chamber and stage
- Computer
- Detectors (one or more) namely Secondary Electron Detector (SED), Backscatter Detector, Diffracted Backscatter Detector (EBSD), X-ray Detector (EDS).

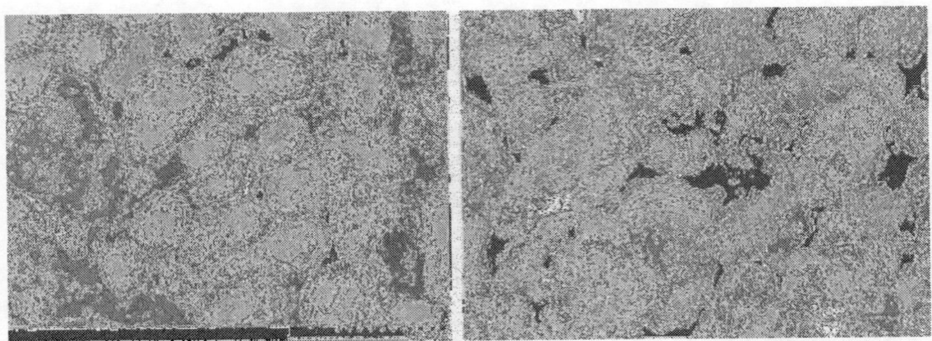

Figure 39: Scanning Electron Microscopy images of Silver (*L*) and Zinc (*R*)

In addition, SEMs require a stable power supply, vacuum and cooling system, vibration-free space and need to be housed in an area that isolates the instrument from ambient magnetic and electric fields.

A Scanning Electron Microscope provides details surface information by tracing a sample in a raster pattern with an electron beam. The process begins with an electron gun generating a beam of energetic electrons down the column and onto a series of electromagnetic lenses. These lenses are tubes, wrapped in coil and referred to as solenoids. The coils are adjusted to focus the incident electron beam onto the sample; these adjustments cause fluctuations in the voltage, increasing/decreasing the speed in which the electrons come in contact with the specimen surface. Controlled via computer, the SEM operator can adjust the beam to control magnification as well as determine the surface area to be scanned. The beam is focused onto the stage, where a solid sample is placed. Most samples require some preparation before being placed in the vacuum chamber.

Of the variety of different preparation processes, the two most commonly used prior to SEM analysis are sputter coating for non-conductive samples and dehydration of most biological specimens. In addition, all samples need to be able to handle the low pressure inside the vacuum chamber. The interaction between the incident electrons and the surface of the sample is determined by the acceleration rate of incident electrons, which carry significant amounts of kinetic energy before focused onto the sample.

When the incident electrons come in contact with the sample, energetic electrons are released from the surface of the sample. The scatter patterns are made by the interaction yields information on size, shape, texture and composition of the sample. A variety of detectors are used to attract different types of scattered electrons, including secondary and backscattered electrons as well as X-rays. Backscatter electrons are incidental electrons reflected backwards; images provide composition data related to element and compound detection. Although topographic information can be obtained using a backscatter detector, it is not as accurate as an SED. Diffracted backscatter electrons determine crystalline structures as well as the orientation of minerals and micro-fabrics. X-rays, emitted from beneath the sample surface, can provide element and mineral information. SEM produces black and white, three-dimensional images. Image magnification can be up to 10 nanometers and, although it is not as powerful as its TEM counterpart, the intense interactions that take place on the surface of the specimen provide a greater depth of view, higher-resolution and, ultimately, a more detailed surface picture.

In addition, SEMs are limited to solid, inorganic samples small enough to fit inside the vacuum chamber that can handle moderate vacuum pressure. Finally, SEMs carry a small risk of radiation exposure associated with the electrons that scatter from beneath the sample surface. The sample chamber is designed to prevent any electrical and magnetic interference, which should eliminate the chance of radiation escaping the chamber. Even though the risk is minimal, SEM operators and researchers are advised to observe safety procedure.

6.2.2: SEM Applications

SEMs have a variety of applications in a number of scientific and industry-related fields, especially where characterizations of solid materials is beneficial. SEMs take incidental electrons and focus them onto a specimen; the electrons that scatter off the surface following this interaction can be analyzed with a variety of detectors that provide topographical, morphological and compositional information regarding the surface of a sample. Although SEMs are large, expensive pieces of equipment, they remain popular among researchers due to their wide range of applications and capabilities, including the high-resolution, three-dimensional, detailed images they produce. In addition to topographical, morphological and compositional information, a Scanning Electron Microscope can detect and analyze surface fractures, provide information in microstructures, examine surface contaminations, reveal spatial variations in chemical compositions, provide qualitative chemical analyses and identify crystalline structures.

SEMs can be as essential research tool in fields, such as life science, biology, gemology, medical and forensic science, and metallurgy. In addition, SEMs have practical industrial and technological applications, such as semiconductor inspection, production line of miniscule products and assembly of microchips for computers. Figure 40 shows the SEM image of nitrogen nanofertilizer. There is no absolute way to eliminate or identify all potential artifacts. To ascertain the surface morphology, the nanoparticles were coated with heavy metal gold and subjected to SEM imaging. A micrograph gives a fairly good idea that the N nanofertilizer particles have a rough outer surface and show a certain degree of agglomeration.

Figure 40: SEM image of nitrogen nanofertilizer

The SEM image (Figure 41) of K nanofertilizer showed different shapes of nanoparticles mainly rectangular and round. Some particles were also square in shape. Similarly, image of B nanofertilizer particles (Figure 42) showed their rough surface with unclear edges.

Figure 41: SEM image of K nanofertilizer particle

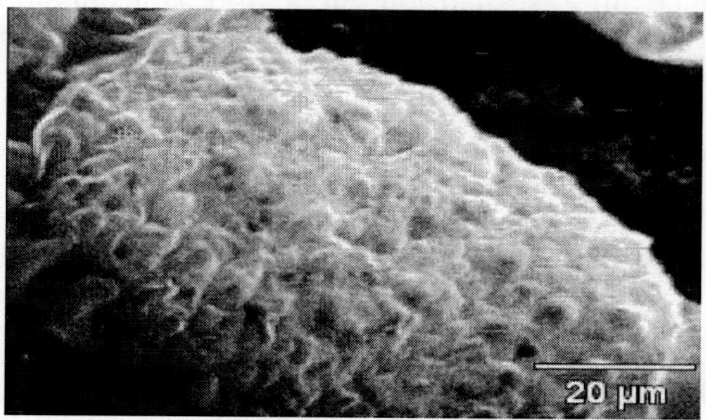

Figure 42: SEM image of B nanofertilizer particle

6.2.3: SEM Advantages

Advantages of a Scanning Electron Microscope include its wide-array of applications, the detailed three-dimensional and topographical imaging and the versatile information garnered by different detectors. SEMs are also easy to operate with the proper training and advances in computer technology and associated software make operation user-friendly. This instrument works fast, often completing SEI, BSE and EDS analyses in less than five minutes. In addition, the technological advances in modern SEMs allow for the generation of data in digital form. Although all samples must be prepared before placed in the vacuum chamber, most SEM samples require minimal preparation actions.

6.2.4: SEM Disadvantages

The disadvantages of a Scanning Electron Microscope start with the size and cost. SEMs are expensive, large and must be housed in an area free of any possible electric, magnetic or vibration interference. Maintenance involves keeping a steady voltage, currents to electromagnetic coils and circulation of cool water. Special training is required to operate an SEM as well as prepare samples. The preparation of samples can result in artifacts. The negative impact can be minimized with knowledgeable experience researchers being able to identify artifacts from actual data as well as preparation skill.

6.3: Transmission Electron Microscopy (TEM)

Transmission electron microscopy is a microscopy technique in which a beam of electrons is transmitted through a specimen to form an image. The specimen is most often an ultrathin section less than 100 nm thick or a suspension on a grid. Transmission electron microscopy (TEM) is a technique used to observe the features of very small specimens. The technology uses an accelerated beam of electrons, which passes through a very thin specimen to enable a scientist the observe features such as structure and morphology. The most important difference between a transmission electron microscope and a scanning electron microscope is: Rather than the broad static beam used in TEM, the SEM beam is focused to a fine point and scans line by line over the sample surface in a rectangular raster pattern.

The intensity of diffraction depends on the orientation of the planes of atoms in a crystal relative to the electron beam, at certain angles the electron beam is diffracted strongly from the axis of the incoming beam, while at other angle the beam is largely transmitted. TEM images revealed the distribution and dispersion with extent of exfoliation, intercalation and orientation of nanofertilizer particles. The micrograph obtained from the TEM, gives a clear indication regarding the shape and size of nanoparticles. A TEM has many uses. Its main purpose is to create high magnification images of the internal structure of a sample. This can be used to gather information on crystalline structures, stress, internal fractures, contamination, and more.

TEM has three essential systems: (1) an electron gun, which produces the electron beam, and the condenser system, which focuses the beam onto the object, (2) the image-producing system, consisting of the objective lens, movable specimen stage, and intermediate and projector lenses, which focus the electrons passing through the specimen to form a real, highly magnified image, and (3) the image-recording system, which converts the electron image into some form perceptible to the human eye (Figure 43).

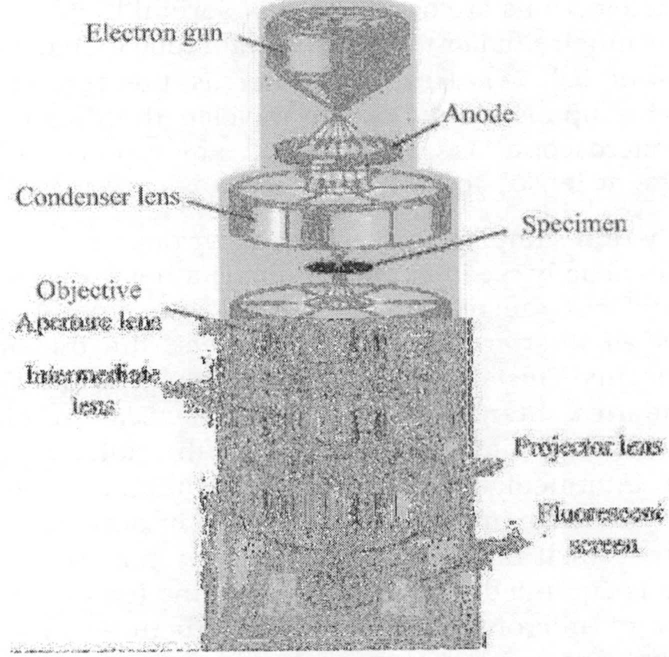

Figure 43: Parts of Transmission electron microscope (TEM)

The image-recording system usually consists of a fluorescent screen for viewing and focusing the image and a digital camera for permanent records. In addition, a vacuum system, consisting of pumps and their associated gauges and valves, and power supplies are required.

Transmission electron microscopy (TEM) is a high magnification measurement technique that images the transmission of a beam of electrons through a sample. TEM is the preferred method to measure the particle size, size distribution and morphology of nanofertilizer particles. In addition, as the technique uses electrons rather than light to illuminate the sample, TEM exhibits significantly higher resolution than light-based imaging techniques. Amplitude and phase variations in the transmitted beam also provide imaging contrast that is a function of the sample thickness and the sample material.

The transmission electron microscope uses a high energy electron beam transmitted through a very thin sample to image and analyze the microstructure of materials with atomic scale resolution. The electrons are focused with electromagnetic lenses and the image is observed on a fluorescent screen, or recorded on film or digital camera. The electrons are accelerated at several hundred kV, giving wavelengths much smaller than that of light (200 kV electrons have a wavelength of 0.025Å). However, whereas the resolution of the optical microscope is limited by the wavelength of light, that of the electron microscope is limited by aberrations inherent in electromagnetic lenses, to about 1–2 Å.

For very thin samples, one is looking through many atoms, and one does not usually see individual atoms. Rather the high resolution imaging mode of the microscope images, the crystal lattice of a material as an interference pattern between the transmitted and diffracted beams. This allows one to observe planar and line defects, grain boundaries, interfaces, etc. with atomic scale resolution. The brightfield/darkfield imaging modes of the microscope, which operate at intermediate magnification, combined with electron diffraction, are also invaluable for giving information about the morphology, crystal phases, and defects in a material. Finally, the microscope is equipped with a special imaging lens allowing for the observation of micromagnetic domain structures in a field-free environment.

High resolution transmission electron microscopy (Figure 44) is an imaging mode of the TEM that allows the imaging of the crystallographic structure of a sample at an atomic scale in which each imaging electron interacts independently with the sample. Due to its high resolution, it is an invaluable tool to study nanoscale properties of crystalline material. Crystal structure can also better investigate by high resolution transmission electron microscope (HR-TEM) also known as phase contrast. When utilizing a field emission source of uniform thickness, the image is formed due to differences in phase of electron waves, which is caused by specimen interactions. Figure 44 showed the HR-TEM micrograph of the single Zn nanoparticle which confirms the crystallity and atomic arrangements in the biologically synthesised Zn nanofertilizer particle. It can be further confirmed by white array of spot found in the selective area electron diffraction (SAED) pattern. It is important to realize that the HR–TEM image may be a direct representation of the samples crystallographic structure.

Figure 44: HR–TEM micrograph of biologically synthesised single Zn nanoparticle

6.3.1: Sample preparation:

Samples are usually prepared by adding a drop of nanoparticle solution (50 μl) on a carbon coated copper grid. Excess of the samples is removed with the help of an absorbent paper and the grid is kept for overnight incubation in a vacuum desiccator at room temperature. TEM imaging can be performed by placing the prepared grid in the specimen holder which is further inserted into specimen port (Figure 45) of TEM instrument until it stops.

The sample for TEM analysis must be able to withstand the electron beam and also the high vacuum inside the chamber that it is put into. For TEM observations, thin samples are required due to absorption of electrons in the material. High acceleration voltage reduces the absorption effects, but can cause radiation damage. At these acceleration tensions, a thickness of about 60nm is required for TEM. The sample preparation is difficult as a thin sample on a support grid must be prepared. The process can also be time consuming and expensive. For biological sample fixation is the first step in sample preparation. This is done to preserve the structure of cell with minimum alteration from living state and to protect them against alterations during embedding and sectional procedures.

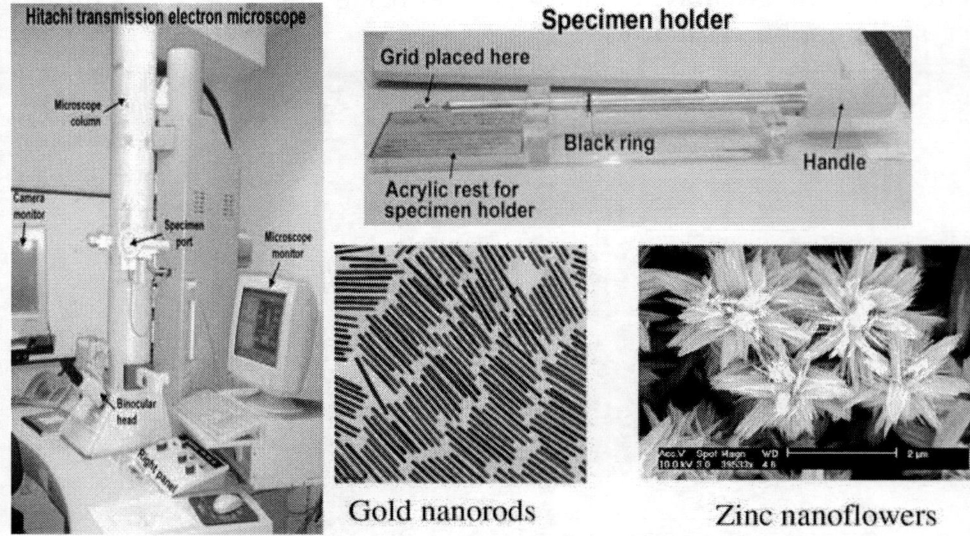

Figure 45: Sample preparation for TEM image

6.3.2: Applications

- Commonly employed to characterize nanomaterials at a very high spatial resolution.
- Can be utilized to gain information about the morphology, crystal structure and defects, crystal phases and composition.
- TEM is almost always the first method used to determine the size and distribution of nanoparticles sample.
- TEM also can be explored for determination of melting points of nanocrystals, measurement of mechanical and electrical properties of individual nanowires and nanotubes.

Insight of Figure 46 shows SAED patterns of a single Mg nanofertilizer particle, where white array of spot supports the crystalline nature of nanoparticle. The high resolution TEM micrograph of a single Mg nano-particle shows the linear arrangements of atoms within the nanoparticle, and the presence of lines also depicts the crystallity of Mg nanoparticles which was otherwise deficient in nonmetallic organic compounds and molecules. The TEM image of single Mg nanoparticles was also presented as Figure 47.

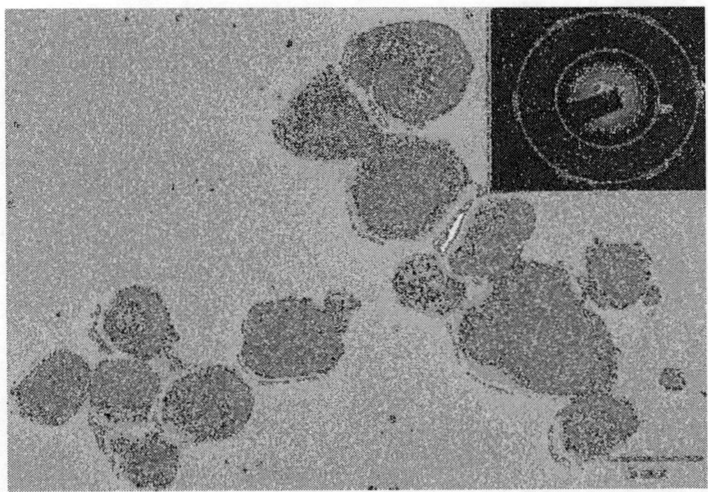

Figure 46: High magnification TEM micrograph of biologically synthesised Mg nanoparticles at 5 nm scale bar, inset showing the SAED pattern of a single Mg nanoparticle

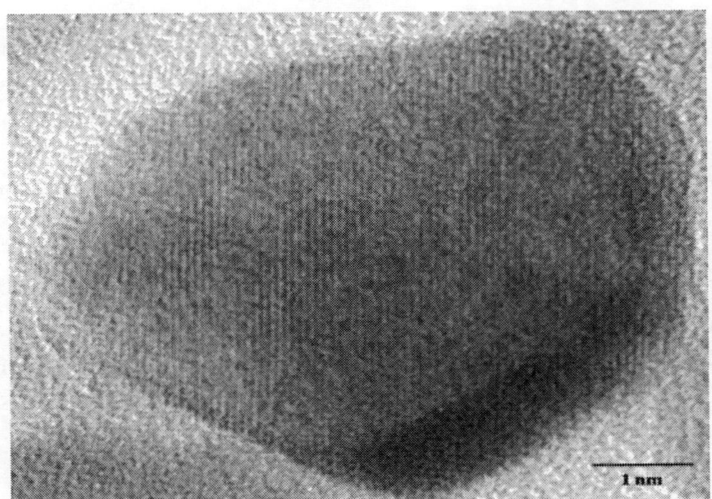

Figure 47: TEM image of single Mg nanoparticle

6.3.3: The Electron gun and condenser system

The source of electrons, the cathode, is a heated V-shaped tungsten filament or, in high-performance instruments, a sharply pointed rod of a material such as lanthanum hexaboride. The filament is surrounded by a control grid, sometimes called a Wehnelt cylinder, with a central aperture arranged on the axis of the column; the apex of the cathode is arranged to lie at or just above or below this aperture. The cathode and control grid are at a negative potential equal to the desired accelerating voltage and are insulated from the rest of the instrument. The final electrode of the electron gun is the anode, which takes the form of a disk with an axial hole. Electrons leave the cathode and shield, accelerate toward the anode, and, if the stabilization of the high voltage is adequate, pass through the central aperture at a constant energy. The control and alignment of the electron gun are critical in ensuring satisfactory operation. TEM image of some of the nanofertilizer particles are shown as Figure 48.

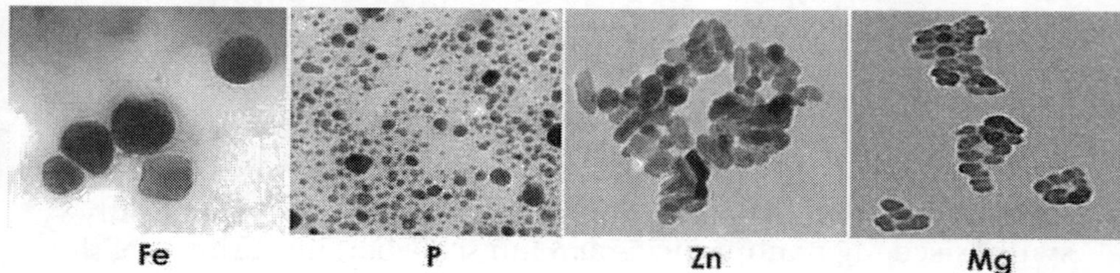

Figure 48: Transmission Electron Microscopy images of different nanofertilizer particles

The intensity and angular aperture of the beam are controlled by the condenser lens system between the gun and the specimen. A single lens may be used to converge the beam onto the object, but, more commonly, a double condenser is employed. In this the first lens is more powered and produces a reduced image of the source, which is then imaged by the second lens onto the object. Such an arrangement is economical of space between the electron gun and the object stage and is more flexible, because the reduction in size of the image of the source (and hence the final size of illuminated area on the specimen) may be varied widely by controlling the first lens. The use of a small spot size minimizes disturbances in the specimen due to heating and irradiation.

6.3.4: The Image-producing system

The specimen grid is carried in a small holder in a movable specimen stage. The objective lens is usually of short focal length (1–5 mm [0.04–0.2 inch]) and produces a real intermediate image that is further magnified by the projector lens or lenses. A single projector lens may provide a range of magnification of 5:1, and by the use of interchangeable pole pieces in the projector, a wider range of magnifications may be obtained. Modern instruments employ two projector lenses (one called the intermediate lens) to permit a greater range of magnification and to provide a greater overall magnification without a commensurate increase in the physical length of the column of the microscope. TEM image of biosynthesised K nanofertilizer is shown below (Figure 49).

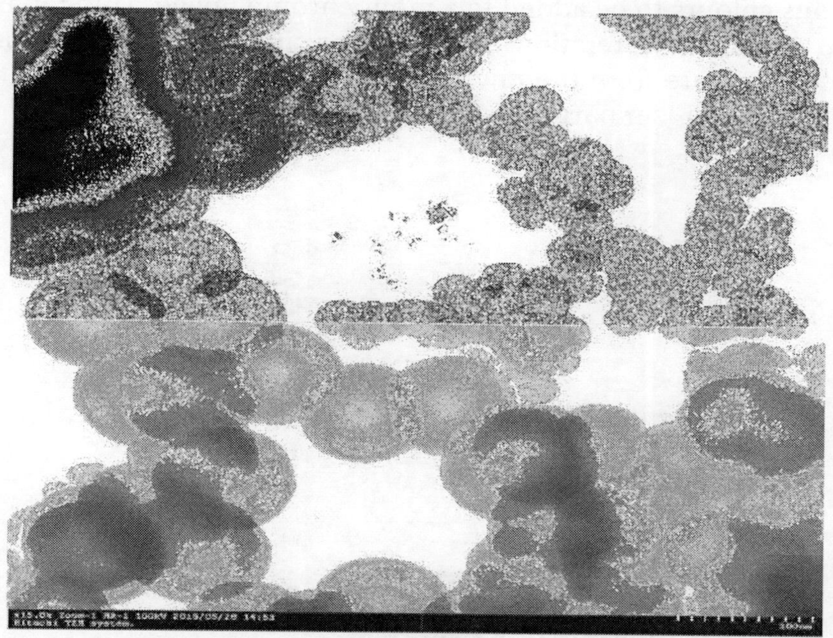

Figure 49: TEM image of K nanofertilizer particle produced by *Trematosphaeria grisea* TFR28 clearly showing their spherical shape

For practical reasons of image stability and brightness, the microscope is often operated to give a final magnification of 1,000–250,000× on the screen. If a higher final magnification is required, it may be obtained by photographic or digital enlargement. The quality of the final image in the electron microscope depends largely upon the accuracy of the various mechanical and electrical adjustments with which the various lenses are aligned to one another and to the illuminating system. The lenses require power supplies of a high degree of stability; for the highest standard of resolution, electronic stabilization to better than one part in a million is necessary. The control of a modern electron microscope is carried out by a computer, and dedicated software is readily available.

6.3.5: Image Recording

The electron image is monochromatic and must be made visible to the eye either by allowing the electrons to fall on a fluorescent screen fitted at the base of the microscope column or by capturing the image digitally for display on a computer monitor. Computerized images are stored in a format such as TIFF or JPEG and can be analyzed or image-processed prior to publication. The identification of specific areas of an image, or pixels with specified characteristics, allows spurious colours to be added to a monochrome image. This can be an aid to visual interpretation and teaching and can create a visually attractive picture from the raw image. The TEM image of some of the single nanofertilizer particles are shown as Figure 50 and 51.

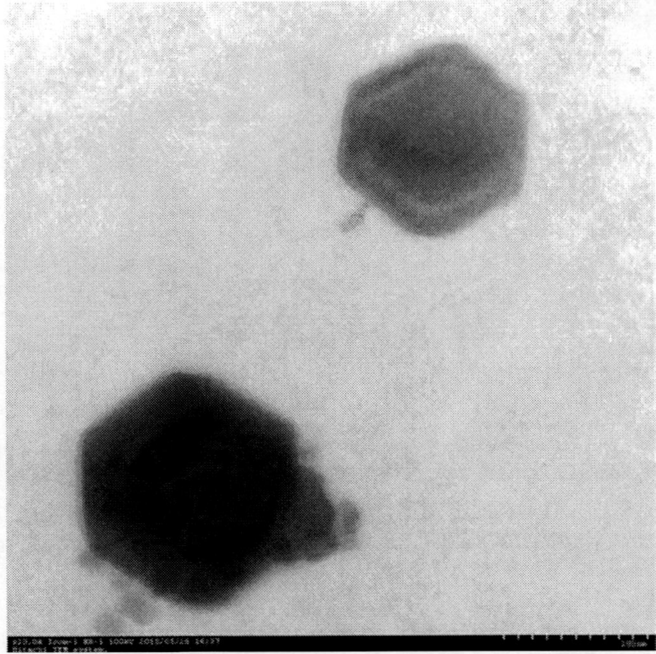

Figure 50: TEM image of Mo nanofertilizer particle produced from $(NH_4)_6Mo_7O_{24}$ by the action of protein released from *Aspergillus tubingenesis* TFR29

TEM Micrographs of B nanoparticles clearly showed (Figure 51) their well distribution and are round to spherical in shape with clear edges of crystal.

Figure 51: TEM image of B nanofertilizer particle produced from H_3BO_3 by the action of protein released from *Aspergillus tubingenesis* TFR29.

6.3.6: Advantages of TEM

- Simple sample preparation on grid
- Good singnal to noise ratio
- Radiation resistant
- Can be determined position of the particle
- High resolution and high magnification

6.3.7: Limitations of TEM

- Sample preparation generally requires more time and expertise.
- Artifacts are common in TEM samples, due both to the sample preparation and to changing the form of the original material.
- The technique involves highly sophisticated instrumentation which requires a rigid maintenance, thus making it a costly affair.

6.4: Atomic Force Microscopy (AFM)

The atomic force microscope (AFM) is one kind of scanning probe microscopes (SPM). SPMs are designed to measure local properties, such as height, friction, magnetism, with a probe. After acquire an image, the SPM raster-scans the probe over a small area of the sample, measuring the local property simultaneously. AFMs operate by measuring force between a probe and the sample. Normally, the probe is a sharp tip, which is a 3–6 μm tall pyramid with 15–40 nm end radius (Figure 52). Though the lateral resolution of AFM is low (30nm) due to the convolution, the vertical resolution can be up to 0.1nm.

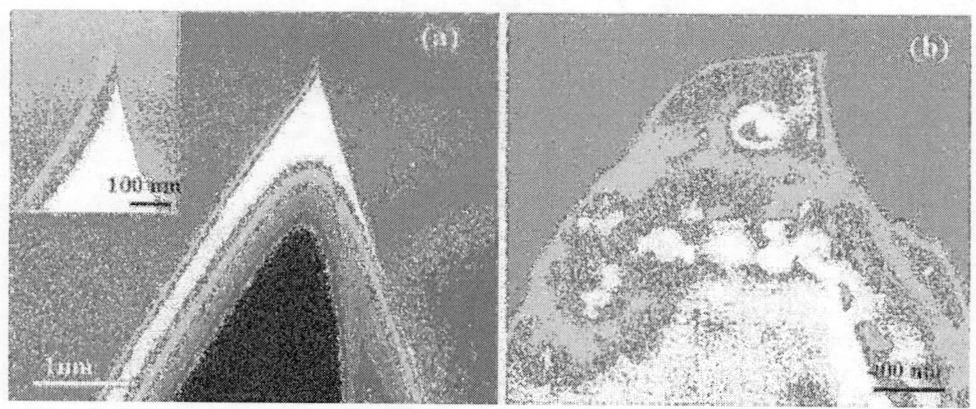

Figure 52: (a) A new AFM tip; inset: The end of the new tip. **(b)** A used AFM tip.

AFM is a mechanical imaging instrument in which a small, <1 μm, probe is scanned over a surface. By monitoring the motion of the probe, the surface topography and/or images of surface physical properties are measured with an AFM. The schematic diagram illustrating the principle and components is shown as Figure 53. The Atomic Force Microscope (AFM) is one of the foremost tools for imaging, measuring and manipulating matter at the nanoscale. Therefore, this technique is ideally suited for nanoparticle characterization. It offers the capability of three dimensional visualization and both qualitative and quantitative information on many physical properties including size, morphology, surface texture and roughness. Advantages of the AFM for such applications are derived from the fact that the AFM is non-destructive technique and it has a very high 3D spatial resolution. In non-contact mode of AFM, the scanning tip hovers at 75Å above the specimen's surface. This particular mode is suitable for biosynthesised nanoparticle. The attractive forces acting between the tip and the specimen were measured and topographic images constructed by scanning, the tip above the surface. In addition to presenting a surface image, AFM can also provide quantitative measurements of feature sizes, such as step height, and sample characteristics, such as capacitance measurements.

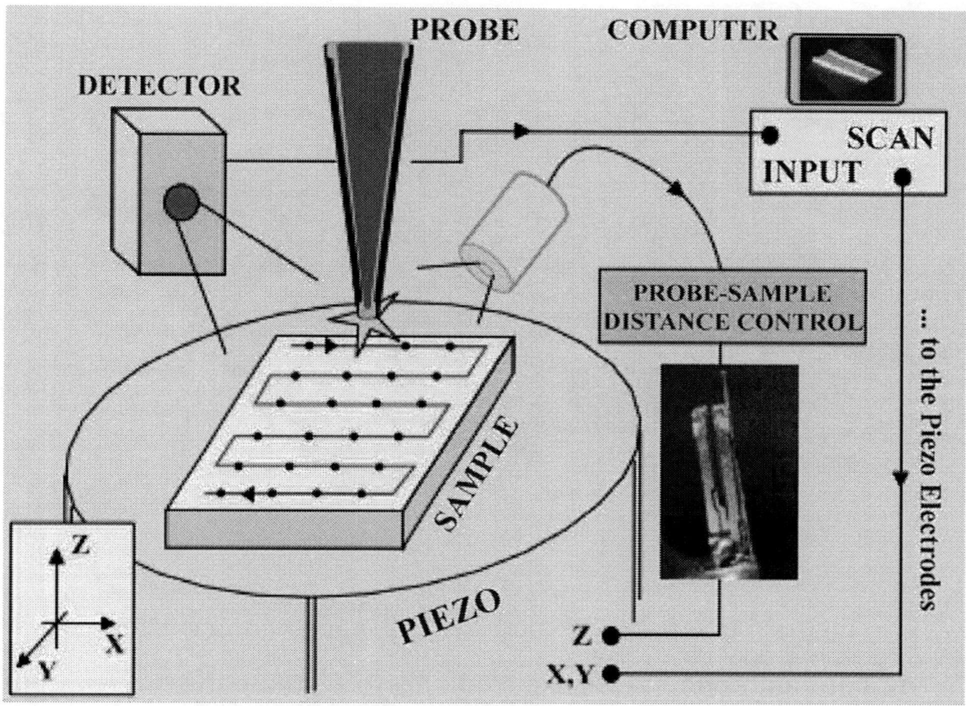

Figure 53: Principle component of AFM

To acquire the image resolution, AFMs can generally measure the vertical and lateral deflections of the cantilever by using the optical lever. The optical lever operates by reflecting a laser beam off the cantilever. The reflected laser beam strikes a position-sensitive photo-detector consisting of four-segment photo-detector. The differences between the segments of photo-detector of signals indicate the position of the laser spot on the detector and thus the angular deflections of the cantilever. Piezo-ceramics position the tip with high resolution. Piezoelectric ceramics are a class of materials that expand or contract when in the presence of a voltage gradient. Piezo-ceramics make it possible to create three-dimensional positioning devices of arbitrarily high precision. The 3–D structure of N nanofertilizer is shown as figure 54. It clearly reveals the rough surface structure and size. The rough surface of biosynthesised particle is important for metal protein interaction, helps in catalyzing enzyme activation in the biosystem.

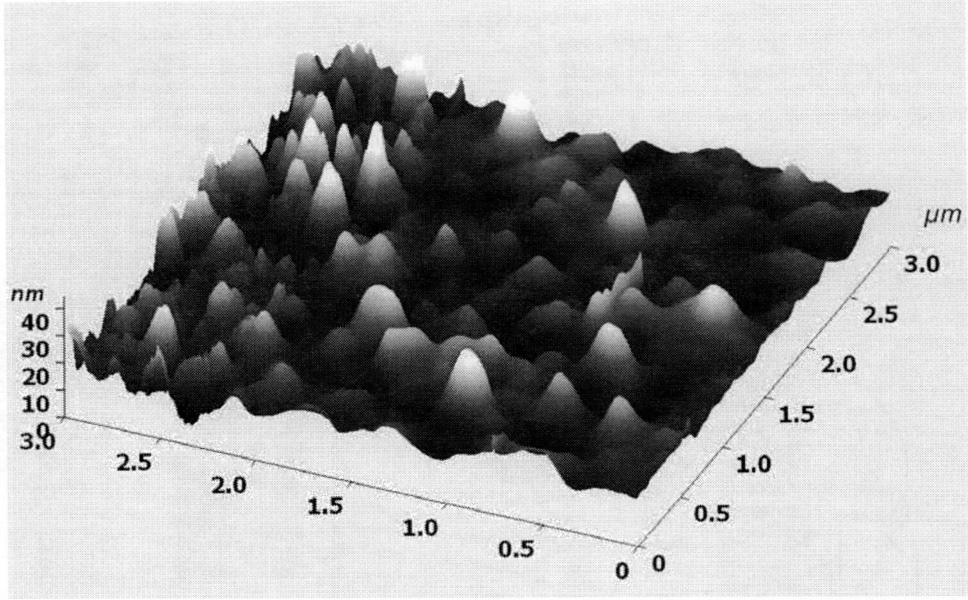

Figure 54: 3–D AFM structure of nitrogen nanofertilizer

In contact mode, AFMs use feedback to regulate the force on the sample. The AFM not only measures the force on the sample but also regulates it, allowing acquisition of images at very low forces. The feedback loop consists of the tube scanner that controls the height of the tip; the cantilever and optical lever, which measures the local height of the sample; and a feedback circuit that attempts to keep the cantilever deflection constant by adjusting the voltage applied to the scanner.

Atomic force microscopy (AFM) is used for quantitative and qualitative data based on different properties like morphology, size, surface roughness, and texture. It is used for the 3D characterization of nanoparticles as compare with other techniques like electron microscopy, dynamic light scattering, and optical characterization methods. It also provides details about the different geometries of nanoparticles, an analysis of hydrated nanofertilizers, and physical properties like magnetic behaviour. It is also used for the study of soft and hard synthetic materials regardless of their conductivity and opaqueness. In AFM, the force between the sharp probe tip (< 10 nm) and the sample surface, with a 0.2–10 nm probe-sample separation, is measured. The probe is attached with a cantilever, which deflects upon interaction; this deflection is measured by the reflection of a laser beam by the "beam bounce" method. Therefore the topography of the surface is directly measured by the deflections of the cantilever (Figure 55). The topography map takes the form of different peaks which are represented by different color gradients (red, orange, yellow, etc.) or grayscale. In this way a multicolor sample surface topology image is produced which can be very helpful for the identification and measurement of parameters under investigation.

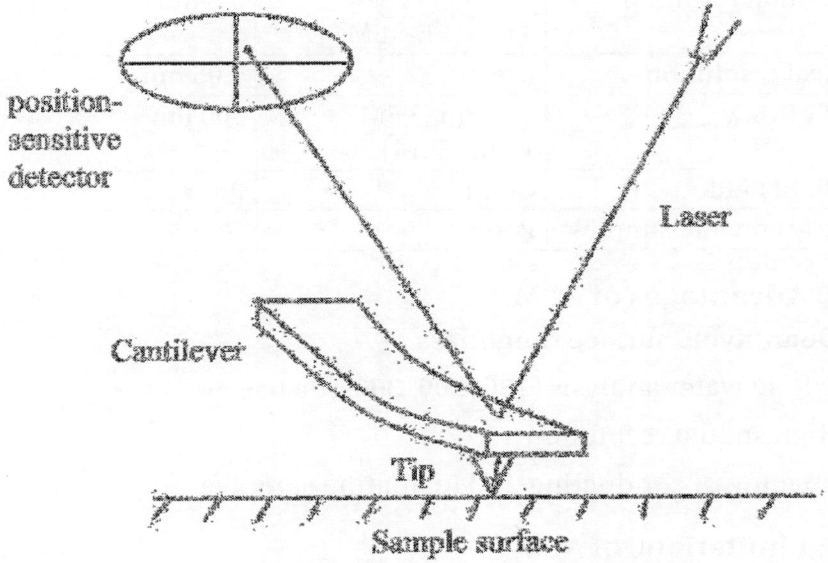

Figure 55: AFM is working with an optical lever.

6.4.1: Applications

- Examining the impact of surface roughness on adhesion and other process challenges
- Determining processing effects (*e.g.,* plasma treatment) used for surface modifications of nanoparticles and other biomedical surfaces
- Assessing trench shape/cleanliness on patterned wafers
- Determining whether morphology is the source of surface hazes.

There are wide differences between SEM/TEM with AFM microscopes on various characters of identification limit. The variations are shown as Table 16.

Table 16: AFM versus SEM/TEM microscope

Character	SEM/TEM	AFM
Sample	Must be conductive	Insulating/Conductive
Magnification	2 Dimentional	3 Dimentional
Time of image	0.1–1 minute	1–5 minute
Environment	Vaccum	Vaccum/Air/Liquid
Horizontal resolution	0.2 nm (TEM) 5 nm (FE-SEM)	0.2 nm
Vertical resolution	n/a	0.05 nm
Field of view	100 nm (TEM) 1 nm (SEM)	100 μm
Depth of field	Good	Poor
Contrast on flat samples	Poor	Good

6.4.2: Advantages of AFM

- Quantifying surface roughness
- Whole wafer analysis (150, 200, 300 mm)
- High spatial resolution
- Imaging of conducting and insulating samples.

6.4.3: Limitations of AFM

- Scan range limits: 100 microns laterally and 5 microns in z direction
- Potential problems with samples that are too rough and oddly shaped
- Tip-induced errors possible.

6.5. UV-VIS Absorption Spectroscopy

Ultraviolet-visible spectroscopy or ultraviolet-visible spectrophotometer refers to absorption spectroscopy in the ultraviolet-visible spectral region. This means it uses light in the visible and adjacent (near UV and near infra-red) ranges. The adsorption in the visible range directly affects the perceived colour of the chemicals involved. In this region of the electromagnetic spectrum, molecules undergo electronic transitions. Spectrophotometry investigates the absorption of different substances between the wavelength limit 160 nm and 780 nm. UV spectroscopy obeys the Beer-Lambert law, which states that: when a beam of monochromatic light is passed through a solution of an absorbing substance, the rate of decrease of intensity of radiation with thickness of the absorbing solution is proportional to the incident radiation as well as the concentration of the substance. The photomultiplier tube is a commonly used detector in UV-Vis spectroscopy. It consists of a photo emissive cathode (a cathode which emits electrons when struck by photons of radiation), several dynodes (which emit several electrons for each electron striking them) and an anode (Figure 56). UV-Vis is often called a general technique because most molecules will absorb in the UV-Vis wavelength range. The UV extends from 100–400 nm and the visible spectrum from 400–700 nm. The 100–200 nm range is called the deep UV. Light sources are more difficult to find for this range, so it is not routinely used for UV-Vis measurements. Typical UV-Vis spectrometers use a deuterium lamp for the UV that produces light from 170–375 nm and a tungsten filament lamp for visible, which produces light from 350–2,500 nm. This technique is complementary to fluroscence spectroscopy, in that fluroscence deals with transitions from the excited state to the ground state, while absorption measures transitions from the ground state to the excited state. The wavelength of light absorbed are determined by the actual electronic transition occurring and hence specific absorption peaks may be related to molecular substructure.

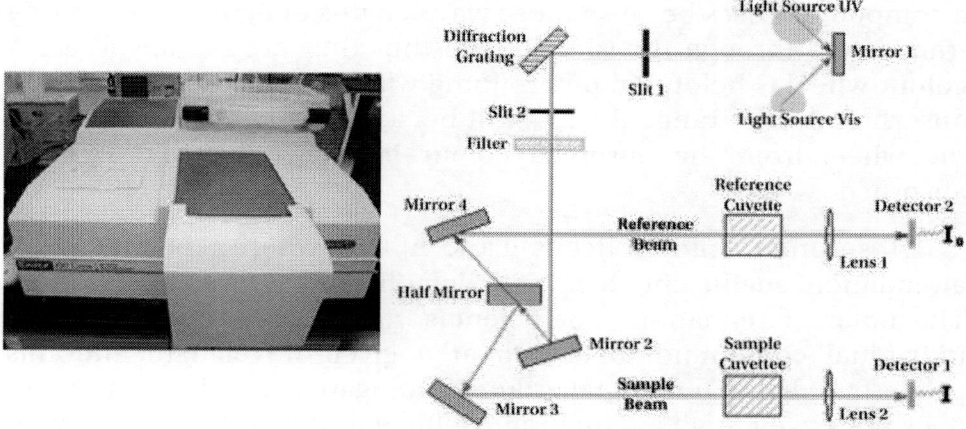

Figure 56: UV-Vis Spectroscopy (left) and the principle involved (right)

When a photon hits a molecule and is absorbed, the molecule is promoted into a more excited energetic state. UV-visible light has enough energy to promote electrons to a higher electronic state, from the highest occupied molecular orbital (HOMO) to the lowest unoccupied molecular orbital (LUMO). The energy difference between the HOMO and the LUMO is called the band gap. Typically, these orbitals are called bonding and anti-bonding. The energy of the photon must exactly match the band gap for the photon to be absorbed. Thus, molecules with different chemical structures have different energy band gaps and different absorption spectra. The most common transitions that fall in the UV-Vis range are π-π^* and n-π^*. Pi orbitals arise due to double bonds, and n orbitals are for non-bonding electrons. Pi star are anti-bonding pi orbitals. Thus, the best UV-Vis absorption is by molecules that contain double bonds. Pi orbitals adjacent to each other that are connected, called conjugation, typically increases absorption. Sigma-σ^* transitions, associated with single bonds, are higher energy and fall in the deep UV, so they are less useful for routine use. The appearance of broad bands or shoulders on the UV-Vis structure is due to the numerous vibrational and rotational states of a molecule, which lead to separate energy band gaps of slightly different energies.

For molecules with absorption in the visible region, the compounds will often appear coloured. However, a common misconception is that the wavelength of peak absorption (λ_{max}) for a compound is the colour it appears. A compound that appears red does not have much absorption in the red region of the spectrum. Instead, the λ_{max} for a compound that looks red is green. The colour of a compound arises because those wavelengths of light are selectively transmitted through the sample, and thus they are not absorbed. A colour wheel is helpful in determining what colour a compound will absorb and what range the λ_{max} will be, as the colour directly across the wheel from the observed colour is the colour that is most absorbed.

Absorption follows Beer's Law, $A=\varepsilon bC$ where ε is the molar attenuation coefficient, b is path length, and C is concentration. The molar attenuation coefficient is the characteristic of an individual compound to absorb at a given wavelength and this property is due to functional groups, conjugation, etc. If a compound does not have a high attenuation coefficient, it could be tagged with an appropriate group to increase its absorbance. Path length is generally related to the size of the cuvette and is 1 cm in standard spectrophotometers. A UV-Vis spectrum is shown as Figure 57.

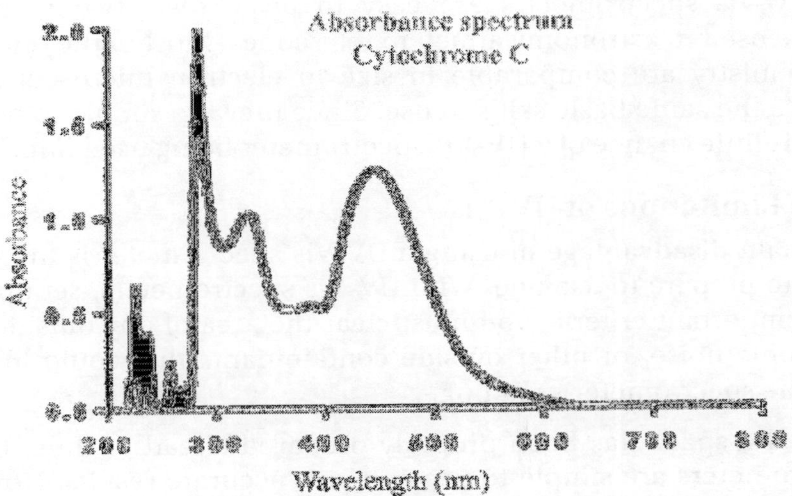

Figure 57: UV-Vis spectrum of a protein. The peak at 280 nm is indicative of a protein. The peak at 450 is due to absorbance of the heme group in cytochrome C.

UV-Vis is performed on a variety of instruments, from traditional spectrophotometers to more modern-day plate readers. The absorbance wavelength must be chosen, either using a filter or a monochromator. A monochromator is a device that separates the wavelengths of light spatially and then places an exit slit where the desired wavelength of light is. Monochromators can be scanned to provide a whole absorbance spectrum. Alternatively, a diode-array instrument allows all colours of light to be transmitted through the sample, and then the light is separated into different wavelengths spatially and detected using photodiodes. Diode-array instruments collect full spectra faster, but are more complicated and more expensive.

UV-Vis spectroscopy is routinely used in analytical chemistry for the quantitative determination of different analytes, such as transition metal ions, highly conjugated organic compounds, and biological macromolecules. Solvent polarity and pH can affect the absorption spectrum of an organic compound.

6.5.1: Advantages of UV-Vis

The biggest advantage for UV-Vis spectrometers is the accuracy of the device. Even small UV-Vis spectrometers can give extremely accurate readings, which is crucial when you are preparing chemical solutions or recording the movement of celestial bodies.

UV-Vis spectrometers are easy to use. Most UV-Vis spectrometers used in astronomy attach to telescopes. Most of the ones used in chemistry are comparable in size to electron microscopes and require the same basic skills to use. Since they are simple to operate, there is little chance of a UV-Vis spectrometer being used improperly.

6.5.2: Limitations of UV-Vis

The main disadvantage of using a UV-Vis spectrometer is the time it takes to prepare to use one. With UV-Vis spectrometers, set up is the most important criteria; you must clear the area of any outside light, electronic noise, or other outside contaminants that could interfere with the spectrometer's reading.

If the space has been properly prepared ahead of time, UV-Vis spectrometers are simple to use and give accurate results. However, if the space has not been properly prepared, even a small bit of outside light or vibration from a small electronic device could interfere with the results you are hoping to achieve in using a UV-Vis spectrometer.

6.6: X-ray Diffraction

X-ray crystallography is the experimental science determining the atomic and molecular structure of a crystal, in which the crystalline structure causes a beam of incident X-rays to diffract into many specific directions. X-ray diffraction (XRD) relies on the dual wave/particle nature of X-rays to obtain information about the structure of crystalline materials. A primary use of the technique is the identification and characterization of compounds based on their diffraction pattern. The XRD image of nano-phosphorus fertilizer is presented as Figure 58.

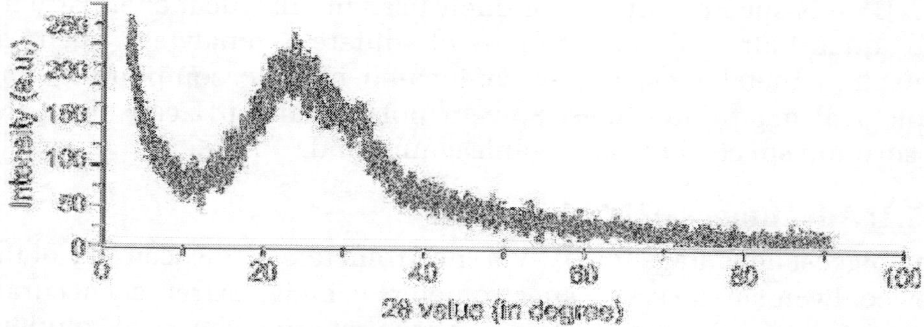

Figure 58: XRD image of biosynthesised nano phosphorus fertilizer

The dominant effect that occurs when an incident beam of monochromatic X-rays interacts with a target material is scattering of those X-rays from atoms within the target material. In materials with regular structure (*i.e.,* crystalline), the scattered X-rays undergo constructive and destructive interference. This is the process of diffraction. The diffraction of X-rays by crystals is described by Bragg's Law, n(lambda) = 2d sin(theta). The directions of possible diffractions depend on the size and shape of the unit cell of the material. The intensities of the diffracted waves depend on the kind and arrangement of atoms in the crystal structure. However, most materials are not single crystals, but are composed of many tiny crystallites in all possible orientations called a polycrystalline aggregate or powder. When a powder with randomly oriented crystallites is placed in an X-ray beam, the beam will see all possible inter atomic planes. If the experimental angle is systematically changed, all possible diffraction peaks from the powder will be detected.

The parafocusing (or Bragg-Brentano) diffractometer is the most common geometry for diffraction instruments. This geometry offers the advantages of high resolution and high beam intensity analysis at the cost of very precise alignment requirements and carefully prepared samples. Additionally, this geometry requires that the source-to-sample distance be constant and equal to the sample-to-detector distance. Alignment errors often lead to difficulties in phase identification and improper quantification. A mis-positioned sample can lead to unacceptable specimen displacement errors. Sample flatness, roughness, and positioning constraints preclude in-line sample measurement. Moreover, traditional XRD systems are often based on bulky equipment with high power requirements as well as employing high powered X-ray sources to increase X-ray flux on the sample, therefore increasing the detected diffraction signals from the sample. These sources also have large excitation areas, which are often disadvantageous for the diffraction analysis of small samples or small sample features.

Polycapillary X-ray optics can be used to overcome many of these drawbacks and constraints to enhance XRD applications. Polycapillary collimating optics converts a highly divergent beam into a quasi-parallel beam with low divergence. They can be used to form a Parallel Beam XRD instrument geometry which greatly reduces and removes many sources of errors in peak position and intensity inherent to the parafocusing geometry, such as sample position, shape, roughness, flatness, and transparency. Polycapillary focusing optics collects X-rays from a divergent X-ray source and direct them to a small focused beam at the sample surface with diameters as small as tens of micrometers for micro X-ray diffraction applications of small samples or small specimen features. Both types of polycapillary optics direct very high X-ray intensities to the sample surface, such that XRD systems employing optics can use low power X-ray sources, reducing instrument size, cost, and power requirements.

X-ray diffraction using X-ray optics has been applied to many different types of applications including thin film analysis, sample texture evaluation, monitoring of crystalline phase and structure, and investigation of sample stress and strain. Figure 59, shows XRD spectra of magnesium nanofertilizer. Determination of unknown solids is critical to studies in nanotechnology, material science, engineering and biology.

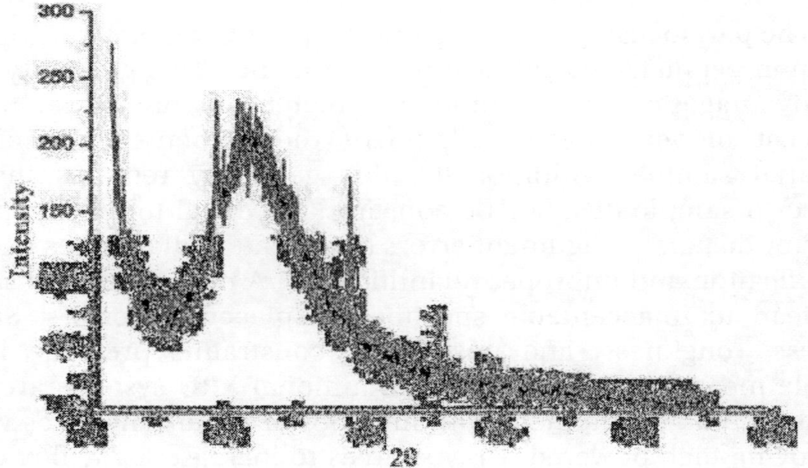

Figure 59: XRD spectra of Mg nanofertilizer

6.6.1: Other applications

- Determination of unit cell dimensions.
- Measurement of sample purity.
- Determine crystal structures and orientation.
- Determine the thickness, roughness and density of the nanostructures.
- Measurement of average crystalline size, strain, or micro-strain effects in bulk and thin-film samples.

6.6.2: Advantages

- Powerful of rapid (< 20 min) technique for identification of an unknown mineral.
- Minimal sample preparation is required.
- XRD units are widely available.
- Data interpretations are relatively straight forward.

6.6.3: Disadvantages

- User must have access to standard database (d-spacings, hkls).
- For unit cell dimensions, indexing of patterns for non-isometric crystal system is complicated.
- Peak overlay may occur and worsens for high angle reflections.

6.7: Fourier Transform Infrared Spectroscopy (FTIR)

FTIR Spectroscopy is a method to determine the chemical functional group present in the sample. Different functional group absorbs different IR frequencies. The identification of functional groups leads to determine the reducing agent and the capping agent responsible for synthesis and stability of nanoparticles. Fourier transforms infrared spectroscopy (Figure 60) is usually better than other methods as it is a non-destructive, environment-friendly technique; it has good speed and more reliable outputs. It is also mechanically easy to operate providing precise measurement without requiring any external calibrations. Further, it gives an indication which functional group is involved in breaking down or absorption of particles. The special advantages of FTIR are identifying changes in protein secondary structures, can resolve between similar components. The FTIR spectra of Zn nanofertilizer is shown as Figure 61.

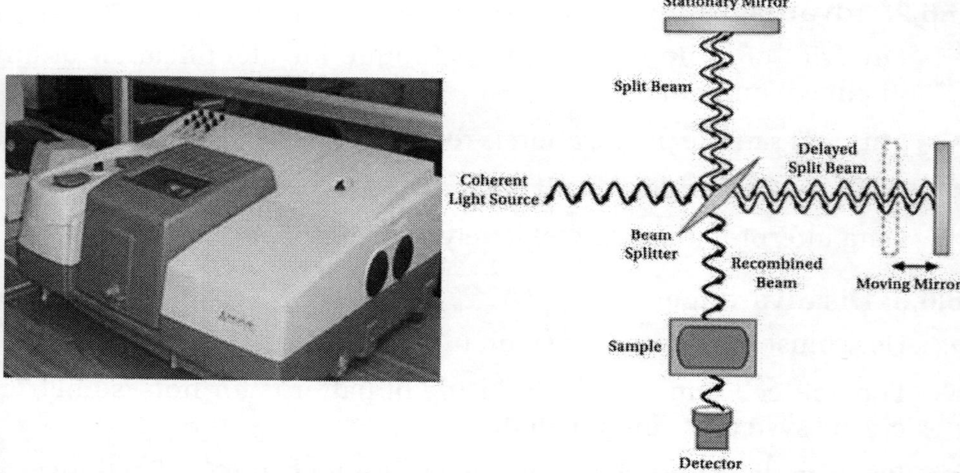

Figure 60: Fourier Transform Infrared Spectroscopy (left) and principle involved (right)

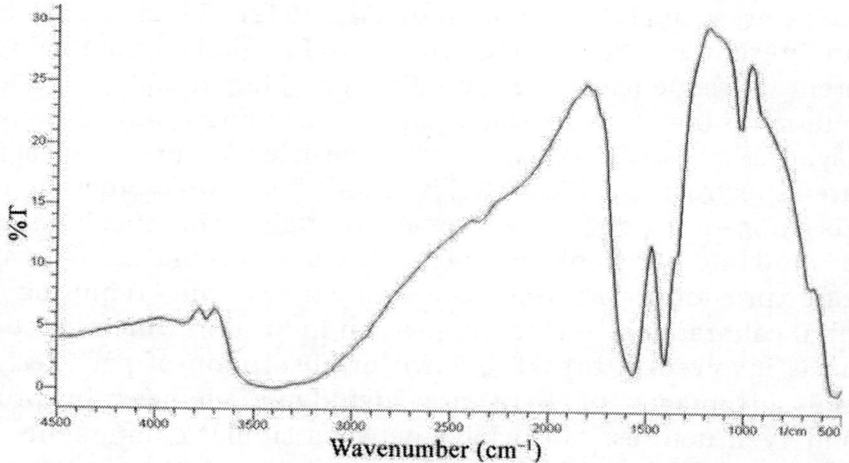

Figure 61: FTIR Spectra of biosynthesised Zn Nanofertilizer

Fourier transform infrared (FTIR) spectroscopy may further be used for quantitative analysis of the reaction products. FTIR is a powerful tool for identification of type of chemical bonds in a molecule by producing an infrared spectrum that is like a molecular fingerprint. The wavelength of light absorbed and transmitted is a characteristic of the chemical bond, as can be seen in the annotated spectrum (Figure 62), for biologically synthesised Mg nanofertilizer particles. FTIR spectra revealed the respective metal as well as peak for fungal protein, which encapsulate metal nanoparticle and make them more stable and eco-friendly. The graphical representation in the figure 62 shows the signature peak for Mg metal marked with arrow and other peaks are due to fungal extracellular protein encapsulating the synthesised nanoparticles.

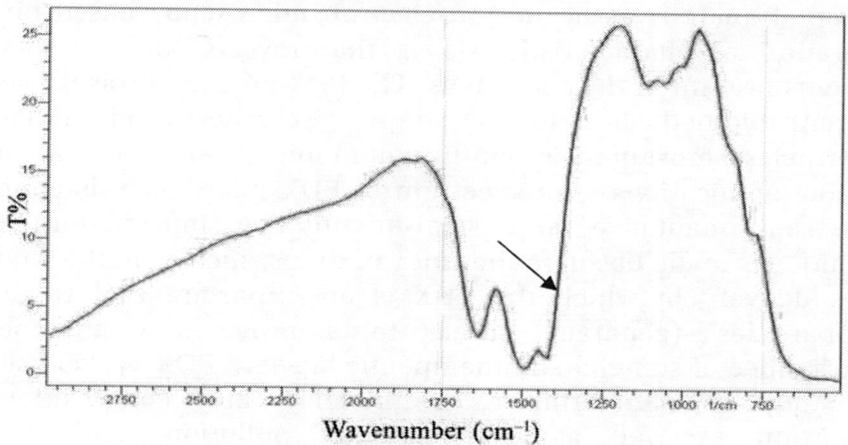

Figure 62: FTIR spectra of biosynthesised Mg nanoparticles

The FTIR measurement can be utilized to study the presence of protein molecule in the solution, as the FTIR spectra in the 1400–1700 cm^{-1} region provides information about the presence of – CO- and – NH – groups. In addition, FTIR spectra also help in identification of presence of metal and their metal oxides.

6.7.1: Application of FTIR

- The most common utilization of FTIR is for the identification of capping molecules. Capping molecules play a critical role in the stability of nanoparticles and also determine their interactions with the environment. Common examples are citric acid, folic acid, polymers, proteins, etc.
- FTIR spectra can also help in qualitative identification of metals and their metal oxides nanoparticles. A broad peak in the vicinity of wavenumbers 400–800 cm^{-1} is observed for the ZnO, Fe_xO_y, etc.

6.7.2: Advantages of FTIR

The method is rapid, cheap, easy to perform, high sensitivity, moderate sample preparation.

6.7.3: Disadvantages of FTIR

Mostly restricted to organic compounds; difficult to analyse liquid samples; data interpretation is bit complicated.

6.8: Energy Dispersive X-ray Spectroscopy (EDS)

The Energy Dispersive X-ray (EDX) microanalysis is a technique of elemental analysis associated to electron microscopy based on the generation of characteristic X-rays that reveals the presence of elements present in the specimens. The EDX microanalysis is used in different biomedical fields by many researchers and clinicians. Nevertheless, most of the scientific community is not fully aware of its possible applications. The spectrum of EDX microanalysis contains both semi-qualitative and semi-quantitative information. EDX technique is made useful in the study of drugs, such as in the study of drugs delivery in which the EDX is an important tool to detect nanoparticles (generally, used to improve the therapeutic performance of some chemotherapeutic agents). EDX is also used in the study of nanofertilizer identification and composition in proportion as well as environmental pollution, and in the characterization of mineral bio- accumulated in the tissues. The results also indicate the purity and nanonutrients content (%) in the sample. In conclusion, the EDX can be considered as a useful tool in all works that require element determination, endogenous or exogenous, in the tissue, cell or any other sample. The Energy Dispersive ray (EDX) microanalysis is involved in different biomedical fields of study due to its high sensitivity in detecting the different elements in tissues. In particular, EDX microanalysis can carry a huge advantage in the detection of heavy metals pollutions.

The EDS technique utilizes X-rays that are emitted from the sample during electron bombardment to determine the elemental composition. When the sample is bombarded by the electron beam of the SEM, electrons are ejected from the atoms comprising the sample's surface. A resulting electron vacancy is filled by an electron from a higher shell, and an X-ray is emitted to balance the energy difference between the two electrons. The EDS detector measure the number of emitted X-rays and plot the value versus their energy. The energy of the X-ray is characteristic of the element from which the X-ray was emitted. A spectrum of the energy versus relative counts of the detected X-rays is obtained and evaluated for qualitative and quantitative determinations of the elements present in the sampled volume. The EDS image of biosynthesised Mg nanofertilizer particle is shown as Figure 63.

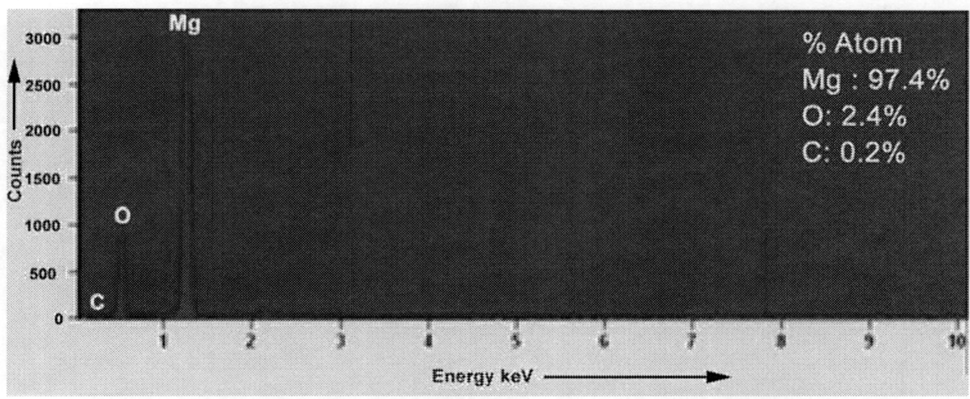

Figure 63: Typical EDX image of biosynthesised magnesium nanofertilizer

Energy dispersive X-ray analysis is a technique to analyze near surface elements and estimate their proportion at different position, thus giving an overall mapping of the sample. Characteristic X-rays are emitted when the electron beam removes an inner shell electron from the sample, causing a higher energy electron to fill the shell and release energy. These characteristic X-rays are used to identify the composition and measure the abundance of element in the sample. The EDS spectrum of figure 63 showed the strong intensity peaks for Mg only, which clearly reveals and confirm the presence of only Mg in the sample. Purity of the product is a necessary prerequisite for nanoparticle synthesis method. There was presence of low intensity peak of oxygen, which confirms that particles were encapsulated by fungal capping proteins entities. Low intensity carbon and oxygen peaks may be due to presence of extracellular fungal protein which encapsulated the metal nanoparticles, and oxygen peaks were shown due to dissociation of precursor compounds.

The EDX detector system performs a simultaneous display of all mid-energy (1–20 keV) X-rays collected during any individual analysis period and the energy of the X-rays is reproduced as a spectrum, which is a histogram plot of number of counts against X-ray energy. The spectrum contains both semi-qualitative and semi-quantitative information. The position of a peak in the spectrum, its energy, identifies the element; the area under the peak is proportional to the number of atoms of the element in the irradiated area. X-rays are also produced when the electron beam is slowed by the electrostatic fields of the atomic nuclei of elements present in the specimen. These X-rays form a continuous radiation that appears below the peaks in that spectrum. Qualitative analysis, *i.e.,* identification of elements in the spectrum, is usually achieved using manufacturer's software. These systems are not fool proof, they are designed for materials applications rather than biology and mis-identification of a peak by the software is a distinct possibility. Conventionally, EDX is a powerful technique allowing an elemental analysis of the surface of the samples (Figure 64).

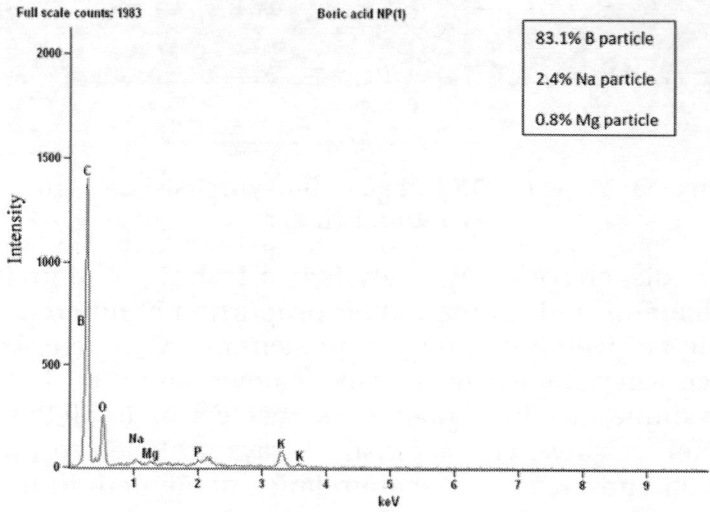

Figure 64: EDS image of boron nanofertilizer

This method implies some limitation in elemental analysis. Firstly, X-ray spectrometry detects elements and it is not capable of distinguishing between ionic and nonionic species. Moreover, in the EDX it is required that all samples are analyzed under relatively vacuum and obviously this has serious implications for the preparation of the specimens because electron and X-rays are strongly absorbed by air molecules. Generally, X-ray detection is not influenced by the chemical state of elements, but it is influenced by inter-elements interference, known in X-ray spectrometry as peak overlap, causing serious problems in the elemental analysis. Therefore, it is possible to detect those elements with atomic number larger than 10. The minimal detectable elemental concentration, which requires some signal averaging, is approximately 0.1 mmol per kg of dry specimen (*i.e.,* 10 ppm), whereas spatial resolution ranges from about 10 nm to a few micrometers. Reducing the detection limit requires more counts, which can be obtained by increasing the counting time and /or the beam current. Values given here for detection limits refer to biological samples for which the mean atomic number (which determines continuum intensity) is generally quite low (*e.g.*, C, H, N, Cl, P, K). Biological samples containing heavy elements give higher detection limits due to the higher background (Figure 65).

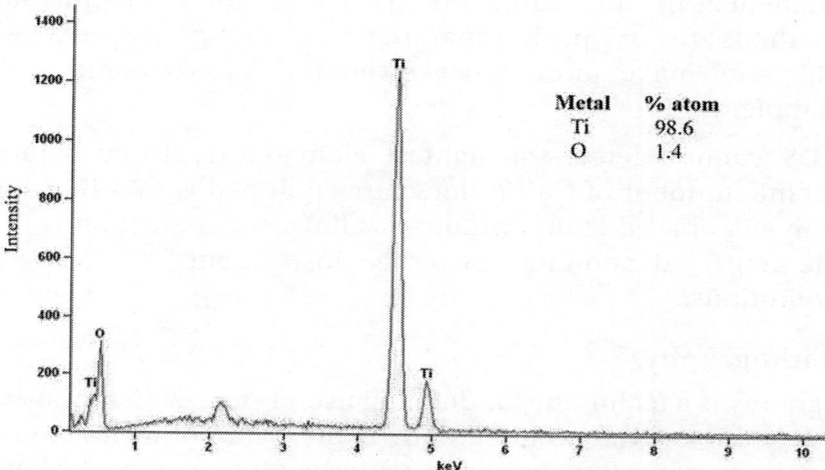

Figure 65: EDS spectra of biosynthesised Ti nanofertilizers

The elemental compositions of biotransformed product containing Ti nanoparticles in the solution were confirmed by TEM equipped with energy dispersive X-ray spectroscopy (TEM-EDX). The results of TEM- EDX clearly showed that the Ti nanoparticles were highly intense and the maximum intensity was found at 4.6keV. The peak at intensity about 360 of oxygen was due to disassociation of Ti and oxygen molecules from the precursor compound TiO_2 used for biosynthesis of Ti nanoparticles. Results clearly depicts that sample contains 98.6% atom of Ti metal only.

6.8.1: Advantages of EDS

- When used in "spot" mode, a user can acquire a full elemental spectrum in only a few seconds. Supporting software makes it possible to readily identify peaks, which makes EDS a great survey tool to quickly identify unknown phases prior to quantitative analysis.
- EDS can be used in semi-quantitative mode to determine chemical composition by peak-height ratio relative to a standard.

6.8.2: Limitations of EDS

- There are energy peak overlaps among different elements, particularly those corresponding to X-rays generated by emission from different energy-level shells (K, L and M) in different elements. For example, there are close overlaps of Mn-Kα and Cr-Kβ or Ti-Kα and various L lines in Ba. Particularly at higher energies, individual peaks may correspond to several different elements: in this case, the user can apply de-convolution methods to try peak separation, or simply consider which elements make more sense given the known context of the sample.

- EDS cannot detect the lightest elements, typically below the atomic number of Na for detectors equipped with a Be window. Polymer based thin windows allow for detection of light elements, depending on the instrument and operating conditions.

6.9: Lithography

Lithography is a technique for defining useful shapes on the surface of a semiconductor wafer. Lithography is the group of techniques used to transfer previously determined patterns on a substrate. There are many lithographic techniques, like photolithography, electron beam lithography, X-ray lithography etc. which are developed by using different lens systems, radiation exposures, but all of them follow a basic and common approach. This technique is used to transfer previously determined pattern on a surface and is a crucial technique in fabrication of nanoparticles. Lithography can be used to print text or artwork onto paper or other suitable material. Lithography originally used an image drawn with oil, fat, or waxes onto the surface of a smooth, level lithographic limestone plate. It is generally used to prepare prototype.

Lithography uses simple chemical processes to create an image. For instance, the positive part of an image is water repelling (hydrophobic) substance, while the negative image would be water retaining (hydrophilic). Thus, when the plate is introduced to a compatible printing ink and water mixture, the ink is adhare to the positive image and the water will clean the negative image. This allows a flat print plate to be used, enabling much longer and more detailed print runs than the older physical methods of printing (*e.g.,* intaglio printing, letterpress printing).

In the field of nanotechnology, lithography represents another route of nano structure by physical method. It is also refer to as photoengraving, and a process of transferring a pattern into a reactive polymer film, termed as resist, which will subsequently be used to replicate that pattern into an underlying thin film.

Lithography is extensively used technique in nanofabrication and nanosynthesis.

6.9.1: Advantage of Lithography

- Clarity is excellent. Offset lithography creates clear, smooth, sharp images and text on a variety of materials.
- Lack of Impressions.
- Low cost.
- Speed and Versatility.

6.9.2: Disadvantages of Lithography

The disadvantages are fewer, but they include the tediousness and cost associated with the production of the printing plates and set up. Also, the printing quality, while high, is not quite as good as photogravure or rotogravure printing.

6.10: Inductively Coupled Plasma Mass Spectrometry (ICP-MS)

It is a highly sophisticated technique uses plasma to dissociate ions contained in a sample. These ions are then extracted from the plasma and analyzed in mass spectrometer for detection. This technique is highly sensitive and capable of detection of nanoparticles (metals/non-metals) with extremely low concentration of order 10^{-12} g. The coupling of plasma with mass spectrometer provides us with an insight structure of nanoparticles. ICPMS is also capable of monitoring isotopic speciation for the ions of choice.

For coupling to mass spectroscopy, the ions from plasma are extracted through a series of cones into a mass spectrometer, usually a quadrupole. The ions are separated on the basis of their mass-to-charge ratio and a detector receives an ion signal proportional to the concentration. The concentration of a sample can be determined through calibration with certified reference material such as single or multi-element reference standard. ICPMS also lends itself to quantitative determinations through isotope dilution, a single method based on an isotopically enriched standard.

6.10.1: Advantage of ICPMS

- The favourable detection limits (0.01 to 0.1 micrograms/L for many elements).
- Simple specimen preparation.

- First detection about 40 specimens per hour.
- The ability to measure more than one element simultaneously.

6.10.2: Disadvantage of ICPMS

A major disadvantage is the high capital cost of the instrumentation.

6.11: Inductively Coupled Plasma Atomic Emission Spectroscopy (ICPAES)

Inductively coupled plasma atomic emission spectroscopy (ICP-AES), also referred to as inductively coupled plasma optical emission spectrometry (ICP-OES), is an analytical technique used for the detection of chemical elements. It is a type of emission spectroscopy that uses the inductively coupled plasma to produce excited atoms and ions that emit electromagnetic radiation at wavelengths characteristic of a particular element. The plasma is a high temperature source of ionised source gas (often argon). The plasma is sustained and maintained by inductive coupling from cooled electrical coils at megahertz frequencies. The source temperature is in the range from 6000 to 10,000 K. The intensity of the emissions from various wavelengths of light is proportional to the concentrations of the elements within the sample (Figure 66). Here the argon gas is ionized in the intense electromagnetic field and flows in a particular rotationally symmetrical pattern towards the magnetic field of the RF coil. Stable, high temperature plasma of about 7000 K is then generated as the result of the inelastic collisions created between the neutral argon atoms and the charged particles.

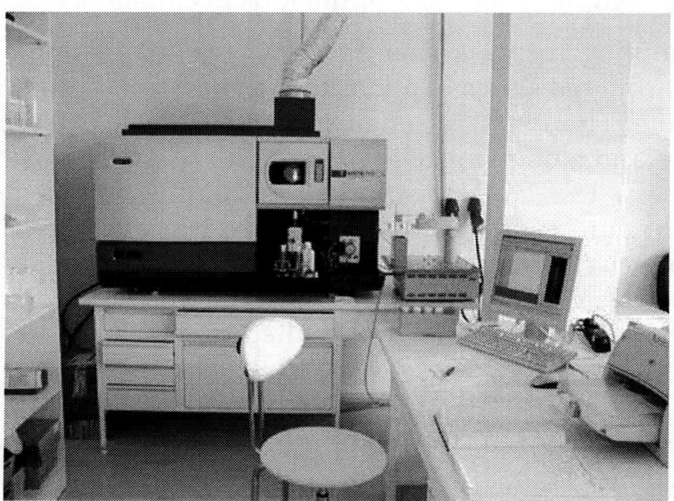

Figure 66: ICPAES or ICPOES

The difference between ICPMS and ICPAES is their detection limit. ICPMS can detect up to 0.6 ppb level of the particle concentration whereas ICPAES detect up to 20 ppb level concentration. Both are becoming the analytical method of choice for the determination of nutrient levels in agricultural soils. This information is then used to calculate the amount of fertilizer required to maximise crop yield and quality besides trace and movement of nanofertilizers.

6.11.1: Advantage of ICPAES

- The possibility of multi elemental analysis.
- Less interference.
- Mostly of spectral nature.
- Good sensitivity.

6.11.2: Disadvantage of ICPOES

The biggest drawback is lack of sensitivity for some elements, and physical and spectral interferences.

Further Studies

Chu, B. (1992). Laser light scattering. Basic Principles and Practice. Academic Press.

Egerton, R. F. (2005). Physical principles of electron microscopy: An introduction to TEM, SEM, and AFM. Springer, 202p.

Griffiths, P. and De Hasseth, J. A. (2007). Fourier transform infrared spectroscopy (2nd edn.), Willey-Blackwell.

Kenkel, J. (1994). Analytical chemistry for techniques. Lewis Publishers, Boca Raton, US.

Reimer, L. (1998). Scanning electron microscopy: Physics of image formation and microanalysis. Springer, 527p.

7

Procedures for Nanofertilizer Application

It is well known that, each of the nutrient element play a major role in growth and development of the plants, and when present in deficient quantities can reduce growth and yields. Soil is the major natural source of plant nutrients. Soil may support growth and development of wild flora just sufficient for their survival and regeneration. However, intensive crop production that aims at high levels of productivity needs supplemental plant nutrition which may be given through soil application and/or foliar application. Nanofertilizer can be applied both on soils and on leaves as foliar. This can also be applied through drip, hydroponic, aqua and aeroponic.

7.1: Foliar Application

Foliar applied fertilizers are facing several structural barriers, because the nutrients are salt based (cations/anions) which may struggle to penetrate the inner plant tissue cells. This is because of pore size of cell wall that ranges between 5 to 20 nm. Hence, nanofertilizers aggregate with diameter less than the pore size of plant cell wall which can easily enter through the cell wall and reach up to the plasma membrane. According to the polar pre-model, for the penetration of the polar and ionic solutes to the cuticle the exclusion limit of pore radius has been estimated as 2 to 2.4 nm, where as for the stomatal diffusion, the pore radius always exceeded 20 nm. The application of nanofertilizer is promising and efficient translocation of nutrients to the desired parts of plant. Engineered nanoparticles can penetrate the stomatal pores with the size of less than 50 nm as observed by many researchers in several plants and the size exclusion limit of stomata in the watermelon is 27.3–46.7 nm. The foliar applied nanoparticles get transported from the site of application to the heterotrophic cells, which carried via the phloem vessels likely through the plasmadesmata (40 nm in diameter). But when applied as foliar or nanofertilizer in suspension then their stability depends on their zeta potential (Table 17) of the particle.

Table 17: Stability of nanoparicles in suspension with respect to zeta potential

Sl. No.	Zeta potential (mV)	Stability behaviour
1	from 0 to ±5,	Rapid coagulation or flocculation
2	from ±10 to ±30	Incipient instability
3	from ±30 to ±40	Moderate stability
4	from ±40 to ±60	Good stability
5	more than ±61	Excellent stability

The uptake of nanoparticles into plant cells via binding to carrier proteins through aquaporin, ion channels and endocytosis. Nanoparticles can also be transported into the plant by forming complexes with membrane transporters that nanoparticles can be entered in to the plant system and move through the vascular system. Nanomaterials are capable to enter from the atmosphere into the leaf stomata and then redistribute to the plant parts as observed in watermelon plant and in cucumber leaves. It has been found that calcium oxide nanoparticles (n-CaO) get transported via phloem tissue of groundnut. In wheat plants nanoparticles were present in phloem tissues which mean that nanoparticles were taken up and transported through phloem route from leaves to stem down to roots, which can be documented with transmission electron microscope. It was general believe that nanofertilizer with the diameter of less than 100 nm can easily penetrate through the stomata of leaves and were redistributed from leaves to stems through the phloem sieve elements. Once the nanoparticle gets entered into the plant system which may be transported from one cell to other cell through plasmodesmata and carried by aquaporins, ion channels, endocytosis or by binding to organic chemicals.

The effect of foliar spray to different nanoparticles on mung bean under green house condition has been shown (Figure 67). In general after foliar spray, the nanoparticles are adsorbed to plant surfaces and taken up through natural nano-or micrometer scale plant openings. Several pathways exist or are predicted for nanoparticles association and uptake in plants (Table 18).

Figure 67: Effect of foliar sprays (optimum concentration) of biologically synthesised nanoparticle *vis-a-vis* control on Mung bean under pot experiment

Table 18: Pathways of nanoparticles entry

Plant Shoot	Plant Root
• Cuticle, epidermis	• Root tips
• Stomata	• Rhizodermis
• Hydathodes	• Lateral root junctions
• Stigma	• Wounding
• Other entry sites	• Cortex

Nanoparticles uptake in the plant body can use different paths. Uptake rates will depend on the size and the surface properties of the nanoparticles. Very small sizes nanoparticles can penetrate through the cuticle. Larger particles can penetrate through cuticle free areas, such as hadathodes, the stigma of flowers and the stomata. Nano-particle must traverse the cell wall before entering the intact plant cell protoplast. Result suggests that only nanoparticles less than 5 nm in diameter will be able to traverse the cell wall of undamaged cell efficiently. The movement of foliar sprayed particle of different sizes on plant leaves at 48h time after application is shown in table 19.

The results clearly showed that smaller particle can move faster than higher particle sizes. The movement inside the plant cell also depends on the shape of the particle. In general, cube shaped particle moves faster than other shapes of the particle. After entering into plant cell the nanofertilizer particle may trigger different enzyme system (Figure 68), enable plants to release more enzymes for native nutrient mobilization.

Table 19: Movement of P nanofertilizers after application as foliage on plants

Nanoparticle size	% present in Leaves	% present in Shoots	% present in Roots
20 nm	36.2	10.4	53.4
40 nm	72.5	5.8	21.7
60 nm	78.3	10.2	11.5
80 nm	85.1	8.0	6.9

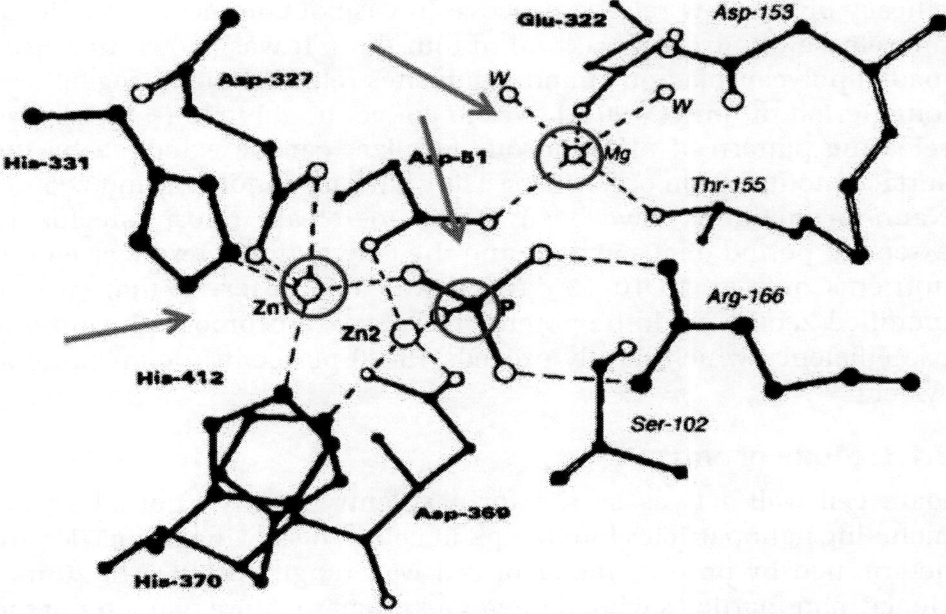

Figure 68: Triggering the phoshphatase enzymes by foliar application of nanofertilizers

Polymer helps to release the nutrients in a controlled manner, a character can be useful in the production of polymer coated nanofertilizer as smart fertilizer. Polymeric nanoparticles have a matrix architecture composed of biodegradable and biocompatible polymers of synthetic or natural origin. The most widely used synthetic polymers are polylactide, polylactide-polyglycolide copolymers, polycaprolactones and polyacrylates. Among the various natural polymers, alginate, albumin or chitosan have been widely explored. The biodegradable, polymeric chitosan nanoparticles with the size of 78 nm were used for the controlled release of NPK fertilizer. It has been found that nano-clay based fertilizers (zeolite and mont-morillonite) released the nutrient for a long period of time (>1000 h) than conventional fertilizer (< 500 h). Zeolite is natural mineral having extensive surface area and can hold wide range of positive and negative nutrient ions after suitable partial modification. The Urea-Hydroxyapatite nanohybrids has been produced for slow release of nitrogen and observed that nanohybrids strongly binds the urea, which released over longer period of time (up to one week) and the release rate was reduced by ~12 times compared to that of pure urea. Urea-encapsulated HA nanoparticles even on the day 60 the nanofertilizer was releasing nitrogen > 10 mg, clearly showing the efficacy of the slow release process. In case of commercial fertilizer, the release of nitrogen was within four days. It was shown that urea loaded polycaprolactone nanocomposites released the nitrogen for a long period of time (> 90 h) over to conventional urea (< 25 h). The releasing pattern of phosphorous fertilizer can be extended by the surface modification of fertilizer using various nanoclays and zeolite. Nano-formulations have been shown to release phosphate for an extended period of 40–50 days and the conventional fertilizer let out nutrients only up to 10–12 days. It was also observed that surface modified zeolite could be potential strategy to promote phosphorus use efficiency which hardly exceeds 18–20 per cent in conventional system.

7.1.1: Mode of entry

Plant cell wall acts as barrier for easy entry of any external agents including nanoparticles into the plant cells. The sieving properties are determined by pore diameter of cell wall ranging from 5 to 20 nm. Hence, nanoparticles with diameter less than the pore diameter of the cell wall could easily pass through and reach the plasma membrane. There is also a chance of enlargement of pores or induction of new cell wall pores upon interaction with nanofertilizer particle which in turn enhance nanoparticles uptake. Further internationalization occurs during endocytosis with the help of a cavity like structure that formed around the nanoparticles by plasma membrane. They may also cross the membrane using embedded transport carrier proteins or through ion channels. In the cytoplasm the nanofertilizer particles are applied on leaf surfaces, they enter through the stomotal openings or through the base of the trichomes and then translocated through various tissues. However, accumulation of nanofertilizer particles on photosynthetic surface cause foliar heatings which results in alterations of gas exchange due to stomotal obstructions that produce changes in various physiological and cellular functions of plants.

7.1.2: Entry of nanocapsules

The nanoencapsulated and the resulted polymer nanocapsules are spread on to the leaf tissue. These encapsules enter the plant through the stomata orifices. During the process, the nanocapsules chemical bonds of the polymer wall can be weakened or broken by a critical amount of stress enzyme present. Plant cell stress enzymes are activated by mechanical, thermal, chemical or biological stress. This stress sensitizes the plant during an attack and infection from fungi and bacteria. These polymer based nanocapsules are able to prevent this infection.

7.1.3: Entry of secondary and microelements

The secondary and microelements, such as Ca, Mg, Fe, S, Zn are encapsulated into microspores and these polymer microcapsules are, with time, incorporated and dissolved into the soil. Once close to the root, the chemical bond of the microcapsules wall polymer is broken down by the organic acids or phenolic substances from the root exudates. These root exudates are typically released to enhance plant feeding during the plant growth process. The possibility of targeting the movement of nanofertilizer particles to specific sites of an organism paves way for the use of nano-biotechnology in the treatment of plant diseases that affects specific parts of the plant. Different procedures have made use of nanofertilizer particles in plants, such as the controlled release of bioactive substances in solid wood and plant transformation through bombardment with gold or tungsten particles coated with DNA.

7.1.4: Effect of foliar application of nanofertilizer on growth parameters

Nanofertilizers have important role in physiological and biochemical processes of crops by increasing the availability of nutrients, which help in enhancing metabolic processes and promoting meristematic activities causing higher apical growth and photosynthetic area. It was documented by some research studies, where foliar spraying of nanoformulations of NPK and micronutrients mixture increased the plant height and number of branches in black gram. Nano NPK increased the growth of leaves in wheat, which was obtained by enhanced availability of nutrients by easy penetration of nano formulation of NPK through stomata of leaves via gas uptake. Foliar applied nitrogen nanofertilizer increased the leaf dry weight of pepper mint by 165 per cent over control. Foliar application of zinc nano-fertilizer on pearl millet (*Pennisetum americanum* L.) significantly increased shoot length, root length, root area and plant dry biomass and on cotton crop increased fresh weight and dry weight have been recorded due to improved physiological processes like chlorophyll content and antioxidant activity. Growth parameters like plant height, leaf number and fresh and dry weight of savory plant get increased by nano-zinc application. Zinc has an effect on synthesizing of natural auxin (IAA) and also can activate many enzymes involved in the biochemical pathways, such as carbohydrate metabolism, protein metabolism, growth regulator metabolism, pollen formation and maintaining the integrity of biological membranes. Thus, the plant growth promoting harmone contents get increased with the application of nano-Zn fertilizer. Foliar application of nano-iron fertilizer increased the growth of forage corn and *Ocimum basilicum* L. because of enhanced production of crude protein and soluble carbohydrates. The foliar spray of nano-TiO_2 increased the total dry matter production of plants by enhancing nitrogen assimilation, photo-reduction activities of photosystem II and electron transport chain and scavenging of reactive oxygen species. It was observed that foliar application of nano-TiO_2 had no significant effect on number of fertile tillers in barley because number of tillers in wheat and barley is mainly controlled by genetic factors and nutrition has a minor effect on this trait.

7.1.5: Nanofertilizer foliar application on physiological parameters

There was a remarkable increase in physiological and biochemical parameters of crops with the application of nanofertilizers as foliage. Biocompatible magnetic nanofluid had positive influence on the total chlorophyll content (a and b) in sunflower leaves. However, with higher concentrations (> 0.75 % MNF) the growth rate of the chlorophyll content is negative. Foliar application of nano-TiO_2 has been recorded significantly higher chlorophyll content, carotenoids and anthocyanins of maize crop, which can facilitate an increase in corn yield. Application of nanosized TiO_2 particles as a foliar spray positively influenced some morpho physiological characters like days to anthesis and chlorophyll content of barley and wheat. In fact, nano TiO_2 can improve structure of chlorophyll and helps better capture of sunlight, facilitates manufacture of pigments, stimulates rubisco activity and also increases photosynthesis. Nano formulation of TiO_2 improved spinach growth and also enhanced nitrogen metabolism, protein and chlorophyll contents. In another study, nano-TiO_2 had significantly increased chlorophyll content on the spinach leaves, and it was 17 times higher than the control plot. The photosynthetic rate gets increased by 29 per cent compared to control. Nano-Mg fertilizer also improves the chlorophyll content of wheat (Table 20).

Table 20: Effect of foliar application of Mg nanofertilizer on sunlight absorption and chlorophyll content in wheat (6 weeks old plant)

Treatment	Light absorption (Lux)	% improvement over control	% improvement in chlorophyll content
Control	1432.8	-	-
Chemical fertilizer	1481.5	3.4	1.9
Nano-fertilizer particle size (< 20nm)	1743.5	21.7	16.7

Nano-chelate zinc fertilizer application proved to enhance the activity of peroxidase, catalase, and polyphenol oxidase enzymes in cotton and soybean crops which increases the shoot and root growth. In pearl millet crop, increased chlorophyll content, total soluble leaf protein and plant dry biomass were obtained with foliar application of zinc nano-fertilizer and in savory plants the contents of chlorophyll, essential oil and phosphours were increased by nano-zinc application. Foliar application of zinc absorption by leaf epidermis and remobilization in to the grain through phloem and several membranes of zinc regulated transporters which might have regulated this process. Nanofertilizer application increased the antioxidant potential in rice. Antioxidants are secondary metabolites produced under unfavourable conditions faced by the plants, such as water stress, salinity and limited nutrients. Moreover, nanofertilizer application was supplemental; its better absorption through plant cells some how provided enough nutrients to enhance antioxidant activities. The cardioprotective, antimutagenic, and anticarcinogenic effects of phenolic compounds are reported to be generally associated with their antioxidant properties that eliminate free radicals and alleviate lipid peroxidation.

The result of the foliar application of Zn, Mg and Ti nanofertilizers on total chlorophyll contents and total soluble protein in leaves of pearl millet were presented (Table 21). The results clearly demonstrated that the total chlorophyll contents increased from stage I to stage II with the application of fertilizer in their ordinary size compound and/or in nanoform. Nevertheless, a significant increase in chlorophyll contents was recorded with the application of fertilizers as nanoparticles as compared to ordinary size compound application, at stage II of plant growth.

Table 21: Effect of application of metal particles on total chlorophyll content and total soluble leaf protein content in pearl millet

Treatment	Total chlorophyll content ($mg\ kg^{-1}$)		Total soluble leaf protein ($mg\ kg^{-1}$)	
	I*	II	I	II
Control	9.1	30.3	28.9	37.7
Mega Zn	9.5	31.5	34.7	43.6
Nano Zn	9.8	37.7	43.3	52.3
Mega Mg	10.7	39.3	34.9	44.6
Nano Mg	12.6	48.9	46.5	57.6
Mega Ti	10.6	46.3	35.5	47.2
Nano Ti	12.3	47.3	48.5	62.3
LSD ($p = 0.05$)	0.41	0.46	0.3	0.49

*I = 2 week (28 days old plant) after foliar spray; II=4 week (42 days old plant) after foliar spray; foliar sprayed at 14 days old plant

The more total chlorophyll contents of 48.9 $\mu g\ g^{-1}$ (61.9 % over control and 24.4% over ordinary Mg particles) were recorded with the application of Mg nanoparticles, whereas, total soluble leaf protein was recorded more of 62.3 $mg\ kg^{-1}$ (65.3% over control and 32% over ordinary Ti particles) with the application of Ti nanoparticles at stage II of plant growth. In another study, recorded 23.3 % increase in grain yield with a foliar application of 2 % Nano-Fe fertilizer over the control. Spraying of manganese nanoparticles has been shown to increase growth, yield and its components compared with manganese sulphate on *Vigna radiata* (L.) Wilczek. Application of 30 mg/L nano iron, manganese and zinc fertilizers produced maximum values of yield and yield attributes of peanut by reason of increased nutrient use efficiency of nano-fertilizers which enhance pigments formation, photosynthesis rate, dry matter production and thus leading to better growth and yield. The tuber yield of potato increased with the foliar application of nanosilver possibly by reason of its antimicrobial effect which might have helped seed tubers to stay healthier for longer time in the soil and subsequently produced more vigorous plants. Foliar applied nano-chelate molybdenum has a significant effect on the traits, such as plant height, number of pods per plant, number of ripe pods per plant, hundred seed weight, seed number per plant, seed length and seed and pods yield and the number of lateral branches and the biological performance of peanut.

Foliar application of Zn, Mg and Ti nanofertilizers particles on microbial population found in the rhizosphere of pearl millet (Table 22). The results clearly exhibited that the additional application of any of the three agriculturally important metals significantly enriched the soil microbial population, in general, from stage I to stage II, and decrease thereafter at the time of harvest of the crop (Stage III). The maximum bacteria and fungi populations of 51 cfu$\times 10^{-6}$ (34.2%) and 23.3 cfu$\times 10^{-4}$ (45.6%) were recorded after application of nano-Mg at stage II, whereas nano-Ti supported the more enhance in actinomycetes population with 17.3 cfu$\times 10^{-5}$ (61.6% over control).

Table 22: Effect of foliar application of fertilizers (mega or nanoparticles) on rhizospheric microbial population of pearl millet

Treatment	Bacteria (cfu$\times 10^{-6}$)			Fungi (cfu$\times 10^{-4}$)			Actinomycetes (cfu$\times 10^{-5}$)		
	I*	II	III	I	II	III	I	II	III
Control	31.0	38.0	21.3	15.0	16.0	9.3	10.0	10.7	6.7
Mega Zn	41.3	44.7	28.3	16.7	18.0	12.0	11.3	13.7	7.7
Nano Zn	42.3	48.7	35.7	21.7	21.0	12.3	15.3	14.3	11.3
Mega Mg	40.7	44.0	31.3	15.0	19.0	9.7	11.7	14.0	8.3
Nano Mg	47.0	51.0	32.7	21.3	23.3	15.3	15.0	16.7	13.3
Mega Ti	36.0	41.3	25.3	17.0	18.7	12.0	12.7	13.7	8.7
Nano Ti	42.7	47.7	32.0	18.3	21.7	13.7	16.0	17.3	14.0
LSD ($p = 0.05$)	0.12	0.18	0.09	0.06	0.08	0.08	0.04	0.03	0.08

*I = 2 week (28 days old plant) after foliar spray; II=4 week (42 days old plant) after foliar spray; III=Crop harvest stage; foliar sprayed at 14 days old plant.

In the last few years, some researchers tried to examine the potential of nanofertilizers to increase the yield of crops after foliar application. Foliar applications of nanofertilizer had reflected in improvement in yield parameters of wheat plants. Foliar spray of NPK nanofertilizers in chickpea increased the yield and yield components as a result of increased growth hormone activity and enhancement of metabolic process, tended to increase in flowering and grain formation. Application of nanofertilizers have greater role in enhancing cotton yield production besides reducing the cost of fertilizer and also minimizing the pollution hazard. Significant increases of total and open bolls per plant, boll weight and seed cotton yield with the foliar nanofertilizers application than soil application. It was suggested that foliar application of zinc, iron and NPK manipulates the growth of chickpea, resulting in beneficial effects on yield and yield components. Moreover zinc nano-fertilizer applied as foliar spray on pearl millet (*Pennisetum americanum* L.) significantly increased the grain yield by 37.7 per cent and also suggested that application of nanochelate of zinc can improve cotton performance by increasing the number of bolls per plant and boll weight. The seed oil content increased with increased concentration of nano-ZnS in sunflower. In groundnut crop pod yield gets increased with the application of nano-scale zinc oxide compared to $ZnSO_4$ application, on account of nano-scale zinc is absorbed by plants to larger extent than its chemical form. Nano-ZnO has proved to be more effective in enhancing productivity and absorption of Zn because of high surface area to volume ratio. The required dosage of nano based Zn fertilizer had 10 folds less than the conventional $ZnSO_4$. It was suggested that nanofertilizer application increased grain zinc content without affecting grain yield, protein content, spikelets per spike, 1000 kernel weight, etc., owing to enhanced enzyme activity and carbohydrate metabolism leading to an increased yield. Nano-scale zinc oxide particle at 40 mg/L treatment was associated with increased rice grain yield and its components in mid tillering stages.

Foliar application of metal oxide nanoparticles *viz.*, MgO, ZnO and CuO recorded more than 22, 33 and 18 per cent of seed cotton yield, respectively than control. In pomegranate, fruit yield and number of fruits per tree get increased with the foliar spraying of nano-scale zinc and boron fertilizers (34 mg B $tree^{-1}$ or 636 mg Zn $tree^{-1}$, respectively). It was reported that foliar application of nano-TiO_2 manipulates growth of barely, resulting in beneficial changes in yield and yield components. These possible reason for, such a beneficial role is due to increase in activity of photosynthesis by promoting cyclic and linear phosphorylation by spraying of nano-TiO_2 and it enhanced the photo assimilates supply in leaves (*i.e.*, increasing source capacity) which ultimately increased the yield attributes. The application of nanofertilizers improved fertilizer use efficiency and significantly increased the grain yield and straw yield of barley. The application of nano-TiO_2 improved the photosynthetic complexes and nitrogen metabolism which led to increase in fresh and dry mass of plant. Nano-TiO_2 applied as a foliar spray in maize crop also enhanced plant growth and grain yield by its efficient photo-catalyst activity, which promoted the manufacturing of pigments and transformation of light energy to active electron and chemical activity. The use of iron nano-fertilizer on soybean crop improved the yield. Iron is a component of ferrodoxin and it may improve photosynthesis. It was suggested that spraying of 0.5 g L^{-1} nano-Fe to the black-eyed pea improved the number of pods per plant, weight of 1000 seeds, yield, and chlorophyll content compared to common Fe. Nano-Fe fertilizer application at tillering and stem elongation did increase the number of seeds per spike, whereas early application of Fe fertilizer decreased the number of seeds per spike in wheat. Hence, the foliar application of Fe was more suitable than seed dressing or soil application attributable to being a suitable time for seed formation. In addition, Fe availability can increase the leaf area index, leaf area duration, and decreased leaves senescence that can increase economic yield.

7.2: Nanofertilizer under Drip

Drip application of micro-nutrients in nanoform is an important method of nutrient feeding and in some cases these applications are more effective than soil applications. Micro-nutrients in nano form are feed through irrigation water under drip because many micro-nutrients are not readily translocated within the plant, and sometimes it is required in midseason to correct deficiency. Drip application of micro-elements in nano size is very helpful when the roots cannot provide necessary nutrients. Although in treatments of nano micronutrient in combinations can also be done in most cases. Once a potential deficiency has been diagnosed, a single compound application with irrigation water in drip can be used to prevent or correct deficiency. Sometimes diagnose in deficiency symptoms is confused then mixture of nano nutrients may be applied in drip. Nano micronutrients are also crucial for achieving higher yields and can be used under drip. Six micronutrients, *i.e.,* Mn, Fe, Cu, Zn, B and Mo are involved in photosynthesis, respiration and other biochemical pathways. However, both macro and micronutrients availability influenced by soil chemical and physical properties. Nanofertilizer applied through drip may also reduces adverse environmental impact. The application of nanonutrients through drip can increase the crop yield by 30–70%. The more yield increase was observed under potato. The small particles of nano-fertilizers and their high surface area make them able to penetrate rapidly into plant tissues. This stimulates the action of plant hormones within the plant that promote the growth of secondary roots, which is reflected in growth and production. The presence of these elements in nano form side by side works to reduce stomatal resistance and increase stomatal conductivity, which provides the plant with enough carbon dioxide and water to continue photosynthesis and withdraw nutrients from the soil leading to an increase in yield. The behaviour of nano-fertilizers once they enter the plant is related to protein carriers, such as Aquaporin and Endocytosis. This leads to the formation of new openings that penetrate cell walls. It then stimulates water absorption and promotes plant growth and production.

7.3: Nanofertilizer for Sprinkler Irrigation

Nanofertilizer can be used as springer, but the doses should be minimized as under sprinkler the environmental losses are higher. Sprinkler irrigation is suited for most row, field and tree crops and water can be sprayed over or under the crop canopy. However, large sprinklers are not recommended for irrigation of delicate crops, such as lettuce because the large water drops produced by the sprinklers may damage the crop.

7.4: *In-vivo* Method (hydroponic, aeroponic) of Nanofertilizer Application

The efficiency of nano-fertilizers and their impact on plant systems is influenced by the method of their application. The delivery of nano-fertilizers to plants can be realized through the *in vivo* application, either as hydroponic or as aeroponic. The principle of aeroponics consists in continuously spraying of a nano-nutrient solution on roots suspended in air. The technique allows strict control of the gaseous environment around the roots. But the technique requires a high level of nutrients to sustain rapid plant growth, thus its application is restricted.

Hydroponics is commonly known as solution culture as the plants are grown with their roots immersed in a solution without soil. Volumes of nutrient solution as well as concentration, maintenance of oxygen demands, and pH are factors that need attention while using this method of nano- nutrient delivery.

7.5: Nanofertilizer Application in Soil

Soil application of nutrients is the most common practice, but it has many limitations with respect to availability of nutrients to the plants.

The inorganic nutrients get fixed in soil as insoluble forms and also subjected to leaching by rainfall or irrigation water. Moreover, anything which restricts root growth reduces the nutrient uptake. Foliar application overcomes these limitations. In addition to that, foliar feeding has proved to be the fastest way of correcting nutrient deficiencies and increasing yield and quality of crop products and it also minimizes environmental pollution and improves nutrient utilization by reducing the amounts of nanofertilizers added to the soil. Even though leaves allow gas exchange, but cuticle present in the leaves restricts the penetration of substances. The nano coated substances enhance the penetration via stomata or cuticle with a size exclusion limit above 10 nm. In addition to this, nanocarriers deliver the nutrients in the right place and right times which reduce the extra amount of active chemicals deposited into the plant system and increase the nutrient use efficiency. Nano-fertilizers have high surface area, sorption capacity, and controlled-release kinetics to targeted sites, and have been considered as smart delivery system. But careful choice of the persistence time of the nano fertilizer in the soil is required besides special attention to the soil texture, salinity, plant sensitivity to salts, and pH of the amendment. Negative soil particles affect the adsorption of mineral nutrients. The anion exchange capacity of most agricultural soils is small compared to the cation one. Among anions, NO_{3-} remains mobile in the soil solution and is susceptible to leaching by water, PO_4^{3-} binds to soil particles containing Al or Fe because the positively charged $Fe^{2+/3+}$ and Al^{3+} exchange OH- groups with phosphates, resulting in tight bounding of the later, whose mobility and availability in soil can limit plant growth.

Metal based nanoparticles possess physical characteristics, particularly size and shape, and chemical characteristics, such as the acid-base character of the surface and the aqueous solubility of the metal. These characteristics in turn will determine the extent to which metal based nanofertilizers undergo transformations that will control their fate, behaviour and ecotoxicity in the environment. Such processes will include aggregation/agglomeration, sorption to surfaces, and dissolution to the ionic metal. Furthermore, metal based nanofertilizers are frequently manufactured with surface coatings, which may modify their intrinsic behaviour. The existence of these multiple characteristics implies their characterization of nanofertilizer form and presentation in the environment will be important to understanding their behaviour, fate, and ecotoxicity.

The assessment of the form and presentation of nanofertilizer particles in environmental matrices and soil specially, is currently hampered by the relative lack of appropriate procedures for the characterization, and development of suitable techniques for greater characterization posses a great challenge. For relatively simple analysis, such as the aggregation state of nanofertilizer in aqueous solution, it is possible to use dynamic light scattering or microscopy based techniques, such as scanning and transmission electron microscopy and atomic force microscopy. However, quantify state of nanofertilizer particle in soils, compared with solutions, is hampered by the need for knowledge of the nanofertilizer presentation within the soil matrix, rather than in aqueous extracts of the soil. The properties like dispersability, agglomeration/aggregation, dissolution rate, size, surface area , charge and surface chemistry are likely to be the key in controlling the nanofertilizer stability, and consequently on their transport through environment and availability to the organisms.

The antifungal activity of zinc oxide (ZnO) and magnesium oxide (MgO) nanofertilizer prepared by bio-safe method was evaluated for *Alternaria alternate, Fusarium oxysporum, Rhizopus stolonifer* and *Mucor plumbeus*. It was observed from the study that all the nanofertilizer particles at different concentrations brought about significant inhibition in the germination of spores of *Alternaria alternata, Fusarium oxysporum, Rhizopus stolonifer and Mucor plumbeus*. However, the highest inhibition in the germination of all the test fungi was observed at higher concentrations followed by lower concentrations of nanoparticles. The nanofertilizer particles of MgO at highest concentration were found most effective in reducing the spore germination followed by nanofertilizer particles of ZnO at the same concentration. It was revealed from the study that different concentration of nanofertilizer particles of MgO and ZnO caused inhibition in the spore germination of *Rhizopus stolonifer*. However, inhibition in spore germination increases with the increase in the concentration of nanofertilizer particles. The maximum inhibition in the spore germination was found at highest concentration '0.5mL'. It was followed by 0.4mL, and 0.1mL concentrations of nanoparticles. The nano-Mgo fertilizer at highest concentration was found most effective in reducing the spore germination followed by nano-ZnO fertilizer.

Naturally-present colloids and macromolecules in soils are nanoparticulate and microparticulate clays, organic matter, iron oxides and other minerals play an important role in biogeochemical processes. Generally, nanoparticles are present in soil in two broader categories like nano-minerals and mineral nanoparticles. Nano-minerals are defined as the minerals that only exist in this size range, *e.g.,* certain clays and Fe and Mn (oxyhydroxides) whereas mineral nanoparticles are defined as minerals that can also exist in larger sizes, *e.g.,* most known minerals. The formation of natural nanoparticles that are present in volcanic dust, most natural waters, soils, and sediments, has been governed by the different soil, geological, and biological processes. Nanosheets, nanorods and nanoparticles are the three classes of natural nanoparticles present in the soil system. Nanosheets in the form of thin coatings are commonly present on surfaces of primary minerals of the soil matrix. Nano-sheets may also occur in the form of thin platelets, which promote accelerated and intensive dissolution of existing minerals and subsequent precipitation of neophases. Nanorods are formed as a result, intense dissolution of primary soil minerals, such as feldspars and mica, which initially occurred, and were followed by precipitation of goethite, sodalite, and other secondary phases. The nanoparticles may also be formed in soils as a result of biotic processes.

7.6: Crop Quality

Quality nutrients are required for improving the quality parameters of crops. In this aspect, nanofertilizer application gave better quality of crop products than the conventional fertilizer, which was supported by some research studies. Fibre quality parameters of cotton like uniformity ratio and fibre strength were improved by the application of metal oxide nanoparticles than control. The application of fertilizer in nanoform is completely controlled and has led to an increase yield and protein content in peanut. Foliar application of nano-forms of iron and zinc fertilizers increased the phosphorus concentration, biomass, and crude protein and soluble carbohydrate concentration in forage corn over chemical forms of fertilizers. It was suggested that a positive close relationship between protein concentration and the concentration of iron and zinc in corn. Report shows that foliar application of ZnO nanoparticles increased the quality parameters like oil content in sunflower. Zinc nanofertilizers increased soluble carbohydrate concentration, probably due to involvement of zinc in photosynthesis, chlorophyll synthesis, starch formation and enzyme carbonic anhydrase, accelerating carbohydrate formation. In peanut, total carbohydrate, total soluble sugars, protein and oil percentages in seeds increased by nanofertilizers. In protein content zinc element had an additive role for protein formation that showed an important role in the protein content of plants. Zinc nanofertilizer plays a positive role in root develepment, which helps plants absorb important nutrients, especially nitrogen responsible for protein synthesis. Additionally, it is involved in the metabolism of carbohydrate, protein and plant hormones especially IAA, and helps in the formation of starch and seed maturity. Nano-Fe appreciably influenced the seed protein content by 2 per cent compared to chemically synthesised Fe fertilizerrs in black-eyed pea. The positive effect in foliar applied nanofertilizers, have enhanced the zinc uptake in upland paddy by 48 per cent over control and beneficial enzyme activity by 53 per cent.

Influential effects of biologically synthesised Zn nanofertilizer on different crops may be due to increase in plant growth hormone like indol-acetic-acid, and more formation of photosynthetic pigment chlorophyll due to high interception of photosynthetically active radiation. The influential growth rate found in *Cicer arietinum* by application of ZnO nanoparticles owing to increase of indol-acetic-acid (IAA), a naturally occurring auxin plant growth regulator contents which is responsible for cellular growth. It is well reported that zinc play very important role in growth regulation, enzyme activation, gene expression, phytohormone activity, protein synthesis, photosynthesis and carbohydrate metabolism, fertility and seed production. Zinc deficiency impairs physiological function, leading to severe reduction in growth, lower yield and poor quality crop products. These reports strongly support the presented result of influential effect by application of nano-Zn particle over the ordinary size Zn particle and control. The effect of nano-ZnO on growth of mung bean (*Vigna radiata*) and chick pea (*Cicer arietinum*) seedlings showed the best response of nano-ZnO on mung bean at 20 ppm and on chick pea at one ppm. Beyond these concentrations the seedling showed the retardation in growth and development. The photosynthetic pigment chlorophyll, total soluble leaf protein content and rhizosphere soil enzymes activity was increase due to more availability of Zn to the plant as nano Zn particle has better penetration power than that of ordinary size Zn particle. It is also well known that zinc act as cofactor for six enzyme classes, *i.e.,* oxidoreductase, transferases, hydrolases, lyases, isomerases and ligases. There are three primary Zn ligand binding sites underpinning structural, catalytic and cocatalytic function of zinc in enzymes, which play major role in regulatory function of enzymes. It was reported that Zn increase the activity of carbonic anhydrase in the plant leaves, which is responsible for photosynthetic electron transfer and consequently chlorophyll contents. The reports from Ho, very clearly displayed that Zn plays a pivotal role in protein synthesis metabolism and being involved in stability and function of genetic materials. .

7.7: Alleviation of Abiotic Stress

Abiotic stresses include salinity, drought, heavy metals, flooding, chilling, freezing, heat, ozone and ultraviolet radiation. Among abiotic stresses salinity and drought represent a great threat to crop production all over the world. These abiotic stresses including drought and salinity may cause a loss in crop production about 50 per cent. Salinity causes nutrient imbalance, membrane damage and enzymatic inhibition in the plants by ionic and osmotic stress. In addition, soil salinity negatively interferes with the water availability to plants, absorption of essential nutrients and crop yields and qualities. One of the most important solutions is to use the nanomaterials in alleviating the harmful effects of these stresses. Nanomaterials may mimic the role of antioxidative enzymes like peroxidase, superoxide dismutase and catalase. These antioxidants and enzymes are continuously scavenging the reactive oxygen species. Soil salinity declines growth parameters, *i.e.,* dry matter, uptake of N, P, K, Na and Ca and yield of cotton. Under saline conditions the growth of roots affected, which reduces the nutrient uptake. Application of nano-Zn enhanced the root penetration and increase uptake of nutrients and it caused significant changes in fresh and dry weight of rice, in biomass production of sunflower, in grain yield of wheat under salinity stress and in yield of maize under drought. The salt stress can be alleviated in Moringa plants using foliar applications of ZnO and Fe_3O_4 nanofertilizers-containing Hoagland solution by enhancing enzyme activity related to salt tolerance. Accumulation of less Na is an important indicator of salt tolerance in plants. Potasium (K) content reflects salt tolerance in plants alleviating the adverse effects of NaCl on nutrients uptake through improving root growth and preventing nutritional disorders and increasing in nutrients uptake by the roots. Due to larger specific surface area and more reactive areas of nanoparticles helped in enhanced enzyme activity related to salt tolerance. The nanofertilizers simplify the absorption of fertilizer and enhance the effect of hoagland solution by increasing the availability of Fe and Zn, which are involved in the salt tolerance mechanism. The foliar application of 200 ppm nano-Zn to the cotton crop under stress condition helps to increase the cotton growth and yield. The application of SiO_2 nanofertilizer can have a positive effect on plant growth and yield of cucumber under salinity condition through improved uptake of nitrogen and phosphorus, and reducing the Na content. The SiO_2 nanoparticle as foliar application avoided leaching loss of N and helped in more accumulation of nitrogen in leaf. Foliar applied SiO_2 might help in increasing cell wall turgidity, strength and also cell wall elasticity during growth extension. It was also found that application of Si counter the negative effects of salinity in *Phaseolus vulgaris* L. by decreasing in stomatal conductance and increasing antioxidative enzymes activity. In general, water deficit stress caused significant decrease in plant growth, yield and yield components. In addition, application of titanium dioxide nanoparticles at 0.02 per cent increased the growth of wheat crop. Thus, the application of titanium dioxide nanoparticles under conditions of water deficit stress is recommended. Titanium improves rubisco activity and increases the CO_2 metabolism, increases photosynthesis and improves yield. Generally under water stress condition gluten and starch content of wheat get reduced, the application of nano-TiO_2 improves these contents may be due to the positive correlation between titanium application and photosynthesis rate. In maize crop with the foliar spraying of nano-Zn fertilizer increased the yield and yield components under water stress condition.

7.8: Alleviation of Heavy Metals Toxicity

Nanofertilizers may be more effective than regular fertilizers in improving plant nutrition, enhancing nutrient use efficiency, and protecting plants from environmental stress. The cadmium (Cd) generally decreases the growth and the contents of Mg, Fe, Zn, chlorophyll a, and glutathione (GSH) particularly when accompanied by a significant increase in Cd accumulation. Foliar application with 2.5 mM nano-silicon alleviated Cd stress in rice seedlings as a result of enhanced the availability of Mg, Fe, and Zn nutrition, and the contents of chlorophyll a and decreased Cd accumulation and translocation of Cd from root to shoot. The Cd treatment produced oxidative stress to rice seedlings indicated by a higher lipid peroxidation level and enhanced antioxidant enzymes activity, such as superoxide dismutase (SOD), peroxidase (POD), and catalase (CAT), and a lower glutothionine (GSH) content. However, those nano-Si-treated plants had lower malondialdehyde (MDA) but higher GSH content and different antioxidant enzyme activities, indicating a higher Cd tolerance in them.

7.9: Smart Delivery of Nutrients

A smart delivery system for agriculture should consider the factors or combination of factors, such as time controlled, specifically targeted, highly controlled, remotely regulated/pre-programmed and multi-function characteristics to avoid biological barriers for successful targeted release of required nutrients. The nanoscience has greatly impacted the conventional delivery systems by eliminating the limitations, such as leaching, degradation by photolysis, hydrolysis and bio-instability in atmosphere. This results in repeated use of pesticides and insecticides causing higher cost of cultivation and environmental pollution. Nano-encapsulated agrochemicals should be designed in, such a way that they possess all necessary characteristics like effective concentration, stability, solubility, time-controlled release in response to certain stimuli, enhanced targeted activity and less eco-toxicity with safe and easy mode of delivery thus avoiding repeated applications.

There are various means through which this delivery system can be put to use.

- Herbicides are chemicals used to kill weeds especially when the moisture content in the soil is not as per required by the crop. Nano-herbicides (usually coated material) provide better penetration in the soil and allow slow and controlled release of active ingredients in reaching the targeted weed causing minimum environmental damage.
- The insecticides are also a concern for agricultural purpose as they induce death to insects and other pests that try to harm the crops. Nano insecticides like surface modified hydrophobic nanosilica have been put to use instead of conventional insecticides as they are safe for plants and cause less environmental pollution.
- Fungal pathogens are a major part of the pesticides which harm the crops hence fungicides are prepared to tackle them separately. The small size of nanoparticles puts them to use in nano-fungicides as they penetrate easily and colonize the fungal spores which are a source of fungal pathogens. For example Ag-nanoparticles are a common source of nano-fungicides but if used in very high concentration can produce chemical injuries in crops, such as cucumber.

Further studies

Abdel-Aziz, H. M. M., Hasaneen, M. N. A. and Aya, M. O. (2018). Foliar application of nano chitosan NPK fertilizer improves the yield of wheat plants grown on two different soils. *Egypt J. Expert. Biol.* **14:** 63–72.

Amin, F. and Mohammad, M. O. (2015). Effect of nano-zinc chelate and nano-biofertilizer on yield and yield components of maize (*Zea mays* L.), under water stress condition. *Ind. J. Natur. Sci.* **5:** 4614–4620.

Hussein, M. M. and Abou-Baker, N. H. (2018). The contribution of nanozinc to alleviate salinity stress on cotton plants. *Royal Soc. Open Sci.* **5:** 171809.

Nair, R., Varghese, S. H., Nair, B. G., Maekawa, T., Yoshida, Y. and Kumar D. S. (2010). Nanoparticulate material delivery to plants. *Plant Sci.* **179:** 154–163.

Tarafdar, J. C., Xiang, Y., Wang, W. N., Dong, Q. and Biswas, P. (2012). Standardization of size, shape and concentration of nanoparticles for plant application. *Appl. Biol. Res.* **14:** 138–144.

8

Effect of Nanofertilizers on Soil and Plants

Nano-fertilizers are nutrient carriers, whose substrates in the range of nano-dimension (1–100 nm), capable of supplying nutrients to the plant system for an extended period of time without associated environmental hazard. Plants provide a potential pathway for the transport of nanoparticles to the environment and serve as an important route for their bioaccumulation into food chain. Various studies are now ascertaining the effects of different nanoparticles on plant growth and metabolic functions. Nanoparticles are adsorbed to plant surfaces and taken up through natural nano-or micrometer-scale plant openings. Several pathways exist or are predicted for nanoparticle association and uptake in plants. Nanoparticle uptake into the plant body can use different paths. Uptake rates will depend on the size and surface properties of the nanoparticles. Very small sizes nanoparticles can be penetrate through cuticle. Larger nanoparticles can penetrate through cuticle-free areas, such as hydathodes, the stigma of flowers and stomata's. Nanoparticles must traverse the cell wall before entering the intact plant cell protoplast. Result suggests that only nanoparticle less than 5 nm in diameter will be able to traverse the cell wall of undamaged cell efficiently.

8.1: The Fate of Nanofertilizer Particles in the Soil Matrix

Nanoparticles encompass many different types of particles resulting in great variations regarding their uptake, bioavailability and toxic effects. Beside plant uptake, the some of the particles may aggregate and sorbed, migrate, bioaccumulate and may be in the dissolve pools. Even nanoparticles with the same core material may exhibit a range of behaviours, depending on factors, such as coating material, shape and size. Trapping of the particles within soil pores (pore straining) is a physical mechanism influencing particle retention in soils. The importance of pore straining as a retention mechanism depends on the particle size and pore size distribution. In general, the mobility of the nanoparticles decreases with the increase in the particle size. Larger particles are more likely to retain in pore straining. Aggregation tends to increase the importance of pore straining as a retention mechanism by producing a range of aggregate sizes that overlap to a greater extent with the size distribution of soil pores. The influence of pore straining in particle retention thus tends to depend on similar factors like pH, ionic strength etc. Increases in the particle size attributable to aggregation may also be associated with chemical conditions favouring sorption to the soil (*e.g.,* pH, iso-electric point). However, the smaller particles are more mobile and are likely to be penetrating the ground water depth. The larger particle tends to retain in the upper layer of the soil, potentially causing soil clogging. The fate of nanofertilizer particles in the soil matrix is shown as figure 69.

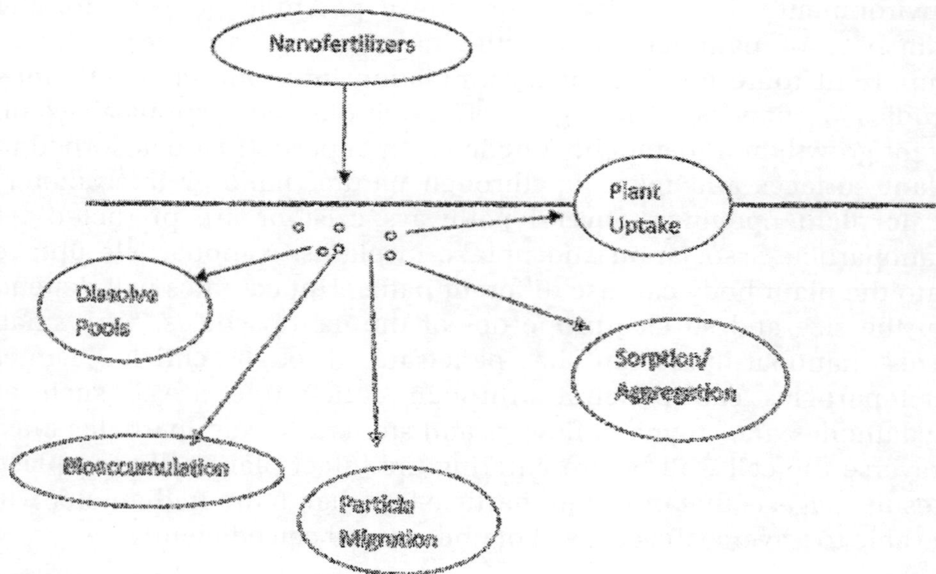

Figure 69: Fate of nanoparticles in soil matrix

8.1.1: Stability

The stability of nanofertilizer particle is a function of their surface energy. In general, the nanofertilizer particles are more stable when they possess a low surface energy. For example, the surface energies of small rutile particles are higher than those for anatase particles of a similar size, consistent with anatase being the more stable phase of nanocrystalline TiO2. Manufactured metal-based nanofertilizer particle possess a number of key characteristics that are believed to exert important controls on their environmental behaviour, fate, and ecotoxicity. These include physical characteristics, particularly size and shape, and chemical characteristics, such as the acid–base character of the surface and the aqueous solubility of the metal. These characteristics in turn will determine the extent to which nanofertilizer particle undergo transformations that will control their fate, behaviour, and ecotoxicity in the environment. Such processes will include aggregation/ agglomeration, sorption to surfaces, and dissolution to the ionic metal. Soil represents a relatively complex medium for the understanding of the physicochemical behaviour of manufactured nanofertilizer particles. In comparison with the dissolved phase, in which behaviour can be understood largely in terms of particle stability against aggregation, soils present a solid matrix with which nanofertilizer particle may interact, as well as an aqueous phase, which may contain appreciable amounts of natural colloidal/ particulate material.

8.1.2: Aggregation and Agglomeration

Aggregation is defined as the association of primary particles by strong bonding, whereas agglomeration is defined as association by weak bonding caused by van der Waals forces. In the environment, physical forces (*e.g.,* Brownian motion, gravity, and fluid motion) and nanofertilizer particle characteristics (*e.g.,* surface properties, particle size) will affect the fertilizer particle agglomeration and aggregation. The particles are constantly colliding with each other because of Brownian motion, and agglomeration will occur when the energy of either motion or attraction exceeds the energy of repulsion. Aggregation may result in the formation of particle flocks that are of sufficient size to sediment out of solution by gravity. The nanofertilizer particle aggregate size in solution depends on properties, such as initial particle size and concentration.

8.1.3: Surface coating

The chemistry of the medium will influence the electrostatic surface charge of the particles, thereby affecting agglomeration, aggregation rates and particle stability. In the absence of a surface coating, metal-based nanofertilizer particle have charged surfaces resulting from the presence of hydroxyl (-OH) groups that can take up and release protons, and can take up dissolved chemical species, such as metal ions and ligands. Surface charging results in the formation of an electrical double layer, comprising the charged surface sites and a diffuse layer containing ions attracted from the solution to the particle surface in response to the charge. The electrical potential at the interface of the diffuse layer and the bulk solution (the zeta potential) can be measured, and its variation with solution chemistry can effectively be used as a surrogate for the variation in particle surface charge with solution chemistry. The zeta potential of uncoated metal or metal oxide particles typically decreases from positive values at low pH to negative values at high pH. As the pH approaches the iso-electric point, or the "point of zero charge," where particle charge/zeta potential approaches zero, the aggregation rate will increase because of the lowering of the electrostatic repulsive forces between particles. A suspension of homogenously charged particles will be stable when the magnitude of the zeta potential is greater than 30 mV. Classical colloid theory can be applied to metal and metal oxide nanofertilizer particle to help explain their stability. The Derjaguin, Landau, Verwey, and Overbeek theory considers stability as a function of the repulsive (*i.e.,* electrostatic) and attractive forces (*i.e.,* van der Waals) to which a particle is subjected. Although the Derjaguin, Landau, Verwey, and Overbeek theory has generally proved unsatisfactory for the quantitative prediction of colloid behaviour in complex natural environments, the conceptual framework that it provides for considering, such behaviour can be useful in explaining trends in observed behaviour in, such systems.

8.1.4: Dissolution

Some types of metal-based nanofertilizer particles are thermodynamically unstable and undergo chemical dissolution if kinetic restrictions are absent. Dissolution occurs when an ion detaches from the particle and migrates through the electrical double layer into the solution. The dissolution behaviour of non-nanosized metals and metal oxides in soils may be of use in prioritizing nanofertilizer particles for study and in choosing which soil properties to consider that may influence their dissolution rates. Behaviour of the nanoparticles within the soil systems will be further complicated by the presence of the solid phase. Soil components, such as clay particles or humic molecules, themselves having the charged surfaces, will influence the association of nanofertilizer particles with the solid phase. Such soil components may also form colloids in the aqueous phase, which will interact with the nanoparticles. For example, humic molecules desorbing into the aqueous phase may sorb to nanoparticle surfaces, and so influence the nanofertilizer particles stability. All of these processes are themselves strongly influenced by overriding characteristics of the soil system, particularly the pH and the ionic strength of the aqueous phase. The nanofertilizer particles were found to quickly equilibrate between the aqueous and solid phases and concentrations in the smaller than 1-mm fraction were stable. The aggregation rate of TiO_2-nanofertilizer particles in soil suspensions has been found to be negatively correlated with soil characteristics, such as dissolved organic matter and clay contents, and positively correlated with the ionic strength, zeta potential, and pH. Nanofertilizer particles are sorbed less strongly to soils of low ionic strength and high dissolved organic matter content. Assuming that bioavailability is related to the particle fraction suspended in the pore water and not the fraction associated with the soil matrix, these factors may be affecting the bioavailability of metal-based nanofertilizer particles in soils.

8.1.5: Ionic strength

The ionic strength of the medium affects the stability of the diffused layer in the electrical double layer. An increase in the ionic strength will lead to a decrease in the electrical double layer thickness, which favours particle association and leads to increasing agglomeration. The effects of ionic strength on the aggregation of Ag-nanofertilizer particles (uncoated, citrate, and sodium borohydride-coated particles) have been observed for suspensions with pH higher than 7. This demonstrates that the particles were stabilized by electrostatic repulsion, caused by the predominance of negative forms (*i.e.,* anions) in the medium and the negative charge of the particles.

8.1.6: Natural organic matter

Natural organic matter has the ability to stabilize engineered nanofertilizer particles. Fullerene was found to be stabilized in the presence of dissolved organic matter in suspension. Generally, after release of the nanofertilizer particle in soil solution, they may interact with dissolved organic matter in particulate form and be suspended, and transported rather than aggregate and deposit quickly. The dissolved organic matter adheres to the surface of nanofertilizer particles and modifies their surface physicochemical properties and enhances their stability.

8.1.7: pH

To understand the fate and behaviour of nanofertilizer particles in soil, it is necessary to understand the effect of pH on the particle suspension. pH is mainly responsible for governing the zeta potential of colloids. The colloidal system exhibits minimum stability or maximum coagulation, when pH is at point of zero charge (pzc). The colloid surface is positively charged when the pH is lower than the pzc value and the zeta potential will increase with decreasing pH below the pzc. Conversely, the colloid surface possesses positive charge at pH above pzc and the zeta potential will be more negative with increasing pH. Nanofertilizer particles with low zeta potential tend to coagulate or flocculate, whereas higher value of zeta potential will impart stability to the nano-particle suspension. The solubility of metal-based nanofertilizer particles and their surface charges are governed by solution pH and consequently the electrostatic interactions between nanofertilizer particles and porous media. Dissolved counter ions in solution determine the long range electrostatic interactions for electronically stabilized nanofertilizer particles, and hence decrease the stability and transport of nanofertilizer particles in the porous media.

Soil pH is a significant factor that directly relates to soil fertility and health. It is an indication of acidic or alkalinity of the soil. Soil pH affects the availability of the nutrients for plants. It is proved that the nature of acidity or basicity of soil depends upon the composition of the soil. In general optimum pH range for plants is 5–7. It is observed that the pH of the soil may be influenced by the accumulation of the different type of nanofertilizer particles like P, Zn, Ag, Cu etc. It was found that the pH of the soil also influences the toxicity of nanofertilizer particles on the microorganisms and nematodes. It was reported that greater toxicity the effect of Zn nanoparticles than dissolved Zn has great toxicity on the bacterial community. Very recently found that copper oxide nanofertilizer particles may able to change the pH of paddy soil. Copper oxide nanofertilizers cause an increase in the pH of soil which ultimately affects soil property. Uptake of silver nanoparticles accumulated in soil by insects may also be influenced by the pH of the soil. Soil pH seems to be the principal factor that governs concentrations of soluble metals. Zn solubility correlates negatively with pH. Moreover, clay content in calcareous soil was higher than in acidic soil, which increase Zn retention in soil and hence, decreases Zn availability to microorganisms, as clays are believed to adsorb zinc ions across both ion exchange and specific adsorption. Zinc availability has also been connected to soil organic matter content.

8.1.8: Zeta Potential

Zeta potential is the potential difference between the dispersion medium and the stationary layer of fluid attached to the dispersed particle. If the dispersion pH is below 4 or above 8 there is sufficient charge to confer stability. However, if the pH of the system is between 4 and 8, the dispersion may be unstable. This is most likely to be the case at around pH 6 (the isoelectric point). Zeta potential value can be related to the stability of colloidal dispersions. The zeta potential indicates the degree of repulsion between adjacent, similarly charged particles in dispersion. For molecules and particles that are small enough, a high zeta potential will confer stability, *i.e.,* the solution or dispersion will resist aggregation. When the potential is low, attraction exceeds repulsion and the dispersion will break and flocculate. So, colloids with high zeta potential (negative or positive) are electrically stabilized, while colloids with low zeta potentials tend to coagulate or flocculate.

8.1.9: Point of zero charge

The potential at the point of zero charge (pzc), describes the condition when the electrical charge density on a surface is zero. Point of zero charge (pzc) is the pH value at which a solid submerged in an electrolyte exhibits zero net electrical charge on the surface. When the pH is lower than the pzc value, the system is said to be "below the pzc." Below the pzc, the acidic water donates more protons than hydroxide groups, and so the adsorbent surface is positively charged (attracting anions). Conversely, above pzc the surface is negatively charged (attracting cations /repelling anions). At pzc, the colloidal system exhibits zero zeta potential (*i.e.,* the particles remain stationary in an electric field) causes minimum stability (*i.e.,* exhibits maximum coagulation/ flocculation rate), maximum solubility of the solid phase and maximum viscosity of the dispersion. Dispersion is stable in the colloid-chemical sense when its particles remain permanently free. In dilute dispersions, it is sufficient to consider only interactions between pairs of particles. In very concentrated systems, it may be necessary to take multi-particle interactions into account.

8.1.10: Behaviour of nanofertilizer particles in the atmosphere

The nanofertilizer particles are ultrafine particles, a category consisting of particles smaller than 10 nm (referred to as nucleation mode) and those from 10 to 100 nm (referred to as Aitken mode). These particles are dominated in the atmosphere in polluted urban, rural, remote continental and marine atmosphere. Their typical concentration is ranges 10^3 to $4x10^6/cm^3$. Nanofertilizer particle in the atmosphere is subject to coagulation, surface coating through condensation of semi, volatile compounds and heterogenous reactions with gaseous pollutants. Coagulation rate is low when the particles are the similar size but increases by orders of magnitudes as the differences in size grow. The time scale of coagulation is dependent on the particle size of interest and the background particle size distribution. For example, in mono disperse population of 20 nm fertilizer particles, the time scale of coagulation in urban and rural setting are around 10 minutes and 10 hours. Particles are removed from atmosphere mainly by Brownian diffusion, gravitational setting (dry deposition) and wet deposition in precipitation to the earth surface. Time scale for wet deposition is highly depends on the frequency of precipitation events. During a precipitation event, atmospheric nanofertilizer particles may be removed via wet deposition mechanism in a matter of minutes to hours. The time scale is for removal by dry deposition in an urban area in 20 h for 40–50 nm nanofertilizer particles.

8.2: Nanofertilizer in Plants

Plant cell wall acts as a barrier for easy entry of any external agent including nanofertilizer particles into plant cells. The sieving properties are determined by pore diameter of cell wall ranging from 5 to 20 nm. Hence, only nanoparticles or nanoparticle aggregates with diameter less than the pore diameter of the cell wall could easily pass through and reach the plasma membrane. There is also a chance for enlargement of pores or induction of new cell wall pores upon interaction with nanofertilizer particles which in turn enhance nanoparticle uptake. Further internalization occurs during endocytosis with the help of a cavity like structure that form around the nanoparticles by plasma membrane. They may also cross the membrane using embedded transport carrier proteins or through ion channels trichomes and then translocated to various tissues.

In the cytoplasm, the nanoparticles may bind with different cytoplasmic organelles and interfere with the metabolic processes at that site. Nanofertilizer particles are when applied on leaf surfaces, they enter through the stomatal openings or through the cuticle or other areas. However, accumulation of nanoparticles on photo-synthetic surface cause foliar heating which results in alterations to gas exchange due to stomatal obstruction that produce changes in various physiological and cellular functions of plants. Studies on the mechanism of uptake and formation of nanofertilizer particles within plants have also led to more investigations on the use of plants as source for nanoparticle synthesis. The potential nanofertilizers effect of soil plant system is shown as figure 70. The main effects are to reduce nutrient volatalisation and leaching. Improve soil quality by increasing microbial activity and water holding capacity. It also leads to the improvement in nutrient uptake.

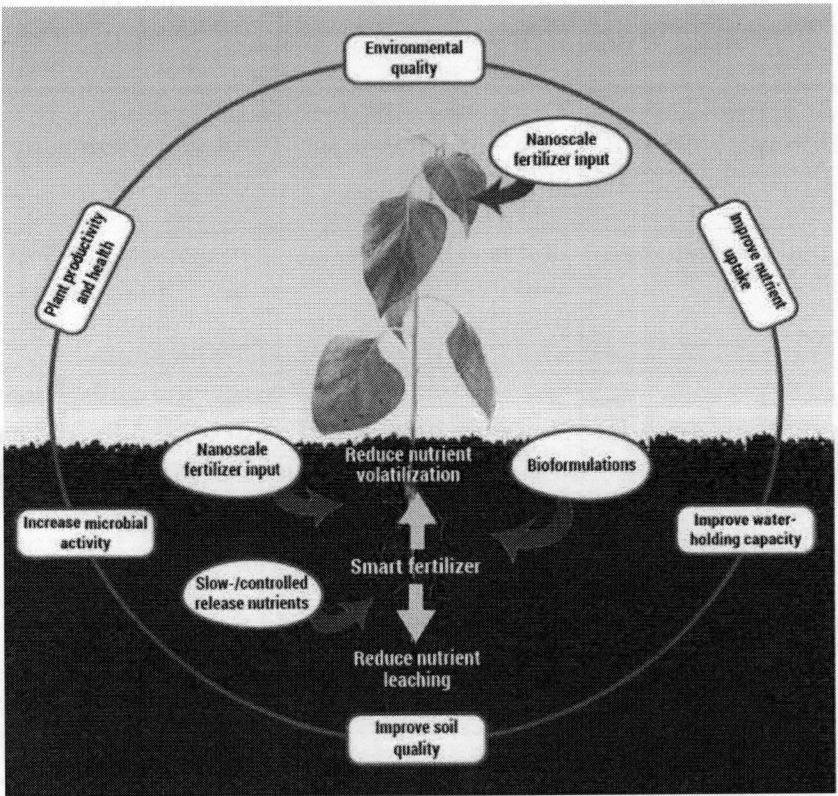

Figure 70: Schematic diagram of potential nanofertilizer effects in the soil-plant system

A comparison was made with the biosynthesised nanofertilizer with the chemical fertilizer used for crop production in India; the effect clearly demonstrated that nanofertilizer is much superior to the chemical fertilizer in terms of crop yield, prize, soil and environmental health, required doses and long term benefits. For example, a comparison was demonstrated between the single super phosphates (SSP) used as chemical fertilizer against biosynthesised nano-P fertilizer use as nanofertilizer (Table 23).

Table 23: A Comparison of phosphorus fertilizer between chemical verses nano-nutrients

Macro (chemical)–P	Nano–P	Other benefits of Nano–P
0.7–3% Native P in plant available form	Native P mobilization increased by 30%	80–100 times less requirement than chemical fertilizer
15% of applied P available to plant	P use efficiency enhanced to 61%	Two times less cost than SSP and 6-times less cost than DAP
Calcium, Iron, Aluminium and Silica inhibits available P in soil	No interference due to nano-size and foliar application	10 times more stress tolerant to the applied plants
Less plant initiative to mobile P	Increase in plant activity due to triggering of enzymes	In general, 23–54% more increase in crop yield than chemical fertilizer of recommended doses
Imported raw material resulting in drain of foreign exchange and decline in government revenue due to subsidy	No requirement of subsidy and import of raw materials	Eco-friendly and can be applied in all types of crops
Adverse effect of higher doses of application	No adverse effect with recommended doses of application	Helping in the improvement of soil health

8.2.1: Entry of nanofertilizer particle in plant

Nanofertilizer particles are adsorbed to plant surfaces and taken up through different plant openings. Several pathways exist or predicted for nanoparticles uptake by plants. Uptake rate will depend on the shape and sizes of the particles. Different entry points of nanofertilizers are shown in Figure 71.

Very small sizes of the nanofertilizer particle can penetrate through cuticle. Larger particles can penetrate through cuticle free areas. Nanofertilizer particles must traverse the cell wall before entering the intact plant cell protoplast. Figure 72 shows how nanofertilizer particles entering the plant cell.

Figure 71: Entry points of nanofertilizers in plants

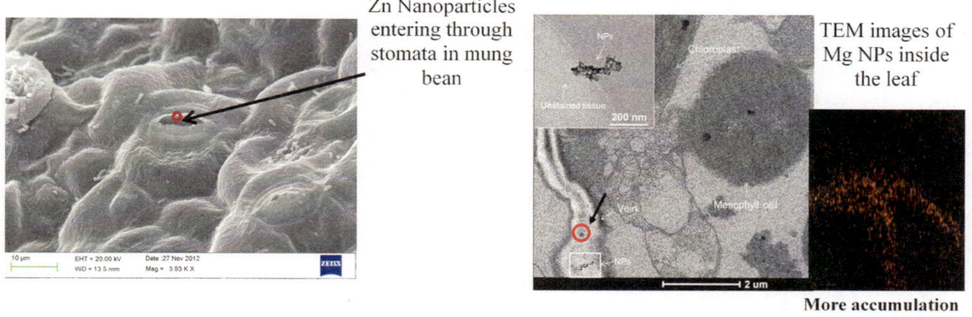

Figure 72: Entry and movement of nanofertilizer particles and their accumulation

After entering the particle through the leaf/plant holes they move through the cell sap and after triggering the co-enzyme system, some of the particles may accumulate in the vacule. The transport of nanofertilizer particle after entering has shown (Figure 73).

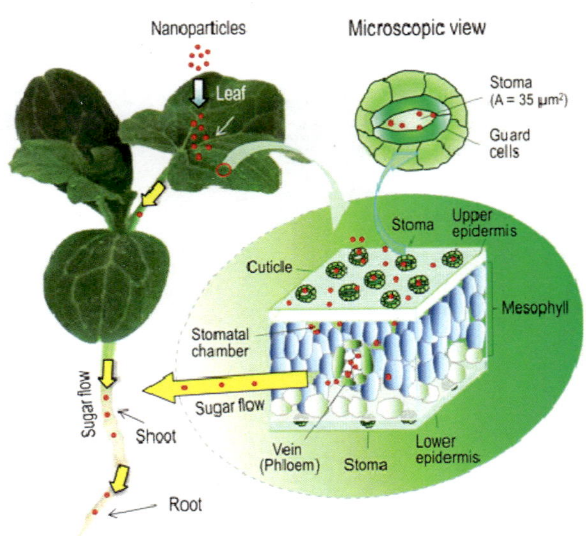

Figure 73: Schematic diagram of nanoparticle transport inside watermelon plants. The stoma openings are 6.7 l nm equivalent in diameter

Plants are an important component of the ecological system and may serve as a potential pathway for nanofertilizer particles transport and a route for bioaccumulation into the food chain. Plant cell wall acts as a barrier for easy entry of any external agent including nano-nutrient particles in the plant cells. The sieving properties are determined by pore diameter of cell wall ranging from 5 to 20 nm. Hence only the nanofertilizer particles or their aggregates with diameter less than the pore diameter of the cell wall could easily pass through and reach the plasma membrane. Certain nanofertilizer particle may increase the permeability of plant cell walls under stress and then penetrate the cells. There is also a chance of enlargement of pores or induction of new cell wall pores upon interaction with nanofertilizer particles which inturn enhance nanofertilizer uptake. The more accumulation of zinc nanofertilizer particles are showing (Figure 74) in vacuole.

Further internationalization occurs during endo-cytosis with the help of a cavity like structure that form around the nanoparticles by plasma membrane. They may also cross the membrane using embedded transport carrier proteins or through ion channels. When nanofertilizer particles are applied on the leaf surface, they enter through the stomotal openings or through the bases of trichomes and then translocated to various tissues. After entering the cells,

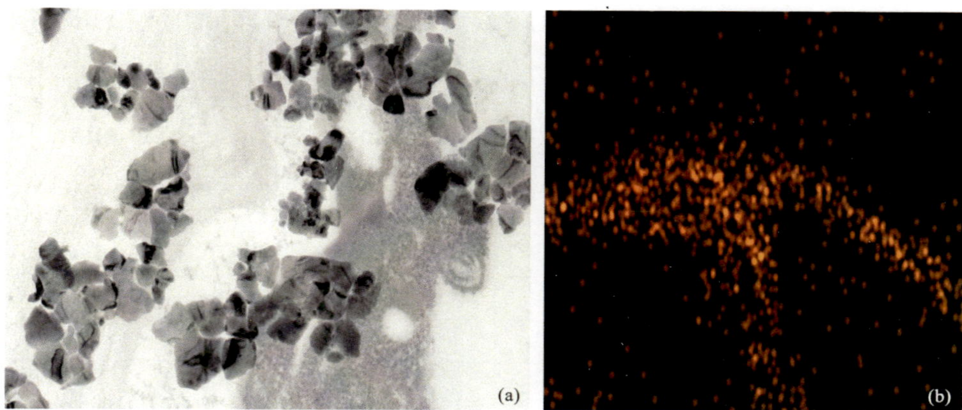

Figure 74: HR-TEM and STEM image of plants which shows more accumulation of zinc nanofertilizer particles in vacuole than other parts

nanofertilizer particle may be able to transport between cells via plasmodesmata, which are microscopic channels of plants traverging the cell walls and enabling transport and communication between cells. Plamodesmata or intercellular bridges were reported to be cylindrical channels about 40 nm in diameter. Thus, nanofertilizer particles less than 40 nm may enter and transport in the plant cell through the plasmodesmata once they are in the plant cells. In the cytoplasm, the nanofertilizer particles may bind with different cytoplasmic organelles and interfere with the metabolic processes at the site. Plant uptake can be a critical transport and exposure pathway of nanofertilizers in the environment. Interactions between plants, and nanofertilizers, such as the mechanism of uptake and the translocation and the interactions between the nanofertilizers and the plant tissues at the molecular and cellular level needs further investigations. But results obtained so far have shown that uptake of nanofertilizer by plants is very much possible. However, more studies are needed in uptake kinetics and toxicity of nanofertilizer doses which are affected by plant type and rhizospheric chemistry.

8.2.2: Effect on plants

The effect on dry matter of six different crops with four different biosynthesised nano-nutrients was reflected (Table 24) an increase in dry matter between 18–39% which was more under pearl millet followed by mung bean and wheat. The more dry matter production may be due to less exudation of carbon from the roots when nano-particles were applied as foliar.

Table 24: Increase in dry matter under different crops with the application of nanonutrients

Crops	Nano-nutrient applied*	% improvement of dry matter over recommended doses of chemical fertilizer
Clusterbean	P, Zn, Fe, Mg	21–33
Maize	P	19–30
Moth bean	P, Zn, Fe,	18–28
Mung bean	P, Zn, Fe, Mg	21–38
Pearl millet	P, Zn, Fe, Mg	23–39
Wheat	P	20–37

* Recommended doses.

Nitrogenases are enzymes that are produced by certain bacteria, such as cyanobacteria. These enzymes are responsible for the reduction of nitrogen to ammonia. Nitrogenases are the only family of enzymes known to catalyze this reaction, which is a key step in the process of nitrogen fixation. Two different proteins, comprise the nitrogenase complex. The FeMo protein binds substrate and reduces H^+ and N_2 to H_2 and ammonia, while the Fe protein receives electrons from ferredoxin, hydrolyzes ATP, and reduces the FeMo-protein. It has been found in several legumes that application of Zn nanofertilizer at 10 ppm concentration enhances nitrogenase activity between 18.9 and 177.5% over recommended doses of chemical fertilizer application (Table 25).

Table 25: Effect of Zn nanofertilizer on nitrogenase activity of some legumes

Crops	% improvement over recommended doses of chemical fertilizer
Clusterbean	38.7
Cowpea	177.5
Horsegram	33.4
Moth bean	41.2
Mung bean	18.9

Foliar application of nano-phosphorus fertilizer enhances the release of more organic acid, as compared to the application of recommended doses of fertilizers, by the plant roots resulted more plant unavailable P mobilization from the soil for plant nutrition

(Table 26). It was noticed that foliar application of 640 mg nano-P ha^{-1} resulted 80 kg P_2O_5 ha^{-1} equivalent yields in many crops.

Table 26: Per cent improvement in organic acid concentration in the rhizospheres and P uptake by plants (Nano-P application 640 mg ha^{-1})

Crops	Organic acid concentration	P uptake
Clusterbean	23.2	27.2
Moth bean	19.5	23.5
Mung bean	20.7	22.7
Pearl millet	15.5	17.3
Wheat	21.8	25.2

Nano-carbon can improve the water environment around the crops, make the nitrogen in the fertilizer more easily absorbed by the roots, increase the root activity of the crop, promote and increase root exudates which helps to increase the microbial population in the soil. The entire root parameters studied for pearl millet showed a significantly higher root activity with the application of different nano-fertilizers (Table 27).

Table 27: Effect on root parameteres of pearl millet with the application of optimum doses of nanofertilizer

Nanonutrients	Length (mm)	Area (mm^2)	No. of Tips	Perimeter (mm)	Average Diameter (mm)
Control**	17996	4402	2313	16082	0.25
N (80ppm)	46560 (158.7)*	13784 (213.1)	2751 (18.9)	18788 (16.8)	0.39 (56.0)
K (40ppm)	19085 (6.1)	9048 (105.5)	2479 (7.2)	16570 (3.0)	0.52 (108.0)
B (2ppm)	26647 (48.1)	13007 (195.5)	2347 (1.5)	16102 (0.1)	0.49 (96.0)
LSD	658.7	855.5	499.4	683.5	0.07

* Figures in parenthesis indicates the percent improvement over control

** Control –recommended doses of chemical fertilizer

LSD - Least significant difference

Soil is the largest receptor for the nanofertilizers. Behaviour of nanoparticles in the soil and their risk evaluation in arable soil ecosystems or other real environmental scenarios is very important.

Soil is the natural matrix and rich in natural nanoparticles both as primary particles and as agglomerates/aggregates. Artificial entry of nanoparticle into the soil may have significant effect, as they may be extremely resistant to degradation and have the potential to accumulate in the soil. The protection of soil microbial biomass and diversity is one of the major issues in the field of sustainable use of soils. The effect of nanoparticle on the soil depends on their concentration, soil type, and enzymatic activity of soil. At high concentrations of nanofertilizer particles, the negative effect on dehydrogenase and esterase enzyme activities (an indicator of soil microbial activities) was observed. Another negative effect caused by nanofertilizer particles is the influence on the rate of soil self-cleaning as well as on the balance of nutrients, which is the basis for the regulation of the processes of plant nutrition and soil fertility improvement. Soil properties, such as pH, texture, structure, and organic matter content, influence the soil microbial community and the ability of pollutants to have toxic effects on microorganism. Nanofertilizer with excess doses could affect mobility of soil pollutants. Therefore, there is a need to compare the toxicity of the nanoparticles in various types of soils. The soil amended with digestate and fly ash reduced the pollutant bioavailability. Intentional influence on soil properties and composition and texture of the soil with various substances can also alter the effects of nanofertilizer particles. The effect of metal nanofertilizer particles under field conditions on clusterbean is shown (Figure 75).

The rapid growth of nanofertilizer production and widespread use in agriculture have increased the quick development in the production of nanofertilizer, its application in farming, and other agriculture practices have expanded the chances of coming and blending to the nanofertilizers in the field's soil. It is presently notable that some nanofertilizer particles influence crop improvement, plant advancement, yield, and huge numbers of them are aggregated in various plant tissues, including the eatable piece of plant tissues. It is difficult to survey the seriousness of the nanofertilizer particles on the eco-frameworks and human well-being at the present moment. The effect of nanofertilizer on crop physiological changes is shown below (Table 28).

Figure 75: Field photograph shows the effect of metal particles *vis-à-vis* control on relative plant growth under field condition in clusterbean (O-Ordinary fertilizer, n-nanofertilizer)

Table 28: Effect of nanofertilizer particles on the physiological changes in plants

Type of nanofertilizer	Concentration applied	Physiological changes
TiO_2	10–200 mg/L	Increased the activities of various enzymes (POD, SOD, and CAT) below concentration of 200 mg/L because of eliminating accumulated reactive oxygen species in plant cells
Al_2O_3	10–1000 mg/L	The activity of SOD and CAT were increased with the treatments of Al_2O_3 nanofertilizers at a concentration of 200 and 500 mg/L
ZnO	10 mg/L	Treatment on clusterbean with foliar sprays of ZnO nanofertilizer caused a significant increase in chlorophyll content (276.2 %), total soluble leaf protein (27.1%), acid phosphatase (73.5%), alkaline phosphatase (48.7%), and phytase (72.4%) over control

Type of nanofertilizer	Concentration applied	Physiological changes
Fe_2O_3	20 mg/L	Increase in root activity, activity of catalase, peroxidase, superoxide dismutase, chlorophyll, malondialdehyde contents, ferric reductase activity, the root apoplastic iron content were recorded by the translocation of the significant amount of Fe_2O_3 nanofertilizers suspended in a liquid medium to various tissues of plants
Sulfur and ZnO nanofertilizer	500–4000 mg/L	Application of various concentrations of both sulfur and zinc oxide nanofertilizers is significantly increase the total lipids, proteins, amino acids, thiol and chlorophyll contents compared to untreated control, but no significance difference has been observed among the treatments with various concentrations for both nanoparticles
TiO_2	200 mg/L	Application of TiO_2 nanofertilizer had a noticeable effect on chlorophyll a and b, and carotenoid contents

Nanofertilizers effect on plants causing different morphological and physiological changes, contingent upon the type and nature of the nanoparticles used. The viability of nanoparticles is controlled by their substance synthesis, measure, surface covering, reactivity, and above all the portion at which they are successful. Scientists from their investigations indicated both positive and negative impacts of nanoparticles on plant development and improvement. The effect of designed nanofertilizer particles on plants relies upon the structure, fixation, concentration, and biochemical and physical properties of nanofertilizer particles just as plant species (Table 29). The results clearly demonstrated that effect on plants depends on the doses of nanofertilizer used.

Table 29: Effect of nanofertilizer particles on plant species and their metabolism

Nanofertilizer particle	Size range/ concentration	Plant	Biological effects
Ag	1–10 mg/L	*Triticum aestivum* L.	Adversely affect seedling growth

Nanofertilizer particle	Size range/ concentration	Plant	Biological effects
Ag	1000–3000 mM	*Pisum sativum* L	chlorophyll fluorescence
Ag	1000–2000 mM	*Allium cepa*	Caused oxidative stress and exhibited toxicity
Ag	5–25 nm, 1–40mg/L	*Phaseolus radiatus*	Higher accumulation in roots and shoots
Au	25–100 mg/L	*Arabidopsis thaliana*	Root length reduced by 75%
CeO_2, ZnO	0.05–0.5 g/kg soil	*Glycine max*	Increased ROS, lipid peroxidation
CuO	200–500 mg/L	*Gossypium hirsutum*	Height and root length decreased

Photosynthesis is the procedure utilized by plants, green algae, and certain other microbes to prepare food. Photosynthesis is a key maker for plants on earth that changes light vitality to concoction vitality. Plants convert just 2–4 % of the accessible vitality in radiation into new plant development. It was reported that nanofertilizer particles in the detached chloroplast enlarged multiple times higher photosynthetic activity than that of controls, and improved most extreme electron transport rates, and nanoparticles empowered the plants to detect nitric oxide, a singling substance. Metal nanoparticles can prompt the productivity of synthetic vitality creation in photosynthetic frameworks. The chlorophyll in photosynthetic response focus binds to the nanocrystals, in this way framing a novel half framework that may create multiple times increasingly energized electrons due to plasmon surface response and quick electron-separation. The improvement components may help in the plant of light-harvesting frameworks.

8.3: Effect of Nanofertilizers on Soil Organic Carbon

With the addition of nanofertilizers in the soil there is hardly any effect was observed on soil carbon content. The natural carbon content in some soils reported to differed between 0.93% and 3.85%, however there was no connection between the natural carbon substance of the soils and the nanofertilizer applied. Sometimes the natural issue can change the impacts of nanofertilizer materials, particularly those which discharge particles. For instance, the poisonous quality of Ag nanomaterials towards the freshwater zooplankton *Ceriodaphnia dubia* and *Daphnia magna* declined altogether as the broke down natural carbon content expanded. Ag nanomaterials are less versatile in soils with a high substance of natural issue in light of the fact that natural materials, for example, sewage slop connect with Ag nanomaterials, with sorption rates more prominent than 90% for uncoated particles and lower rates for functionalized particles. Concentrates with biofilm networks and unadulterated societies have additionally appeared natural issue, for example, exopolysaccharides and humic acids can secure microorganisms and diminish the poisonous quality of Ag nanomaterials. For instance, the nearness of Suwannee River humic acids made it difficult to decide an impact on the development of *Pseudomonas fluorescens*. In any case, these discoveries are for the most part from investigations in which the impact of natural carbon was tried on amphibian life forms though comparable information for soil life forms is increasingly constrained. It is already demonstrated that Ag nanomaterials and the particles discharged by them communicate with natural issue added to soil, diminishing the long haul harmful consequences for the aggregate of microbial exercises bringing about nitrate gathering. The proposed mechanisms for the restraint of Ag nanofertilizer disintegration incorporate surface adsorption onto common natural issue (which thus forestalls Ag nanofertilizer oxidation and the arrival of silver particles) and the decreasing action of humic/fulvic acids which changes over oxidation items once more into silver metal.

8.4: Effect of Nanofertilizers on Microbial Diversity

TiO_2 and CuO nanofertilizer particles reduce the microbial biomass of the soil. Nanofertilizer like ZnO, TiO_2, CeO_2, and Fe_3O_4 influences on enzymatic activities of soil (invertase, urease, catalase, and phosphatase) and bacterial networks of the saline-soluble base and black soils. The outcomes demonstrated an effect on soil chemical exercises and changes in soil bacterial network and danger on natural nitrogen obsession. Results so far have demonstrated that a high convergence of Fe_3O_4 nanofertilizers altogether diminished the substance of microscopic organisms in the soil. Zinc oxide and CeO_2 nanofertilizers influenced the plate checks of *Azotobacter*, P-solubilizing, microscopic organisms and restrain enzymatic exercises. Titanium dioxide nanofertilizer particles diminished the plenitude of utilitarian soil microscopic organisms and enzymatic activities and hindering impact on microbial action, occurrence, and diversity. The information provided by Maliszewska reported that the biogenic Au nanoparticles up to the grouping of 33 mg kg^{-1} do not influence the soil properties, their processes and can be delegated not destructive. Antibacterial exercises (development of a zone of inhibition) were seen on clinical disconnects (*Bacillus subtilis, Escherichia coli, Klebsiella pneumonia, Pseudomonas aeruginosa, Salmonella typhi,* and *Staphylococcus aureus*) treated with ZnO nanofertilizers incorporated by organic and concoction techniques. Copper ions, released from the Cu nanofertilizer particles, were considered as a major cause of lethality to both the pathogenic and the beneficial bacteria. It was investigated the effect of CuO and Ag nanofertilizer particles on leaf microbial decomposition showing that exposure to these nanoparticles led to a decrease in leaf decomposition rate (Table 30).

Table 30: Effect of nanofertilizers on soil microbial community and enzymes

Nanofertilizer	Major effect
Au	Hinderance in enzymatic activities of soil
TiO_2	The loss in diversity of the bacterial community
Cu	Decreases C, N Biomass
Ag	Loss of microbial community, hindrance in enzymatic activities
Mixture of Cu and Fe	Loss of microbial community
Mixture of Co, Fe, Ni and Ag	Not much effect on the soil community
Ag	Effect on β-Glucosidase, acid phosphatase, dehydrogenase, Urease enzymes
C_{60}	Effect on eucine-aminopeptidase, β-cellobiohydrolase, acid phosphatase, β-Glucosidase, chitinase, xylosidase enzymes

The researchers observed that TiO_2, at high concentrations, can disturb the symbiosis relationship between plant and bacteria, and have a negative effect on the growth of clover. In natural soil, the mobility of the nanofertilizer particles was weak and no increase in the uptake of titanium dioxide by the plants was found. The mobility of MWCNT (Multi walled carbon nano-tubes) in soils was also weak;

the experiments showed a concentration-independent uptake of MWCNT by the plants and a reduction in the number of red clover blossoms in the case of high concentrations. Both TiO_2 and MWCNT at a very high concentrations resulted in a change in the composition of the microbial communities that interact with the plants. The effect on soil nematode species are presented (Table 31).

Table 31: Effect of nanofertilizer particles on soil nematodes community

Nanofertilizers	**Affected species**
MWCNT, SWCNT (Single walled carbon nanotubes)	*Eisenia foetida*
Tio_2, ZnO	*Eisenia foetida*
Al_2O_3	*Eisenia foetida*
Pt	*Caenorhabditis elegans*
Ag	*Caenorhabditis elegans*
ZnO	*Caenorhabditis elegans*
CeO_2, TiO_2	*Caenorhabditis elegans*

8.5: Nanobiotechnology

It accelerated almost every sector of agriculture sciences and crop research. It is well understood that the use of nanofertilizer material for the scientific and agriculture field enables a lot of advantages over conventional methods. Biofertilizers based on nanotechnology is quite useful for plant growth and developments. Biopesticides and no fungicides are more effective against disease and harmful insects. DNA delivery in crop research has become very easy with the help of nanobiotechnology. Uncontrolled and unregulated accumulation of nanomaterials in soil is a matter of concern. Since toxicity of nanomaterial is not unknown from late last decades. Investigations indicate the toxicity of nanomaterials is the emerging serious issue. Microbial community and soil health which directly or indirectly related to human health and well-being must be protected. Therefore, nanofertilizer must be applied at right doses and concentration to get maximum benefit as well as maintain soil health and protect environment.

Further studies

Cho, K. H., Park, J. E., Osaka, T. and Park, S. G. (2005). The study of antimicrobial activity and preservative effects of nanosilver ingredient. *Electrochem. Acta* **51:** 956–960.

Kumar, N., Shah, V. and Walker, V. K. (2012). Influence of a nanoparticle mixture on an arctic soil community. *Environ. Toxicol. Chem.* **31:** 131–135.

Mousavi, S. R. and Rezaei, M. (2011). Nanotechnology in agriculture and food production. *J. Appl. Environ. Biol. Sci.* **1:** 414–419.

Tiede, K., Boxall, A. B., Tear, S. P., Lewis, J. and David, H. (2008) Detection and characterization of engineered nanoparticles in food and the environment. *Food Additives Contaminants* **25:** 795–821.

You, T., Liu, D., Chen, J., Yang, Z. and Dou, R. (2018). Effects of metal oxide nanoparticles on soil enzyme activities and bacterial communities in two different soil types. *J. Soils Sediments* **18:** 211–221.

9

Physiological Effects of Nanofertilizers

Nanofertilizers stimulate a number of vital facets of plant biology as plant root and leaf surfaces are the main nutrient gateway of plants which are highly porous at the nanoscale. Consequently, the application of nanofertilizers can improve plant nutrient uptake through these pores, or the process can facilitate complexation with molecular transporters or root exudates through the creation of new pores, or by the exploitation of endocytosis or ion channels. Moreover, it has been unequivocally observed from an extensive research that the size reduction of nanomaterials facilities the increase of surface mass ratio of particles; as a result, a bountiful amount of nutrient ions get adsorbed and desorbed slowly and steadily for an extended period of time. Thus, the nanoformulations of fertilizers ensure a balanced nutrition of crops throughout the growth cycle that eventually improve agricultural production. It is to be noted that the increased efficiency of a product may encourage the farmers to use the product more profitably.

The application of nanostructured materials, designed for sustainable crop production, reduces nutrient losses, suppresses disease and enhances the yields. Nanofertilizers, with a particle size less than 100 nm, influence key life events of the plants that include seed germination, seedling vigor, root initiation, growth and photosynthesis to flowering. Additionally, nanofertilizers have been implicated in the protection of plants against oxidative stress as they mimic the role of antioxidative enzymes, such as superoxide dismutase (SOD), catalase (CAT) and peroxidase (POX). However, besides their beneficial effects on plants, applications of nanofertilizers at higher rate have been proved to be phytotoxic too as they enhance the generation of reactive oxygen species (ROS). The elevated level of ROS may damage the cellular membranes, proteins and nucleic acids. Therefore, in, such a conflicting and ambiguous nature of nanofertilizer in plants, it is necessary to decipher the mechanism of cellular, biochemical and molecular protection render by nanomaterials under stressful environmental conditions. The systematic studies on the role of nanofertilizers under abiotic stresses, such as drought, salt, temperature, metal, UV-B radiation and flooding is necessary. Furthermore, suitable strategies adopted by plants in presence of nanomaterials under challenging environments also need more attention.

9.1: Effect on Seed Germination and Growth

The germination of seeds is a sensitive phase in the life cycle of plant, which facilitates seedling development, survival and population dynamics. However, seed germination is largely affected by different parameters including environmental factors, genetic trait, moisture availability and soil fertility. In this concern, an extensive number of studies have shown that the application of nanofertilizers has positive effects on germination as well as plant growth and development. For example, the application of multiwalled carbon nanotubes (MWCNTs) positively influences seed germination of different crop species including tomato, corn, soybean, barley, wheat, maize, peanut and garlic (Table 32). Similarly, nanofertilizer SiO_2, TiO_2 and Zeolite application positively stimulate seed germination in crop plants. It has also found that Fe/SiO_2 nanomaterials have significant potential to improve seed germination in barley and maize. Despite a significant group of research on nanofertilizer-induced positive effects on germination, the underlying mechanisms how nanomaterials could stimulate germination still not very clear. A few studies have demonstrated that nanomaterials have the potential to penetrate the seed coat and enhance the ability of absorption and utilization of water, which stimulates enzymatic system and ultimately improves germination and seedling growth. Nonetheless, the mechanism of nanomaterial-induced water uptake inside the seed is still largely unknown.

In addition to germination, nanofertilizers, such as ZnO, TiO_2, MWCNTs, FeO, ZnFeCu-oxide, and hydroxyfullerenes are reported to increase crop growth and development with quality enhancement in many crop species including peanut, soybean, mung bean, wheat, onion, spinach, tomato, potato and mustard. For example, carbon nanomaterials fullerols, as OH-functionalized fullerenes have commonly exerted positive effects on plant growth. It was demonstrated that fullerenes enhanced hypocotyl growth in Arabidopsis by stimulation of cell divisions. It has also been found that seed dressings with Fullerol not only increase fruit number, fruit size, and final yield by up to 128%, but also stimulate the content of bioactive compounds, such as cucurbitacin-B, lycopene, carotein and inulin in fruits of bitter melon (*Momordica charantia*). The application of nano-iron fertilizer not only increased the agronomic traits of *Dracocephalum moldavica* with sowing density, but also improved essential oil contents of plants. Similarly, foliar application of nano-zinc and boron fertilizers was found to increase fruit yield and quality, including 4.4–7.6% increases in total soluble solids (TSS), 9.5–29.1% decreases in titratable acidity (TA), 20.6–46.1% increases in maturity index and 0.28–0.62 pH unit increases in juice pH on pomegranate (*Punica granatum*) without affecting any physical fruit characteristics. These findings revealed perspectives of nanofertilizers to improve crop yields and product quality. Although the exact mechanism behind promotion of plant growth and enriched quality is not clear, it may be at least partially explained by the potentialities of nanomaterials to absorb more nutrients and water that in turn helps to enhance the vigor of root systems with increased enzymatic activity. Moreover, the studies of nutrients on slow/controlled release or control loss of nanofertilizers carried out in water and soil have confirmed that the long term availability of all the doped nutrients to the plant over the full crop period of cultivation is crucial for promoting germination, growth, flowering and fruiting. For example, hydroxyapatite nanomaterial-coated urea fertilizer releases nitrogen slowly and uniformly over up to 60 days, whereas the traditional bulk fertilizer losses only within 30 days with uneven release that reduced the nutrient efficiency of plants and adversely affects crop growth. Conversely, the examination in different studies offers conflicting evidence about the positive nexuses of nanomaterials on seed germination and growth of crops. Such variability may arise due to a number of factors related to nanomaterial characteristics, such as size, shape, surface coating and electronic properties, dose as well as mode of application and the plant species studied. For instance, it was demonstrated that the application of TiO_2 at 2.5% level increased photosynthesis in spinach about 3.13%; however, photosynthesis decreased beyond 4% concentration of TiO_2. In another studies, demonstrated that 15 mg kg^{-1} of nanofertilizer Fe/SiO_2 increased shoot length of barley and maize seedlings about 8.25% and 20.8%, respectively; however, the shoot length was negatively affected when concentration reached 25 mg kg^{-1}, meaning that crop growth

Table 32: Effect of nanofertilizers on crop physiology and plant protection

Nanofertilizer	Crop species	Mode of application	Concentration used	Duration of treatment	Responses
MWCNTs	*Hordeum vulgare, Glycine max, Zea mays*	Seed priming	100 mg/L	24 h	Enhanced germination and growth of seedlings
MWCNTs	*Triticum aestivum, Zea mays, Arachis hypogaea, Allium sativum*	Seed priming	50 mg/L	Over night	Improved and rapid germination, increased biomass accumulation and water absorption potential of seeds
ZnO	*Coffea arabica*	Foliar spray	10 mg/L	45 days	Enhanced growth, biomass accu-mulation and net photosynthesis
ZnO	*Triticum aestivum*	Mixed with growth substrate	20 mg/L	Growth cycle	Increased grain yield and biomass accumulation
ZnO	*Cyamopsis tetragonoloba*	Foliar spray	10 mg/L	6 weeks	Improved plant growth, biomass accumulation and nutrient content
FeS_2	*Cicer arietinum, Spinacia oleracea, Daucus carota, Brassica juncea, Sesamum indicum*	Seed priming	80–100 mg/L	12–14 h	Increased germination and crop yield

Nanofertilizer	Crop species	Mode of application	Concentration used	Duration of treatment	Responses
CuO	*Spinacia oleracea*	Mixed with soils	200 mg/kg	90 days	Improved photosynthesis and biomass production
ZnO	*Nicotiana tabacum*	Hydroponics	0.2 µM and 1 µM	21 days	Positively affected growth physiology, increased metabolites, enzymatic activities and anatomical properties of plants
Fe/SiO_2	*Arachis hypogaea*, *Zea mays*	As fertilizers	15 mg/kg	3 days	Enhanced plants growth and biomass accumulation
TiO_2	*Spinacia oleracea*	Seed priming and foliar application	0.25% suspension	48 h and 35 days	Increased biomass accumulation, chlorophyll, nitrogen and protein content
Ag nanofertilizer	*Triticum aestivum*	Mixed with pot soils	50 mg/L and 75 mg/L	Trifoliate stage	Improved growth and tolerance to heat stress
Ag nanofertilizer	*Vigna sinensis*	Foliar application	50 mg/L	40 days	Enhanced growth and biomass by stimulating root nodulation and soil bacterial diversity
TiO_2 and SiO_2	*Oryza sativa*	Foliar application	20 and 30 mg/L	55 days	Mitigated Cd toxicity and improved growth by stimulating antioxidant potential and inhibiting Cd translocation

Nanofertilizer	Crop species	Mode of application	Concentration used	Duration of treatment	Responses
SiO_2 nanofertilizer	*Oryza sativa*	Foliar application	2.5 mM/L	70 days	Alleviated heavy metal toxicity and improved growth by decreasing bio-concentration and translocation in plants
ZnO, CuO and Ag nano-fertilizers	*Prunus domestica* fruits	Fruit spray	100 and 1000 µg/mL	4 days	Suppressed grey mold symptoms caused by *B.cinerea* and soil borne diseases
Al_2O_3 nano-fertilizer	*Solanum lycopersicum*	Foliar application	400 mg/L	20 days	Successfully controlled Fusarium root rot in tomato
Ag nano-fertilizer	*Vigna unguiculata*	Foliar application	50–100 mg/L	7 days	Showed no phytotoxicity, but could inhibit growth of *Xanthomonas axonopodis* pv. *malvacearum* and *Xanthomonas campestris* pv. campestris (*in vitro*)
CuO	*Solanum lycopersicum*	Foliar application	150–340 mg/L	11 days	Effectively controlled late blight disease caused by Phytophthora infestans
MgO	*Solanum lycopersicum*	Drenching	7–10 mg/L	7 days	Controlled bacterial wilt disease by suppressing pathogen *Ralstonia solanacearum*

depends on the concentration of nanomaterial application. It was found that foliar application of nanofertilizer Fe_3O_4 could significantly enhance total chlorophyll, total carbohydrate, essential oil levels, iron content, plant height, branches/plant, leaves/plant, fresh weight, and dry weight of *Ocimum basilicum* plants compared to that of soil application. Accomplishing this task requires a multi-pronged strategy, such as activation of plant enzymatic system, hormonal regulation, stress gene expression, regulation of toxic metal uptake and avoiding water deficit stress or flash flood through shortening plant life cycle. Various efforts have been made by researchers to develop technologies and practices towards sustainable agricultural systems by avoiding adverse effects on environmental compartments.

9.2: Effect on Adverse Environment

Advances in nanomaterial engineering suggest that nanofertilizers can increase crop production in existing adverse environments. Salinity stress seriously limits crop production in about 23% of the cultivated lands worldwide. In contrast, it has been reported that application of nanofertilizer-SiO_2 improves seed germination; increases plant fresh weight, dry weight and chlorophyll content with proline accumulation in tomato and squash plants under NaCl stress. Likewise, it was demonstrated that foliar spray of nanofertilizer, iron sulfate ($FeSO_4$), show positive response to salinity stress tolerance in sunflower cultivars. Nanofertilizer $FeSO_4$ application not only increased leaf area, shoot dry weight, net carbon dioxide (CO_2) assimilation rate, sub-stomatal CO_2 concentration (Ci), chlorophyll content, maximum photochemical efficiency of photosystem II (Fv/Fm) and iron (Fe) content, but also decreased significant amount of sodium (Na) content in leaves. Recently, it has also been explored that silicon nano-particles as nanofertilizer (SiNPs) could effectively alleviate UV-B induced stress in wheat. Nano-zeolite can improve the long-term availability of nutrients and enhance the germination and growth of plants. The life cycle of nanofertilizer-applied wheat plants was 23.5% shorter (130 days compared with 170 days) for yield production from the date of sowing compared to conventional fertilizer-applied plants. Such an acceleration of plant growth and productivity by the application of nanofertilizers demonstrate their potential as effective tools in agricultural practices, especially in drought-prone, or even in sudden flash flood-prone areas where the early maturity of crops is an important aspect for sustainable crop production. In addition, nanomaterials are found to be effective in detoxification or remediation of harmful pollutants like heavy metals.

For example, foliar application of nanofertilizer-Si at 2.5 mM concentration significantly improves Cd stress tolerance in rice plants by regulating Cd accumulation. Nanofertilizer-Si is also effective against Pb, Cu, and Zn with Cd. It appears that nano-Si fertilizers may putatively have an advantage over traditional fertilizers in reducing heavy metal accumulation.

9.3: Effects on Secondary Metabolites

Plants produce several important secondary metabolites which are primarily used as their defence compounds. The medicinal value of secondary metabolites to humans is also well established. The medicinal value of these metabolites has called for the need for their enhanced production in plants. Some of the strategies used for improving the content of secondary compounds through plant cell and organ cultures include altering the culture medium composition, providing precursors and elicitors and altering the environmental conditions. Nanofertilizers among other strategies can also be employed for the increased production of these compounds. It has been suggested through various studies that the application of nanoparticles as fertilizer to plants and their cultures *in vitro* can elicit the production of important secondary metabolites. For instance, the addition of aluminium oxide (Al_2O_3) nanofertilizers significantly enhanced the phenolic content in cell suspension cultures of tobacco. Various types of nanofertilizers have been shown to enhance the quantity of secondary metabolites. In a study, the quantities of essential oils increased in the callus cultures of *Calendula officinalis* when exposed to Ag nanofertilizers. Similarly, the addition of titanium oxide (TiO_2) nanofertilizers enhanced the quantity of chlorogenic acid, cinnamic acid, gallic acid, o-coumaric acid, and tannic acid in the callus cultures of *Cicer arietinum*. Furthermore, combined application of different nanofertilizers has also been proven to be effective in enhancing the metabolite content in plants. For instance, maximum accumulation of total phenolic compounds was observed in *Prunella vulgaris* calli established in response to the addition of Au-Ag nanoparticle combinations (1:3).

The types of nanofertilizer particles employed for the enhancement of secondary metabolites production in plants include Ag, ZnO, $CuSO_4$, Fe and TiO_2 nanoparticles as fertilizer. Specific secondary metabolites enhancement through application of these nanoparticles include anthocyanins in *Arabidopsis thaliana* seedlings , steviol glycosides in shoot cultures of *Stevia rebaudiana*, rosmarinic acid and caffeic acid in *Satureja khuzestanica* shoot cultures, artemisinin in hairy root cultures of *Artemisia annua*, taxol in *Corylus avellana* cell suspension culture , atropine in hairy root cultures of *Datura metel*, aloin in Aloe vera suspension cells, hyoscyamine and scopolamine in hairy root cultures of *Hyoscyamus reticulatus*, and capsaicin in *Capsicum frutescens* cell suspension cultures.

The exact mechanism of speeding up the secondary metabolism by nanofertiliers is yet to be established. One possible mechanism includes the generation of reactive oxygen species (ROS) by the nanoparticles which ultimately trigger the secondary metabolism to cope with the stress induced by ROS. Another mechanism of enhancing their production is by regulating the expression of various genes involved in the biosynthesis of secondary metabolites. The upregulation of genes involved in the thalinol biosynthetic pathway in Ag nanofertilizer treated plants. Conclusively, nanofertilizer particles have been shown to significantly affect the production of bioactive compounds in different plants and plant cell cultures. Hence, can be employed as standard elicitors in plant cell, tissue and organ cultures to produce valuable metabolites.

9.4: Biochemical Parameters

The changes brought about by the application of nanofertilizers sometimes may not be apparent in terms of physiology, but they may translate into the enhanced biochemical features including the antioxidant. For instance, magnetite nanofertilizers (Fe_3O_4) enhanced the activity of different antioxidant enzymes, such as catalase (CAT), superoxide dismutase (SOD) and peroxidase (POD) in wheat plants. These antioxidant enzymes form the defence system of plants and are used to quench ROS generating free radicals. The activation of these enzymes by nanoparticles defines their capacity to enhance antioxidant activities in plants.

Application of CeO_2 nanofertilizer at higher concentration resulted in the increased CAT activity in leaves of *Prosopis juliflora* velutina. Furthermore, incorporation of Ag nanofertilizer into MS basal medium at a concentration of 50mg L^{-1} decreased the production of hydrogen peroxide, proline, and Malondialdehyde (MDA) in seedlings of *Brassica juncea* by causing the activation of antioxidant enzymes. Another study reported that Ag nanofertilizer treatment altered the antioxidant enzymes content and expression profile of different genes that are involved in the biosynthesis of auxin, abscisic acid and ethylene. The ZnO nanofertilizer application can mitigated the effects of NaCl in tomato plant by upregulating antioxidant enzymes, such as glutathione peroxidase (GPX) and SOD. ZnO nanofertilizer particles have been reported to alter the genetics of the plant, thus enhancing the antioxidant enzymes, which imparts antimicrobial potential and therefore has a role in disease protection. ZnO nanofertilizer application in clusterbeans led to alterations in the total protein, lipid and chlorophyll contents as compared to plants treated with ZnO chemical fertilizer. Similarly, ZnO nanofertilizer exposure increased the protein content, photosynthetic pigment levels, and it upregulated the expression level of SOD and peroxidase (POX) isoenzymes. It is thus increasing the activity of antioxidant enzymes in the cotton plant.

9.5: Applications of Nanofertilizers during Plant *in vitro* Cultures

Positive effect under *in vitro* cultures of different plants due to application of nanofertilizers at various stages was found.

9.5.1: Towards callus induction and organogenesis

Nanofertilizers have been shown to positively affect the production of callus, shoots, roots including adventitious roots and hairy roots in *in vitro* cultures of plants. Various studies have shown that nanofertilizer have a positive impact on callus induction, somatic embryogenesis, shoot regeneration, shoot multiplication, and root growth. This effect is supposed to be associated with the altered gene expression, ethylene inhibition and ROS generation in response to the application of nanofertilizer particles. For instance, supplementation of growth medium with Ag nanofertilizer particles increased percentage of shoot induction, and callus formation from stem explants of *Tecomella undulata*. The increased number of shoots has been attributed to the inhibition of ethylene by Ag nanofertilizers. Nonetheless, Ag nanoparticles have been proven effective *in vitro* through different effects on growth parameters, such as shoot regeneration of *Vanilla planifolia*, high frequency callus formation (89%) in *Solanum nigrum*, callus proliferation in *Prunella vulgaris*, and increased embryogenesis (50%) in *Linum usitatissimum*. Similarly, Au nanoparticles have also been shown to positively affect the *in vitro* cultures of *Linum usitatissimum* in terms of increased embryogenesis (70%). Figure 76 presents the overall impact of *in vitro* nanofertilizer particles application.

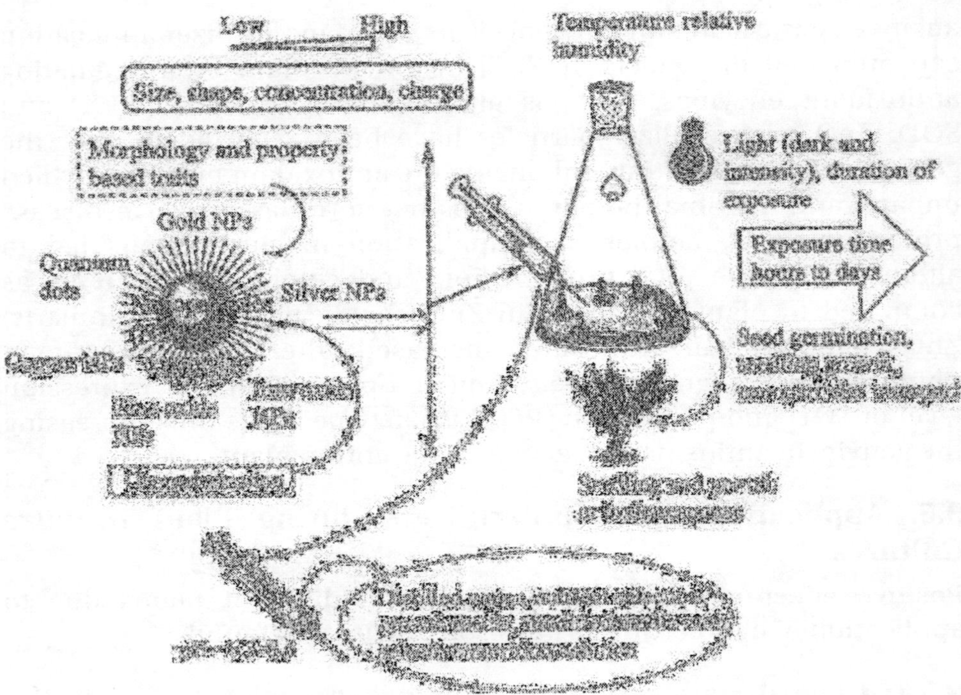

Figure 76: *In vitro* application of different types of nanofertilizer particles in plant cell cultures for production of biomass and important secondary products. (*Source:* Redrawn from S.K. Verma, A.K. Das, M.K. Patel, A. Shah, V. Kumar, S. Gantait, Engineered nanomaterials for plant growth and development: a perspective analysis, Sci. Total Environ. 630 (2018) 1413–1435.)

9.5.2: Nanofertilizer induce somaclonal variations

The application of certain external agents during the growth of plant *in vitro*, can give rise to various changes as compared to the wild type plant. These changes in the plant biochemistry and physiology are called as somaclonal variations. *In vitro* derived plantlets may have altered genetics, such as mutated DNA and changed chromosome structure. This results in varied phenotypes, such as altered plant size, flower colour, resistance and secondary metabolite production in plants. Nanofertilizers have been shown to alter plants in a way to induce somaclonal variations. For instance, supplementation with carbon nanoparticles resulted in a higher number of tetraploid cells with high level of methylated DNA in callus cultures of *L. usitatissimum*. Similarly, Ag nanofertilizer particles were shown to induce changes in protein and DNA profile of *Solanum nigrum* calli that resulted in altered appearance of callus cultures. These variations, like all other somaclonal variations, may be advantageous and disadvantage.

9.6: Effect of Nanofertilizers on Biotic Factors

Crop productivity is also largely influenced by biotic factors including pests and diseases. To minimize crop losses, farmers have been heavily reliant on pesticides, which adversely affect human health and environmental sustainability. However, recent studies have revealed that nanomaterials as fertilizer could successfully reduce the risks of pests and diseases, thereby minimizing the severity of yield losses and environmental hazards. For example, biosynthesised Ag nanoparticles obtained using stem extract of cotton plant (*Gossypium hirsutum*) possess a strong antibacterial activity as evidenced by the zone of inhibition for *Xanthomonas axonopodis* pv. malvacearum and *Xanthomonas campestris* pv. campestris, two major bacterial pathogens of Malvaceae and Brassicaceae family crops, respectively. Metal oxide nanofertilizers, such as CuO, ZnO, and MgO could also effectively control many plant and soil borne diseases caused by *Botrytis cinerea, Alternaria alternate, Monilinia fructicola, Colletotrichum gloeosporioides, Fusarium solani, Fusarium oxysporum* sp, *Radicis lycopersici, Verticillium dahliae, Phytophthora infestans* and *Ralstonia solanacearum* in many plant species. Hence, the judicious use of nanofertilizer materials can increase crop productivity without disturbing the environmental health. In recent years, researches on the use of nanocomposites have greatly expanded in the field of plant protection due to their high efficacy and eco-friendly nature. For example, Ag-incorporated chitosan nanocomposites (Ag@CS) with fungicides antracol show an increased antifungal activity compared to that of each component alone. Again, the development of nanocomposites of *Bacillus thuringiensis* (Bt) containing active Bt has further increased the efficacy and shelf life of pesticides and reduced the dosage required previously. However, the mechanisms behind, such actions of Bt-based nanocomposites are still to be elucidated.

9.7: Uptake and Translocation of Nanomaterials

Nanofertilizers can be absorbed by crops through the roots or leaves. Nanoparticles can penetrate root epidermis and endodermis reaching the xylem vessels, allowing it to be transported to the aerial part of the plant. Moreover, nanofertilizers can be absorbed by leaf stomata and transported to other plant parts through the phloem. In both cases, nanofertilizer particles must penetrate the cell wall by pores, and pore sizes may range 3–8 nm. Therefore, only nanofertilizer particles which are smaller than 8 nm could pass through pores and reach the plasma membrane. Accordingly, nanomaterials or aggregates bigger than 8 nm cannot enter into cells. It was recently shown that tomato roots can absorb 3.5 nm Au nanoparticles, but not 18 nm Au nanoparticles. It has been shown that CeO_2 nanofertilizer can be absorbed by cucumber leaves and consequently transported to different plant tissues and Ag nanofertilizer particles could be absorbed and distributed throughout the plant after foliar exposure in lettuce plants.

The uptake and translocation of nanofertilizer particles may vary from plant to plant depending on its particular physiology and several mechanisms of their uptake, transport and distribution within the plant. In several cases, plants activate defense responses against nanoparticles. This appears to be particularly true for the metallic oxide-based nanofertilizers, in which the plant faces the parent nanomaterials effects as well as the metal ions produced by the dissolution of engineered nanofertilizers. An experiment with carrots comparing metal oxide nanoparticles (ZnO, CuO, or CeO_2) and metal ions uptake demonstrated that uptake of both nanometal oxides and metal ions occurred. Such uptake and accumulation in edible parts may impose not only problems to the physiology of plants, but also pose serious risks to human health. It was shown that metal oxide nanoparticles accumulated in the outer layer of carrot and did not enter the fleshy part, while metal ions entered the fleshy edible part and were potentially more toxic to human health. It was suggested that this outer layer acts as a barrier to restrict the inward penetration of engineered nanoparticles in the edible tissues. Therefore, peeling the outer layer of root and tuberous vegetables will be required to reduce the toxic exposure to these metal nanoparticles.

9.8: Transformation of Nanofertilizer Particles

Nanomaterials are highly reactive. Therefore, when nanofertilizers are used in crops, and they interact with different components in the environment, they are subject to transformations and changes in their physicochemical properties. Nanofertilizers interact with organic and inorganic substances in the soil, as well as with various plant components that may alter the behaviour, fate, and toxicity of nanofertilizer particles. When nanofertilizers are used in roots, nanoparticles are exposed to root exudates in rhizosphere that can determine the behaviour and toxicity of heavy metals. For example, mesquite plants treated with Ni-$(OH)_2$ nanoparticles had Ni $(OH)_2$ in the roots, but contained Ni nanoparticles in the shoots and leaves, demonstrating the biotransformation process of this nanoparticle inside the plant. Figure 77 shows zinc nanofertilizer sprayed mung bean leaf surface. It clearly stated that nanoparticles are evenly distributed on plant leaf upper surface upon spray and they might enter to the plant.

Other cases of biotransformation have been observed using CeO_2 nanoparticles in cucumber plants. Results showed 15% of Ce (IV) being reduced to Ce(III) in the roots and 20% to Ce(III) in the shoots . However, another study showed that CeO_2 nanoparticles biotransformation did not happen in cucumber roots, suggesting that nanoparticles biotransformation requires specific conditions in the plant rhizosphere. Biotransformation was also demonstrated in Ag nanoparticles applied on lettuce plants. Ag nanoparticles were oxidized and Ag^+ ions formed Ag complexes with thiol-containing molecules. Also, in another experiment using Ag nanoparticles in lettuce, it was found that 33% of the nanoparticles was transformed to Ag-glutathione.

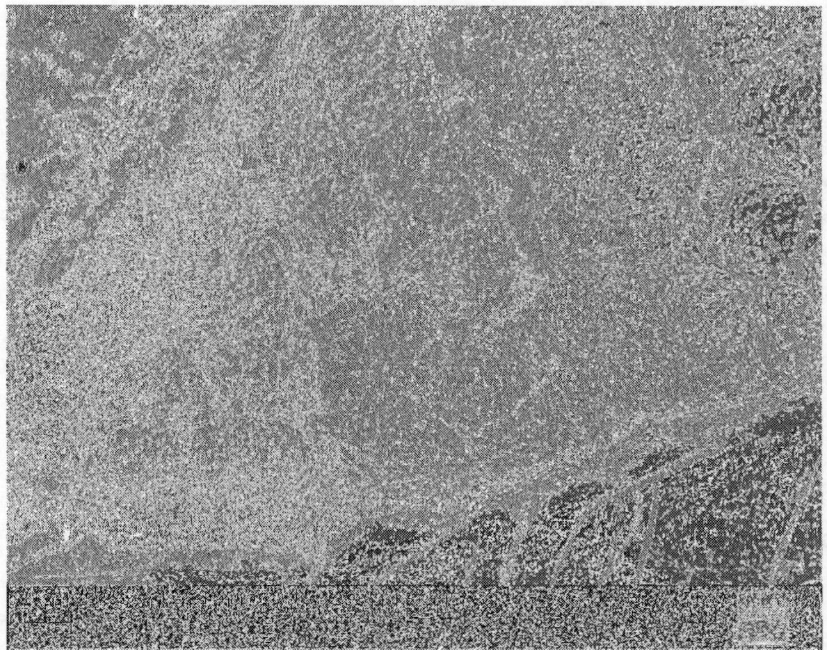

Figure 77: Zinc nanoparticle sprayed mung bean Leaf surface

9.9: Accumulation of Nanofertilizer Particles

Among the various issues, the most important might be the accumulation of nanoparticles in plants and their food parts. The accumulation of nanoparticles depends on many factors, but mainly on plant species, tissue/organ that will be used directly as food or for food processing, and nanoparticle type and size. Due to the variability in interactions between nanoparticles and plants, nanomaterials used in nanofertilizers can accumulate in plants and, in some cases, they can cause toxicity problems, not only to plants but also to humans. For example, multi-walled carbon nanotubes induce phytotoxicity in red spinach (*Amaranthus tricolor* L.) causing growth inhibition, production of reactive oxygen species and cell death. CeO_2 nanoparticles can accumulate and shut down the nitrogen fixation potential of soybean, thus not only threatening the future of leguminous crops in agriculture but also can cause human health problems. Further, in another experiment, application of C_{60} (fullerene) to zucchini, soybean, and tomato plants increased the accumulation of dichlorodiphenyldichloroethylene (DDT).

9.10: Plant Growth Promotion and Stress Tolerance

Despite numerous studies on nanomaterials induced plant growth promotion and stress tolerance, the underlying mechanisms still remain largely uncovered. The influential effects of nanomaterials on crop growth under unfavourable conditions can, at least in part, be explained by the increased activity of enzymatic system. For example, nanomaterials like nano-SiO_2 or nano-ZnO application increases the accumulation of free proline and amino acids, nutrients and water uptake, and activity of antioxidant enzymes including superoxide dismutase, catalase, peroxidase, nitrate reductase, and glutathione reductase, which ultimately improve plant tolerance to extreme climatic conditions. In addition, nanomaterials could also regulate stress gene expression. For example, a micro array analysis showed that a number of up-ragulated or down-regulated genes by the application of Ag nanoparticles in Arabidopsis. Among the up-regulated genes, a major part is associated with the response to metals and oxidative stress (cation exchanger, cytochrome P450-dependent oxidase, superoxide dismutase, and peroxidase). In contrast, down regulated genes are related to response to pathogens and hormonal stimuli including systemic acquired resistance, ethylene signaling, and auxin-regulated gene involved in growth and organ size. Such nanomaterials-induced responses are directly involved in plant protection against stresses. However, the response of plants to nanofertilizers varies with the plant species, their growth stages and the nature of nanomaterials used. Therefore, further work is needed to identify the signaling cascades and the genes regulated by specific nanomaterials in different plant species before the technology reaches the farm. The effect of nanofertilizers on abiotic stresses is shown (Table 33).

Table 33: Alleviating effects of nanofertilizers on abiotic stresses in plants

Nanofertilizer	Plant	Abiotic stress	Effect
TiO_2	*Triticum aestivum* L.	Drought	Increased growth, yield, gluten and starch content
ZnO	*Glycine max*	Drought	Increased germination percentage and germination rate, decrease in seed residual fresh and dry weight
Analcite	*Triticum aestivum* L., *Zea mays* L.	Drought	Increased activation of CAT, accumulation of flavonoids and carotenoids, decreased proline accumulation, improved seed germination, photosynthetic pigments and biomass accumulation
Fe	*Carthamus tinctorius* L.	Drought	Enhanced chlorophyll and carotenoids content, improved growth and yield attributes, decreased H_2O_2 and MDA content
TiO_2	*Ocimum basilicum* L.	Drought	Improved the negative effects of drought stress
SiO_2	*Crataegus* sp.	Drought	A positive significant effect on photosynthetic rate, stomatal conductance and plant biomass, non-significant effect on chlorophyll and carotenoid content
Fe	*Arabidopsis thaliana*	Drought	Activation of plasma membrane Hþ-ATPase, stomatal opening, increased chlorophyll content and plant biomass, maintained normal drought sensitivity, increased CO_2 assimilation
Maghemite (g-Fe_2O_3)	*Helianthus annuus* L.	Drought	Countered drought stress with no effect on proline, total amino acids and mobilization of trace elements

Nanofertilizer	Plant	Abiotic stress	Effect
Ag	*Carum copticum*	Drought	No significant effect on yield attributes and water use efficiency
SiO_2	*Lycopersicum esculentum*	Salinity	Lower levels of nano-SiO_2 enhanced seed germination potential, root length and dry weight. Higher levels suppressed seed germination characteristics
SiO_2	*Solanum lycopersicum* L.	Salinity	Alleviated the effect of salinity on fresh weight, chlorophyll concentration, photosynthetic rate, and the leaf water content
SiO_2	*Ocimum basilicum*	Salinity	Increased fresh and dry weight, chlorophyll content and proline content
SiO_2	*Lens culinaris* Medik	Salinity	Enhanced seed germination and seedling growth
SiO_2	*Cucurbita pepo* L.	Salinity	Improved seed germination and growth characteristics, reduced levels of malondialdehyde, hydrogen peroxide and electrolyte leakage, reduced chlorophyll degradation and oxidative damage, enhanced photosynthetic parameters and antioxidant enzymes
SiO_2	*Vicia faba* L.	Salinity	Enhanced seed germination, increased growth, activities of antioxidant enzymes relative water content and total yield
ZnO and Fe_3O_4	*Moringa peregrina*	Salinity	Reduction in Na^+ and Cl contents, increased N, P, K^+, Ca^{2+}, Mg^{2+}, Fe, Zn, total chlorophyll, carotenoids, proline, carbohydrates, crude protein, and enzymatic and non-enzymatic antioxidants

Nanofertilizer	Plant	Abiotic stress	Effect
ZnO	*Helianthus annuus* L.	Salinity	Increased growth, net CO_2 assimilation rate, sub-stomatal CO_2 concentration, chlorophyll content, Fv/Fm and Zn content, and decreased Na^+ content in leaves
SiO_2	*Solanum lycopersicum* L.	Salinity	Salinity up-regulated the expression profile of four salt stress genes AREB, TAS14, NCED3 and CRK1, and six genes RBOH1, APX2, MAPK2, ERF5, MAPK3 and DDF2 were down-regulated, suppressed the effect of salinity on seed germination rate, root length and fresh weight
Se	*Cucumis sativus* L.	Cold	Modified the physiological response of plants to short-term chilling stress, increased proline content in leaves, reduced lipid peroxidation in roots, no significant resistance of plants to low temperature
TiO_2	*Cicer arietinum* L.	Cold	Enhanced activities of antioxidant enzymes, decreased H_2O_2 content and electrolyte leakage, enhanced accumulation of TiO_2 by sensitive genotypes than tolerant one
TiO_2	*Cicer arietinum* L.	Cold	Enhanced expression of Rubisco and chlorophyll binding protein genes, decreased in H_2O_2 content, enhanced activity of phosphoenolpyruvate carboxylase
SiO_2	*Agropyron elongatum* L.	Cold	Overcome seed dormancy, enhanced seed germination and seedling weight
Ag	*Arabidopsis thaliana*	Cold	Enrichment of antioxidant activity for the genes, 35% of similar genes were regulated by both Ag nanoparticles and cold stress

Nanofertilizer	Plant	Abiotic stress	Effect
MWCNTs	*Lycopersicum esculentum*	Heat	Up-regulated the expression of various stress-related genes including HSP90. It was suggested that plants sense nanoparticles as a stress factor
TiO_2	*Lycopersicum esculentum* Mill	Heat	Enhanced photosynthesis by regulating energy dissipation, caused cooling of leaves through inducing stomatal opening
Na_2SeO_4	*Lycopersicum esculentum* Mill.	High and low temperature	Improved plant growth, chlorophyll and leaf relative water content
TiO_2	*Chlamydomonas reinhardtii*	Cd	Reduced Cd toxicity

Apart from their beneficial effects several nanofertilizers sometimes show toxicity symptoms. For example, presence of nanofertilizer in the growth medium induces oxidative stress and causes reduction in germination rate, root and shoot length, and loss of photosynthesis, chlorophyll (Chl), biomass, and nutritive value of crops. The nanofertilizer can also alter gene expression involved in biotic and abiotic stress responses.

9.10.1: Nanofertilizer for stress tolerance

Nanofertilizer enhances the water stress tolerance via enhancing root hydraulic conductance and water uptake in plants and showing differential abundance of proteins involved in oxidation-reduction, ROS detoxification, stress signaling, and hormonal pathways. Concerning the plant abiotic stresses, they include salinity, drought, flooding, chilling, freezing, ultraviolet radiation, etc, which causes a great loss in crop production worldwide. Due to the multiple abiotic stresses and the increasing chances of global climate change, an urgent need for mitigation and adaptation of these plant abiotic stresses. Therefore, the investigation of different effects of abiotic stresses on plant growth and development is crucial issue at different levels including biochemical, physiological and molecular levels. Different plant mechanisms in counteracting the abiotic stresses and maintaining their growth should be a great significance. The using of nanomaterials is an emerging solution among the accepted solutions of plants towards abiotic stresses.

Due to their unique characteristics, nanomaterials already have been used in several applications including medicinal, industrial and agricultural sectors. Concerning the agricultural application of nanomaterials, there is an increasing and unlimited uses in agriculture and food systems including nanofertilizers, nanopesticides, soil nano-reclaimants, nanosensors for nano-farming, soil and water nanoremediators, etc. These nanomaterials may help in reducing the consumption of different agro-chemicals (fertilizers, pesticides, etc.) leading to minimizing the environmental pollution and hence the sustainable agriculture. Regarding the relationship between nanomaterials and plant stress, many studies have been published explaining more details about this relation. The main distinguished effect of plant abiotic stress is to cause in general the oxidative stress. Under this oxidative stress, nanomaterials may help stressed-plants in enhancing their defense system including the antioxidative enzymes mainly peroxidase, superoxide dismutase and catalase. On the other hand, these nanomaterials under higher concerntration may also cause oxidative stress on plants due to the accummulation of reactive species (oxygen and nitrogen) leading to a damage in proteins, nucleic acids and cell membrane. Therefore, the interaction between nanomaterials and plants under abiotic stress should be investigated under different levels including the physiological, biochemical and cellular levels.

Under normal conditions, all living organisms can ideally grow and develop but under undesirable conditions several problems can happen. These unfavourable conditions include stresses and/or environmental constraints. The plant stresses could cause losses in crop production. Thus, crop productivity faces a lot of environmental stresses including biotic and abiotic ones. Concerning the abiotic stresses, they include salinity, drought, heavy metals, flooding, chilling, freezing, heat, ozone and ultraviolet radiation. Whereas, the biotic stress includes different pathogens like bacteria, virus, fungi, etc. It is well known that, plants may be exposed to different combinations of abiotic and biotic stresses at the same time under field conditions. Concerning stress combinations, it is reported about common stress combinations including drought and salinity, salinity and heat, drought and pathogen. This enforced many researchers recently to focus on different interactions among combined abiotic and biotic stress under molecular and other levels. Therefore, different tailored molecular and physiological responses by plants have been recorded under these combined stresses. These responses may occur in case of plants exposed to simultaneous stresses, but can not deduce in case of different individual stresses. Thus, the responses of plants to abiotic and biotic stresses are dynamic and complex. Concerning multiple and integrated omics studies, it could be used them in discovering new areas of interactions and regulation as well as response of stress kinetics and identification of multiple response phases

Among abiotic stresses salinity and drought represent a great threat to crop production all over the world. This threat will be accelerated in frame of the global climate change as well as the frequency and severity of different stresses. Furthermore, these abiotic stresses including drought and salinity may cause a loss in crop production about 50 %. Nanofertilizers can help plants to synthesise and then accumulate more osmolytes or osmoprotectants including amino acids (proline, glycine, glutamine, etc.), quaternary ammonium compounds (*e.g.,* glycine betaine, β-alanine betaine, proline betaine), tertiary sulphonium compounds (dimethyl sulphoniopropionate or DMSP), sugars (sucrose, trehalose, fructose, maltose, etc.) and sugar alcohols (pinitol, mannitol, myoinositol, ononitol and sorbitol) under stress conditions. Therefore, under drought and salinity stresses plants have a hormonal regulation system including abscisic acid (ABA), salicylic acid (SA), melatonin (N-acetyl-5–methoxytryptamine) and nitric oxide (NO). These molecules are important in regulating plant status during drought, controlling the opening of stomata, photorespiration and antioxidant defenses.

A strong relation among drought, the global climate changes and greenhouse emissions leading to an expansion in the arid land mass as well as great damages to agriculture due to elevated temperatures. So, the plant abiotic stress is one of the most important challenges facing several countries all over the world. The main effect of drought and salinity stresses can be represented in the plant metabolic functions, and maintenance of turgor pressure as well as disrupt the redox balance by aggravating the production of reactive oxygen or nitrogen species causing oxidative injury. Nanofertilizer may increase the fertility of soils as well as crop production. Moreover, salt affected soils management also can be achieved using these nanomaterials. Agroecosystem environments include soils, sediments, plants, water, air etc. These environments may suffer from several stresses even abiotic or biotic. These stresses represent the vital pressure or main problem in deterioration of the agricultural production. There are many case studies regarding these stresses, such as salt affected soils, polluted soils and water, global climate changes, marginal lands, wetlands, bushfires and agroforests, etc. Nanofertilizers have found to serve as great role in minimizing these problems considerably.

9.11: Plant Productivity

Plant productivity depends on a combination of various vital factors, such as soil fertility, good quality irrigation water, an appropriate light intensity and temperature, among other environmental factors, so that any deviation in one or more of these factors causes adverse effects on plant productivity. Unfortunately, during the crop growth cycle, a plant has to constantly face several biotic and abiotic stresses. Drought, heat, salinity, water logging, and cold, among others, are major abiotic stresses that cause huge losses to agriculture globally by reducing yield and product quality. Being sessile in nature, plants must combat these stresses *in situ*. These stresses either individually or in combination, negatively affect the morphological, physiological, biochemical and molecular changes in plants that ultimately decrease productivity. In recent years, in view of climate change the severity of these abiotic stresses is expected to increase in the upcoming decades that clearly depict a threat for crop production. In this context, there is an increased interest towards the use of nanofertilizer and nanopesticides in the agriculture sector. Different types of nanomaterials have been evaluated for their possible role in managing different abiotic stresses, and a summary of their positive, negative and neutral effects on various crops under various abiotic stresses (Table 34) have been presented. Numerous studies have been performed for the endurance of abiotic stresses with different nanomaterials resulting both in positive or negative impacts on plant growth under abiotic stress conditions. Furthermore, evaluation of salt stress-related genes using semi-quantitative RT-PCR showed that exposure to Ag nanoparticles caused upregulation of four genes (AREB, P5CS, MAPK2 and CRK1) and down regulation of three genes (TAS14, ZFHD1and DDF2) under salt stress.

Table 34: Impact of some nanofertilizers on various crops under non-stressful condition

Nanofertilizer	Concentration	Crop	Effect
Fe	30–60 ppm	Garden pea	Improved seed mass and chlorophyll content
	50–500 ppm	Cucumber	Dose dependent effects on biomass and antioxidant enzymes
Zn	10–50 ppm	Coriander	Improved pigment content and defence responses
	10 ppm	Clusterbean, moth bean, pearl millet	Improved antioxidant properties and stress tolerance
Cu	130 ppm	Lettuce	Increased shoot/root ratio
	100 ppm	Bean	Growth inhibition and nutrition imbalance
	50–500 ppm	Tomato	Improved fruit firmness and antioxidant content

Some fertilizer companies now-a-days are producing nanofertilizers commercially. In cotton, the growth and yield gets increased with the foliar application of commercial nanofertilizer nanomol than the other commercially available nanofertilizers *viz.*, Richfield, Agriklik and Nualgi. Unlike combined nutrient nanofertilizers, single micronutrient nanofertilizers *viz.*, Nanobor, Nanomag and Nanozinc also showed the positive results on number of bolls, boll weight and seed cotton yield. Moreover, foliar spraying of nano-iron oxide recorded significant increase in boll weight and seed cotton yield compared to magnetite nanoparticles, normal iron oxide and iron sulphate fertilizers. In another trial, 10 per cent nano-NPK fertilizer which is commercial one, treated as a foliar spray on wheat crop and improved the quality of wheat grain. Foliar application of commercial nano-fertilizer (potacrystal and phosphoone) at two or three stages (vegetative, flowering or filling) increased yield and yield components of faba bean in both seasons and also in maize crop. Few studies have reported negative effects of nanoparticles on higher plants. The deposition of nanoparticles like titanium oxide, zinc oxide, cerium oxide, and silver nanoparticles on the surface of cell and in the organelles can cause oxidative stress to the cell by the stimulation of oxidative stress signaling. Foliar application of Fe_3O_4 nanoparticles on maize crop at high or low concentration affects the photosynthesis as well as interferes with the plasma membrane ion channels. Some research suggested that with the application of Ag nanofertilizer particles, which interfered membrane proteins and activates signaling pathways that led to inhibition of cell proliferation. A separate study suggested that higher concentration of Ag nanofertilizer application causes the accumulation of Ag in shoots. Increasing of Ag ions produces reactive oxygen species (ROS) that leads to oxidative stress, which reduces the root and shoot growth of wheat. Nano-Mg demonstrated to give the greatest PMDI (plasma membrane damage index), suggesting that the application of nano-Mg would be harmful to the membrane of black-eyed pea. In drought conditions, applied nanofertilizers in rice alter the gene expression for protein synthesis during grain development. An investigation revealed that foliar application of nanofertilizer on wheat decreased the protein content and increased the fat content, while total phenol content and antioxidant activity has been increased in rice. Enhancement in protein content was observed with the application of Zn and Fe nanofertilizer on chickpea as compared to same doses of chemical fertilizer application (Table 35). The result shows nano Zn and Fe fertilizer application was associated with high protein content and less semi-oxide dismutase activity resulting in more stress tolerance by the plants.

Table 35: A comparison of chemical and nanofertilizer on protein and SOD content of chick pea

Treatment	Protein content (mg/mL)	SOD (Unit/mg protein)
10 ppm ZnO	+ 13.5	-28.7
10 ppm nano ZnO	+ 30.4	-38.7
1.5 ppm ZnO	+ 29.5	-37.7
1.5 ppm nano ZnO	+ 41.4	-43.2
1.5 ppm Fe_2O_3	+20.6	-33.2
1.5 ppm nano Fe_2O_3	+ 32.9	-39.7

Influence in plant phenological character as well as protein and chlorophyll content in leaf were recorded by foliar application of biologically synthesised Mg nanoparticles. The Mg nanoparticles were synthesised using soil fungi and their effect on arid crops *i.e.,* pearl millet, mung bean, moth bean and clusterbean were recorded. The effect of Mg nanoparticle on these crops was observed, might be due to its essentiality for plants growth, development and physiological regulations. Mg plays an important role in photosynthesis and 15–30% of total Mg is found in chloroplast. As it is central atom for photosynthetic pigment chlorophyll, plays a pivotal role in capturing of light energy during photosynthesis. Increasing in photosynthetic rate is primarily responsible for higher root exudates secretion, which may create chemical attraction for plant growth promoting microbes, resulted in their higher population in the rhizosphere of nano-Mg sprayed plants. In addition to chlorophyll structural component, Mg is essential for photosynthesis during the light reaction in stroma, where Mg dissipates the electrical component of the proton motive force generated by light perception. In this manner, Mg ensures the continued influx of protons into the thylakoid lumen of the chloroplast.

Nano-magnesium fertilizers has a definite role in enzyme regulation more particularly those involved in CO_2 fixation as Mg act as cofactor for ribulose 1,6-bis phosphate carboxylase (RuBP Carboxylase), the first enzyme of the Calvin cycle and for fructose-1,6-bis phosphatase, which regulates the partitioning between starch synthesis and export of trios phosphate to the cytosol might be responsible for more exudates secretion from roots and indirectly responsible for increased carbon and microbial population in the rhizosphere soil. Mg nanoparticles also act as a cofactor and promoter for numerous cellular enzymes including DNA and RNA polymerases, ATPases, phosphatses, kinases, carboxylases, particularly Phospho-enol Pyruvate (PEP) carboxylase and synthetase mainly glutathione synthetase. Such a pivotal role of Mg increased the activity of these enzymes when it provides in nanoscale form due to its high surface area to volume size ratio as compared to regular size Mg particles, resulted in more nutritional availability of NPK in the soil. ATP-dependent proteins, including the plasma membrane H^+ ATPases and the vacuolar pyrophosphatases (PPi) use Mg-ATP as a substrate. During the reaction cycle of these enzymes, Mg bridges the oxygen atom of ATP phosphate groups with nitrogen from the protein and as, such promotes catalysis. This might be one of the reasons for higher availability of NPK in soil when biologically synthesised Mg nanoparticle sprayed on plants, which has better penetration than regular size particles. Within plants, Mg is transported either in ionic form or complexes with organic acids. As Mg is most mobile mineral nutrient in plants, both within the xylem and phloem, it is easily translocated within plants.

It was noticed that growth, development and yield of arid crops, *i.e.,* pearl millet, moth bean, mung bean and clusterbean, was increased when biologically synthesised Ti nanoparticles sprayed by foliar mode. In leaves, total chlorophyll and protein content was influenced might be due to photocatalytic activity of Ti, when it exposed under sun light. In general, more photosynthetic capability was noticed when TiO_2 nanoparticles were applied. It was also observed that TiO_2 nanoparticles significantly reduced the incidence of blast and mold with a correspondent control plants in which no TiO_2 nanoparticles were applied. A combination of TiO_2, aluminum (Al) and silica (Si) was effective in controlling downy and powdery mildew of grapes by stimulating plant physiological defenses. Increasing chlorophyll contents in the leaves responsible for higher photosynthetic activity in plants, that is directly or indirectly responsible for increasing microbial population and carbon contents in the rhizosphere zone, as they may act as chemo-attractant for microbes. Increasing microbial population might be responsible for increasing NPK nutrients availability by increasing soil enzymatic activity, and nutritional availability promote the plant growth, development and biomass-yield of the plants. Increased growth and yield by increasing photosynthetic rate and reducing the severity of foliar and pod disease by application of TiO_2 nanoparticles was reported. The effects of nano-TiO_2 on the photochemical reaction of chloroplasts of *Spinacia oleracea* were recorded. The obtained results are due to increase in Hill reaction and the activity of chloroplasts, which accelerated fe-cy reduction and oxygen evolution by application of TiO_2 nanoparticles.

It was demonstrated that, application of biologically synthesised Zn, Mg and Ti nanoparticle improve growth, development, yield of arid crops, and also increase soil microbial enzymes, total carbon, NPK nutrients in the rhizosphere in comparison to regular size Zn, Mg and Ti particles, and control in which no metal particle was applied. Phytotoxicity of five types of multi–walled nanoparticles at the level of seed germination and root growth in six higher plant species (*Raphanus sativus, Brassica napus, Lolium multiflorum, Lactuca sativa, Zea mays* and *Cucumis sativus*) were reported. Seed germination was not affected except for the inhibition of nanoscale zinc on *Lolium multiflorum* and nanoscale zinc oxide on *Zea mays.* Inhibition of root growth varied greatly among nanoparticles and plants and it is partially correlated to nanoparticles concentration. The inhibition was only occurred during the seed incubation process rather than seed soaking stage. In nano-anatase TiO_2 treated *Spinacia oleracea* showed that Rubisco carboxylase activity was 2.67 times that of control Rubisco. Since in the photosynthesis, the molecular mechanism of carbon reaction promoted by nano-anatase TiO_2 was still not clearly understood, by reverse transcription PCR and Northen blotting experiments, evidenced that Rubisco small subunits and Rubisco large subunit messenger RNAs were promoted in the nano-treated plants. Accordingly, the protein expression of Rubisco from the nano-anatase treated spinach was increased by 40% compared with the control. The influence of metal nanofertilizer particles on the soil microbial community and germination of *Lactuca* seeds were recorded. The rhizosphere is the zone of soil surrounding a plant root where the biology and chemistry of the soil are influenced by the roots. As plant roots grow through soil they mostly release water soluble compounds, such as amino acids, sugars, phenolic compounds that supply food and nutrition for microorganism to a larger extent than elsewhere in the soil. Root exudates content increased with plant age of up to six weeks and thereafter decrease due to de-pigmentation in the leaves followed by lowering the photosynthetic activity, results in the decreasing microbial population. On the contrary metal nanoparticles influence the growth of *Lactuca* seeds, this influence was tested by measuring the length of the root and shoot of the plant after 15 days of incubation. An increase in the shoot/root ratio compared to that of the control was evidenced.

9.12: Effects on Human

Nanoparticles can easily enter into the biological systems which is not possible for the larger particles. Nanoparticles may cross cell membrane which depends on the dimension of nanoparticle. There is a possibility to inhale the nanoparticles by farm workers while spraying, which can reach the blood. A preliminary study was reported of the incorporation of NPK fertilizer into chitosan nanoparticles which can reach other target sites, such as brain, liver or heart. These nanoparticles might affect the regulatory mechanisms of enzymes and other proteins. There is a chance of nanoparticles entering into the food chains which ultimately affects the human health system. Hence, there is an urgent need to develop human resources with an understanding of the complexities of the agricultural production system to serve nanotechnology applications in agriculture successfully. The deposition of nanoparticles like titanium oxide, zinc oxide, cerium oxide, and silver nanoparticles on the surface of cell and in the organelles can cause oxidative stress to the cell by the stimulation of oxidative stress signaling. Foliar application of Fe_3O_4 nanoparticles on maize crop, at high or low concentration, affects the photosynthesis as well as interferes with the plasma membrane ion channels. In *Oryza sativa* with the application of multi-walled carbon nanotubes (MWCNTs), which increased the production of reactive oxygen species (ROS) and oxidative stress caused by Fe_3O_4 nanoparticles in the shoots and roots of cucurbits.

9.13: Effect of Nanofertilizers on Growth Hormones

It was found that nanofertilizers (Zn, Fe, Mg, K etc.) may influence different plant growth hormones which ultimately affect the plant growth and development.

9.13.1: Indolebutyric Acid (IBA)

In a study, it was investigated that indolebutyric acid (IBA) increases the marigold (*Tagetes erecta*) flower size, leaves per plant, plant hight, root size, and non-bloom flower at lower concentration (100 mg/L). Flower size increased at optimum concentration (300 mg/L). Leaf size and roots per plant increased at higher concentration (400 mg/L). Maximum branches were observed at medium (200 mg/L) concentration. The appropriate concentration of IBA and Naphthaleneacetic acid (NAA) before cutting, improve the survival rate of cuttings of Chinese rose. In production, 250 mg/L solution of equal amounts of NAA and IBA dipped the cuttings for 2–3 seconds greatly improve the survival rate and the production efficiency.

9.13.2: Gibberellic Acid (GA_3)

Lentil (*Lens culinaris*) plants development expanded 43%, by Gibberellic acid (GA_3) while development retardants (Prohex-Ca and Topflor) repressed it by 34% as estimated. Gibberellic acid (GA_3) as well expanded the lentil plant height, while diminishing the seed weight per plant and seeds weight. Indole-acetic acid (IAA), Prohex-Ca or Topflor diminished lentil per plant dry weight. Predominantly GA_3, IAA and kinetin broadly improved specific phenolic compounds (gallic acid and rutin) which are fundamentally utilized as a primary source of naturally occurring antioxidants. Different growth parameters of African marigold (*Tagetes erecta*) considerably enhanced by different concentrations (50, 100 and 200 mg/L) of Gibberellic acid (GA_3) for example height, number of fundamental branches, basal measurement of plant, number of flower per plant, weight of flower, size of flower and flower yield per plant over control. Additionally early flowering and long duration of flowering also occurred at all concentration of GA_3.

9.13.3: Abscisic Acid (s-ABA)

Abscisic acid (s-ABA) was constantly valuable at reducing water loss and extending shelf life for all species treated (impatiens, seed geranium, petunia, marigold, salvia, and pansy). The treatment of bedding plants before transport and transaction with abscisic acid maintain high-gradeability even under harsh deficiency stress environment. Reduction in transpiration and water use in tomato (*Solanum lycopersicum*) significantly enhanced by abscisic acid drenches (125–2000 $mg{\cdot}L^{-1}$). So abscisic acid treated plants leaves had lower relative water content (RWC) as compared to control plants at 24h after the abscisic acid drench. The leaf relative water content of control plants is 97%, while plants treated with abscisic acid had relative water content 57% to 62%. Abscisic acid induced wilting which was due to the reduction in root conductance and theorized that Abscisic acid affects aquaporins in the roots, preventive water uptake.

9.13.4: Salicylic Acid (SA)

It is reported that, biomass accumulation, number of inflorescences and flavonoid content of marigold plant was significantly expanded due to linear utilization of Salicylic acid (SA). Before the reproductive stage exogenous use of salicylic acid brought about higher biomass production of marigold plants and by expanding complete flavonoids content in the inflorescences increased the value of the raw material. Salicylic acid increased the antioxidant activity of the leaves of *Alternanthera tenella* and can also be recognized to the increase in betacyanin content, which are compounds with standard antioxidant action.

9.13.5: Indoleacetic Acid (IAA)

Ripening of avocado fruits (*Persea americana*) had clearly effected by Indoleacetic acid (IAA). Indoleacetic acid invigorated respiration and prompted preclimacteric ethylene generation, at higher concentrations (100 and 1000 μm), which bringing about quickened ripening of the fruits. However, the ripening of fruits was delayed, the climacteric respiration and ethylene creation was disguised at the low concentrations (1 and 10 μm). *Pythium ultimum* symptoms on tomato plant were controlled by the Indoleacetic acid (IAA). The harshness of the side effects brought about by *Pythium ultimum* inside the rhizosphere of plantlets expanded by the low concentrations of IAA (0– 0.1μgml^{-1}), while higher concentrations (10μgml^{-1}) applied either by drenching to the developing medium or by showering on the shoot, decreased the side effect caused by this pathogen.

9.13.6: N-Acetyl Thiazolidine 4–Carboxylic Acid (NATCA)

Forchlorfenuron (CPPU) (5 and 10 ppm) and Nacetyl thiazolidine 4–carboxylic acid (NATCA) (50 and 100 ppm) and their combinations were applied to Apricot (*Prunus armeniaca* L.) at pink bud and petal fall stage. Minimum (27.03 %) fruit drop was observed in tree receiving 50 ppm NATCA at petal fall stage and NATCA at 100 ppm at pink bud stage decreased titratable acidity and increased the total sugars, reducing sugars and non-reducing sugars content of fruit. Foliar spray of CPPU at 10 ppm enhanced the vegetative quality like tree height (17.40 %), tree spread (22.17 %), and tree volume (27.82 %) over control. However, NATCA (100 ppm) at petal fall stage increases the fruit set (52.71%) and fruit retention (38.12%). Foliar spray of (NATCA) on apple cv. Starking Delicious at petal fall stage (1.0 mL/L followed by 1.5 mL/L) recorded maximum fruit set (22.73% and 28.68%), L/D ratio (0.93 and 0.98) and colour of fruit (62.50% and 76.58%). Lowest fruit drop (17.78% and 16.36%), highest yield (102.50 and 89.13kg/tee) and fruit weight (137.62g and 144.15g) was recorded in trees sprayed at petal fall stage (1.0ml/L followed by 1.5ml/L) and 30 days after petal fall stage.

9.14: Effect of Nano-fertilizer on Nutrient Recovery

Nutrient recovery by crops remains relatively low. Since, a nanoparticle is a particle with at least one dimension below 100 nm; nanofertilizers could soon offer a technological solution to the nutrient-loss problem, thereby aiding technologically-minded farmers and subsistence farming. Nanofertilizers refer to nanoscale dimension products that deliver nutrients to crops. These nutrients can be i) encased inside nanomaterials for example nanotubes or nanoporous materials, ii) covered with a thin defensive polymer film, or iii) conveyed as particles or emulsions. Germination and seedling development, physiological activities including photosynthetic movement, nitrogen metabolism, mRNA articulation, protein level, and positive changes in genes articulation, have improved by nanoparticles and nanotubes in numerous crops (sunflower, common bean, and maize, among others) which indicating their potential use for increasing crop yields. Nevertheless, additional research is necessary in sort to realize the effect of nanofertilizers on the genetic, physiologic, and morphologic changes in crops, as well as their effect on soil microbial communities, symbiosis, physicochemical soil properties, and pollution. One may speculate that the creation and improvement of fertilizers at nanoscale could have a deep impact on energy, the economy and the environment. Assumption although, the scientific, technical, and agricultural projects linked with nanoferti-lizers should be including side effects in order to accurately determine progress and shape a sustainable future. Plant Growth regulators are substances which can manipulate growth and development in plants. Plant height, leaf number, leaf area index, dry mass, chlorophyll content, photosynthetic parameters, seed yield, oil yield, nutritional status etc. can be influenced by plant growth regulators. The spectacular achievements in sex modification of plants have been reported from several quarters by the application of auxins. Ethylene can regulate ripening, senescence, abscission, epinasty, swelling and elongation, hypertrophy, dormancy, hook closure, leaf expansion, flower induction, sex expression and exudation. In cases, where auxins promote femaleness, an application of GA will promote the formation of male flowers. It was reported that application of GA_3 induced early flowering and more inflorescence per plant. It is evident that Plant Growth Regulators exhibit a positive impact on different vegetable crops, specifically those belonging to the family cucurbitaceae. Hence, they can be utilized commercially for enhancing productivity of, such crops and thus fulfill the demand of the market. However, there is much scope to investigate the effect of these phytohormones on other economic plants.

9.15: Conventional Fertilizers versus Nano-fertilizers

Conventional or traditional fertilizers are generally applied on the crops by either showering or spreading. One of the fundamental factors that choose the method of utilization is the final concentration of the fertilizers that activate the plant. In the practical circumstance, because of draining of synthetic compounds, float, overflow, dissipation, hydrolysis by soil moisture, photolytic and microbial degradation the much lower concentration (well underneath the desired concentration) achieves the objective site. Feasible and economic losses happens, because it had been evaluated that around 40–70 % of nitrogen, 80–90 % of phosphorus, and 50–90 % of potassium substances of applied fertilizers was lost in the environment and could not be taken by the plants. These issues have been caused because of steady utilization of fertilizer and pesticide which destructively influences the natural nutrient balance of the soil. Environmental pollution brought about by extreme utilization of chemicals as fertilizers and pesticides irritating ordinary flora and fauna. That is why, to satisfy the crop nutrient necessities and to lessen the risk of environmental pollution, it is very essential to upgrade the utilization of chemical fertilization. Therefore, it tends to be invaluable that different methods of fertilization be also tested and used to supply essential nutrients for plant development and yield production, as maintenance the soil structure in good shape and the environment clean. To expand the nutrient use competence and decrease the cost of environmental contamination, nanofertilizer has given the possibility of investigating nanoscale or nanostructure materials as fertilizer transporter or controlled-discharge vectors for working of the smart fertilizers as new facilities. Nanomaterials coatings on the surface of fertilizer particles grasp the material more intensely because of higher surface pressure than the conventional surfaces which help in controlled discharge. Utilization of nanofertilizer in agriculture is a significant characteristic to release of agrochemical for example, fertilizer supplying macronutrients and micronutrients to the plants. Nanofertilizers show controlled discharge of agrochemical site targeted delivery, decline in poisonous quality, and improved nutrient utilization of delivered fertilizers. These attributes of nanoparticles are because of their high surface area to volume proportion, high solubility, and specific focusing because of small size, high versatility, and low toxicity, targeted delivery.

Nano-SiO2 improved defense mechanisms of plants against salt stress toxicity by augmenting net photosynthetic rate, stomatal conductance, transpiration rate, water use efficiency, total chlorophyll, proline, and carbonic anhydrase activity. Moreover, protecting effects of nanofertilizers on photosynthetic parameters against heat, drought and salinity have been extensively studied.

Photosynthesis is highly susceptible to changes in light and continuous exposure of plants to UV-B radiation causes excessive accumulation of ROS in chloroplast which damages PS II and decreases electron transfer rate and thylakoid membrane stability. Damaging effects of UV-B radiation on photosynthesis can be alleviated by nanofertilizers which improve the rate of photosynthesis by limiting oxidative stress, enhancing chlorophyll biosynthesis, Rubisco activity, light absorbance, transport and transformation of light energy, absorbance of UV-radiation without scattering the useful visible one. Overproduction of ROS by various cell organelles is the signature effect of abiotic stress-induced oxidative stress. Apart from damaging effect, ROS are also known to trigger various defense systems through activating cell signaling cascade and inducing or suppressing the expression of many genes. Nonetheless, plants are equipped with enzymatic and nonenzymatic system of antioxidants which continuously scavenge harmful ROS. Regardless of, such defense systems, plants face detrimental effects of abiotic stresses when the cellular defense system lags behind the destructive capacity of abiotic stresses. Therefore, accurate temporal and spatial response of stress stimulus by the defense system before the onset of stress-induced damage is crucial for the survival of plants under stressful environmental conditions.

Nanofertilizers play important role in the protection of plants against various abiotic stresses by stimulating the activities of antioxidant enzymes, accumulation of osmolytes, free amino acids and nutrients. On contrary, it was suggested that extremely high concentrations of nanofertilizer doses were used in most of those toxicological tests, while application of nanofertilizer at lower concentrations was found effective in alleviating various abiotic stresses and enhanced plant growth and development conditions. Among various strategies adopted to counter drought-induced destruction in plants, use of nanofertilizers has been proved more promising. It was concluded that exogenous application of lower dose of nanoTiO_2 performed better in alleviating drought stress-induced damage to linseed compared to higher doses. It was found that suitable concentration of applied TiO_2 increases absorbance of visible light and improved energy transfer among amino acids within

PSII and improved energy transport from tyrosine residue to Chorophyll a (Chl a) and oxygen evolution in PSII. Application of nanofertilizer particles of analcite to wheat and corn significantly alleviated drought stress by enhancing photosynthetic pigments and accumulation of protective antioxidants. However, a decrease in proline content was recorded in drought stressed seedlings treated with analcite, which reflects that analcite induced protection mechanism against drought was not related with proline.

The micronutrient iron is constituent of various enzymes and its deficiency leads to chlorosis. Foliar application of iron nanofertilizer on safflower reduced the adverse effects of drought stress and improved yield components of safflower. Under the conditions of water scarcity stomatal closure is the prime response that the plants adopt to preserve water. During the opening of stomata, CO_2 is taken up at the cost of water transpired, however, when the stomata are closed water is conserved, but uptake of CO_2 is compromised. Perception of drought stress signal triggers the activation of signal transduction cascades leading to stomatal closure and thus, conservation of water through reduced transpiration coupled with reduced CO_2 uptake. Iron nanofertilizers activate stomatal opening through activating plasma membrane Hþ-ATPase in *Arabidopsis thaliana*. It was reported that nanosized maghemite (g-Fe_2O_3) significantly countered the symptoms of drought stress in *Helianthus annuus*.

Soil salinity is mainly caused by excessive accumulation of NaCl which poses osmotic and ionic stress in plants. Osmotic stress reduces the ability of plants to take up water and nutrients, whereas, ionic stress due to hyper accumulation of Naþ in cytosol leads to low Kþ/Naþ ratio and excessive generation of ROS which causes severe damage to biological molecules that ensures leakage of electrolytes and affects metabolic activities in cytosol. In recent years nanofertilizers are being used as a vital tool for improving growth and productivity of crop plants under adverse environmental conditions including salt stress. For example Si, the second most abundant element in the soil, significantly alleviates salt stress and enhances seed germination and activities of antioxidative enzymes, photosynthetic rate and leaf water content. The treatment with 1 mM nano-Si significantly alleviated 25 mM salinity; however nano-Si at 2 mM could not alleviate 50 mM salinity. High concentration of Naþ and Cl, the dominant ions in the saline soil, are commonly responsible for salt induced damages in plants. Therefore, reduced uptake from soil or sequestration of Naþ and Cl ions into the vacuoles by some means would be an important defense strategy against salt stress. It has been found that foliar spray of ZnO and Fe_3O_4 as nanofertilizers at the level of 60 mg/L proved best and alleviated the effect of salt stress on *Moringa peregrina* plants. The alleviation of salt stress on various parameters was greater in sunflower plants supplied with nano-sized ZnO as compared to the plants treated with bulk ZnO as chemical fertilizer. Salt alleviating effect of nanofertilizers was further confirmed with the report that nano-Si differentially regulated the expression of salt stress genes.

Temperature stress is a significant factor influencing the sustainable growth and production of crop plants. Temperature stress includes both the high temperature (HT) stress, and low temperature (LT) stress (cold stress or chilling stress and freezing stress). High temperature stress is the rise in temperature beyond a critical threshold for a period of time sufficient to cause irreversible damage to plant growth and development. However, cold stress is caused by the temperature cool enough to produce injury without forming ice crystals in plant tissues, while freezing stress. Although, plants are equipped with a network of defense units, but the precise perception and transduction of stress stimulus to the defense system and accurate temporal and spatial activation of these defense units in response to stress stimulus before the onset of stress-induced damage is crucial for the protection of plant's cellular machinery. While going through the available information on nanofertilizer plant interaction under abiotic stresses, it became evident that generation of ROS is a common response of plants to almost all the stresses. These ROS act as a stress signal to activate plant's defense system as well as aggravate cellular damage. Although, detailed mechanism of action of nanofertilizer is not completely understood, however, using proteomic and genomic approaches several attempts have been made to propose possible mode of action of nanofertilizers in plants under abiotic stress. Perception of stress stimulus elevates the cytosolic Ca2þ concentration ([Ca2þ]cyt). Elevated [Ca2þ]cyt is sensed by Ca2þ binding protein (CaBP). Binding of Ca2þ activates CaBP that directly binds to the promoters of specific genes and induce or repress their expression leading to plant tolerance to the stress. The nanofertilizer particles cause over-expression of CaBPs and bind to CaBPs that trigger downstream signaling and finally the expression of stress related genes and activation of plant's defense system. Similar to [Ca2þ]cyt, nanoparticles also induce NO (Nitric Oxide) synthesis that acts as stress signaling molecule and induces expression of stress related genes that activate the defense system and ultimately tolerance to stress. Optimum concentration of nanofertilizer particles, through regulating antioxidant defense system, also maintains ROS at the level sufficient enough for stress signaling leading to activation of plant's defense system and stress tolerance.

Interaction of nanofertilizer particles with miRNAs also sheds light on the mechanism of action of nanomaterials. Exposure of tobacco plants to nano-TiO_2 and nano-Al_2O_3 upregulates miRNA expression and play vital role in response to metal stress, however increasing levels of these nanomaterials exhibited yellowing, wilting, reduced leaf sizes and leaf counts, reduced root growth, and a decrease in biomass. It has been noticed that the treatment with nano-ZVI improved drought tolerance by up-regulation of the expression of AHA2, a gene related to the stomatal opening process in Arabidopsis. It was observed that application of nanofertilizer particles of TiO_2, Ag and MWCNTs to Arabidopsis under various stresses repressed expression of phosphate starvation and root-development genes, however, supplementation with salicylic acid prevented the reduction in root hair density caused by exposure to nanoparticles. Activation of defense system is primarily carried out by a signaling network which triggers the molecular machinery against a particular stress. Calcium (Ca2þ) acts as a second messenger and plays important role in signaling responses to various stresses. Perception of stress stimulus causes mobilization of Ca2þ from its stores to cytosol through Ca2þ channels leading to elevation in cytosolic Ca2þ concentration, which is sensed by Ca2þ-binding proteins (CaBPs) that initiate downstream events leading to changes in gene expression and plant adaptation to stress. It has been shown that nitric oxide (NO) induces elevation in cytosolic Ca2þ in plant cells exposed to biotic and abiotic stresses and consequently Ca2þ induces NO synthesis, adopting proteomic approach noticed that Ag-nanoparticles-responsive proteins in *Oryza sativa* roots were primarily associated with oxidative stress response pathway, Ca2þ regulation and signaling, transcription, protein degradation, cell wall synthesis, cell division, and apoptosis. Ag nanofertilizer particles or released ions interfere cell metabolism by binding to second messenger Ca2þ receptors, calcium ion channels, and Ca2þ/Naþ pumps. The interaction of water-suspended fullerene C_{60} nanocrystals caused functional modulation of the Ca2þ/calmodulin-dependent protein kinase II (CaMKII). It was noted that exposure of *Arabidopsis thaliana* to cadmium sulfide quantum dots (QDs) caused over-representation of calcium binding protein CML45 and calcium-dependent protein kinase 23. These CaBPs have been shown to regulate stress responses, and over-expression results in enhanced resistance to various abiotic stresses in plants. Moreover, it has already been reported that metals effectively substitute for Ca2þ in CaBP. Nanofertilizer particles have been shown to enhance the activity of nitrate reductase, an enzyme responsible for NO synthesis in plants that elevate the NO level. However, NO has been reported to counteract nanoparticles-induced phytotoxicity and triggers the expression of antioxidant genes and suppresses the ROS and lipid peroxidation. Furthermore, the application of nanofertilizers elevates expression of genes related with cell division, cell elongation and stress.

Considering the cell wall penetrating property of MWCNTs, it can be hypothesised that plants can sense carbon nanotubes as stress factors similar to pathogen attack. However, these associated suppositions are nascent and therefore, the entire mechanism of nanoparticle-induced signal transduction in plants that starts from sensing to the gene expression needs further in-depth investigation. In addition, phytotoxicity studies on nanofertilizers exhibited increased production of ROS that play dual role in plants as toxic compounds and signaling molecules. The dual role of ROS delicately depends on their generation and scavenging; any imbalance between these two processes will lead to either excessive accumulation or reduced availability of ROS that may cause oxidative stress or signaling failure, respectively. However, this equilibrium is maintained by continuous generation and scavenging of ROS. As mentioned earlier that most of the toxic effects of nanofertilizers were noticed at their higher concentrations. Whereas, lower concentrations of nanofertilizers show beneficial or no effect on the cellular processes of plants which lead us to conclude that lower concentration of nanofertilizer maintains normal functioning of antioxidant defense system that facilitates controlled production of ROS to a specific level sufficient enough for signaling, but unable to cause damage. However, these observations need extensive and careful investigations, so that the status of nanofertilizer in biological system could be ascertained. Table 36 showed the effect of different nanofertilizers against stress.

Table 36: Effect of nanofertilizers on plant growth/physiology/ tolerance against stress

Nanofertilizer particles	Plant	Impact on plant parts/process
Al_2O_3	*Lemna minor*	Increased root length, photosynthetic activity, biomass accumulation
TiO_2	*Triticum aestivum*	Increased root length
CeO_2, ZnO	*Zea mays*	Reduced yield
CuO	*Brassica napus*	Increased plant growth
$FeCl_3$	*Lepidium sativum, Sinapis alba, Sorghum saccharatum*	Increased seed germination, seedling length, biomass
$AgNO_3$	Lentil seed	Seed germination/elongation of root and shoot
Fe_2O_3	Soyabean	Increased root length, regulated the enzyme
Cu, Zn	Wheat	Increased RWC and stabilized photosynthetic pigments
$Ca_3(PO_4)_2$	Rice	Increased growth, microferti-lizer and promoter of growth
Fe_3O_4, TiO_2	Soya bean	Enhanced plant growth, crop yield, effect on leaf carbon and phosphorus
CeO_2	Soya bean	Stimulated plant growth, rubisco carboxylase activity, relative water content
ZnO	Chickpea	Effect on root, accumulation of biomass in seedlings, lowered ROS, promoted antioxidant activity
Ag	Wheat	Increased shoot fresh and dry weight, enhanced salt tolerance ability of crop
SiO_2	*Zea mays* L., *Phaseolus vulgaris* L	Increased in seed germination, root and shoot length, photosynthetic pigments, total protein and total amino acids

Nanofertilizer particles	Plant	Impact on plant parts/process
SiO_2	*Nigella sativa* L.	Increased dry weight, photosynthetic pigments, total protein and total amino acids

9.16: Effect of Nanofertilizers on Photosynthesis

Photosynthesis is the key mechanism that transforms light energy into chemical energy. SiO_2 nanofertilizers increased the photosynthesis rate by increasing the activity of carbonic anhydrase and the formation of photosynthetic pigments. Carbon anhydrase acts as a supplier for CO_2 to the rubisco enzyme, which enhances photosynthesis. TiO_2 nanofertilizer particles has photocatalytic properties that not only increase the efficiency of light absorbance but also increase the conversion of light energy into chemical energy. TiO_2 prevented the plant from aging, and ultimately enhanced the photosynthesis process. TiO_2 nanofertilizer particles have potential to increase the activity of rubisco and ultimately improving plant growth. It can also enhance the net rate of photosynthesis, water conduction, and plant transpiration. ZnO nanofertilizer particles showed a positive effect on the growth of cotton (*Gossypium hirsutum* L.). The growth (130.6%) and biomass (131%) of cotton were signficantly enhanced by ZnO nanofertilizer application. It also increased the level of chlorophyll a, b, and carotenoids (141.6%, 134.7%, 138.6%, respectively) and increased soluble protein (179.4%), but reduced malondialdehyde (MDA) level in plant leaves. Various enzymatic activities of catalase, superoxide dismutase (264.2%), and peroxidase (182.8%) were also increased and improved the growth of cotton plants.

9.16.1: Effect of CNTs on photosynthesis mechanism

Sunlight is the most available source of energy, which is conserved in many ways in an ecosystem. One of the most efficient methods for the conservation of sunlight is photosynthesis. For this purpose, the higher green plants, algae, and bacteria contain special pigments that use water and CO_2 to form organic molecules. These photosynthetic organisms contain the photo-elements chlorophyll a, b, d, and f, and a series of electron carrier redox reactions. The thylakoid membrane of plastids acts as a photo-current producer in the presence of potassium ferrocyanide. The cell surface (1 cm^2) produced maximum electric power, 24 mW, at 625 nm of red light. The thylakoid membrane immobilized with MWCNTs (Multi-walled carbon nanotubes) acts as an anode with MWCNTs as a cathode, which produced the maximum current density, 38 mA/cm. The maximum electric power produced at this current density is 5.3 mW/cm^2. The effect of CNTs on chlorophyll f and d was more than that on chlorophyll a and b: it enhanced the absorption ability of far-red and infrared light (700–750 nm) and also enhanced the ability of photo-convertors. The CNTs were synthetic nanoparticles that penetrate into the biological matrix and have multifunctional properties, such as water uptake and conduction for electricity in biological systems. MWCNTs were most electro-conductive and have a combined effect on thylakoid, the multi-protein complexes PS I and II, and photo-electrochemical properties. SWCNTs (Single Walled Carbon Nano-Tubes) enhanced immobilization of the reaction centre of the bacterium *Rhodobacter sphaeroides* and also enhanced the photo-electrochemical activity. MWCNTs significantly enhanced direct transfer of electron in the thylakoid of spinach and of the cyanobacterium *Nostoc* sp. CNTs enhanced the expression of *Arabidopsis aquaporin* in tobacco plant and enhanced photosynthetic activity by production of the photo-electric current. It was observed that CNTs activate the gene and protein expression of aquaporin in tobacco cells. MWCNTs enhanced activity of anaerobic ammonium oxidation bacteria, high carbohydrate and protein. It also enhanced soil enzyme activity, soil microbial biomass. CNTs can manipulate the composition of soil microbes as well as growth of gram-negative bacteria.

9.17: Effect of Nanofertilizers on the Soil Microbial Community

Soil microbes have a signifcant role in soil health, plant growth, productivity, and biological and chemical reactions within soil and plants. Nanofertilizer particles enter into the soil through several ways including human activity, sewage, and industrial waste. Nanoparticles of silica, palladium, gold, and copper have benefcial effects on soil microbes and seed germination of lettuce. Biological and physicochemical properties determined their health and increased soil productivity. Zinc oxide and copper nanoparticles did not show harmful effects on soil microbes although silver and titanium nanoparticles showed an adverse effect on the microbial biomass richness. It was noted that a low concentration of Au nanoparticles (0.1 mg/kg) reduced the size of PVP. Au nanoparticles stimulate soil enzymatic activity; the Au nanoparticle size and soil enzymatic activity showed no correlation at a high dose (100 mg/kg). Citrate-coated Au nanoparticles signifcantly increased soil enzymatic activity as compared to PVP-coated Au nanoparticles at 50 nm size of both particles.

Nanoparticles, such as Fe_3O_4, TiO_2 changed the soil microbial community, influenced the colonies of nitrifying bacteria associated with roots. CuO Influenced the composition and activity of the bacterial community decreased the oxidative potential of the soil. ZnO nanofertilizer enhances ammonifcation, increases dehydrogenase, and hydrolase activities. TiO_2 nanofertilizer influenced carbon mineralization, pH of soil, organic matter; identified soil type and moisture status. CeO_2, Fe_3O_4, SnO nanofertilizer particles has no effect on microbial biomass C and N. Ag nanoparticles have an impact on ion release shape and function of the natural soil microbes. TiO_2, ZnO nanofertilizer particles can altered soil microbes, enhanced the degradation of organic pollutants. Ag nanoparticles also influenced soil microbial diversity and functional bacterial diversity. It can help to increase biomass of *Aspergillus niger* and *Penicillium chrysogenum.*

9.18: Effect of Nanofertilizers on the Physiological Changes in Plants

Use of various nanofertilizers has a significant role in the physiology of crop plants. It may directly or indirectly influences the physiological parameters by the alteration in the formation of reactive oxygen species, catalase, peroxidase, superoxide dismutase activities, chlorophyll, phenol and leaf protein contents. Ag nanofertilizer treated *B. monnieri* plants have higher protein and carbohydrate contents and lower total phenol contents, catalase and peroxidase activity. It is also obvious that the application of TiO_2 nanofertilizer at lower concentration (200 mg/mL) increases the chlorophyll, peroxidase catalase, superoxide dismutase activities, and malondialdehyde contents on *Lemna minor* compared to bulk by the elimination of reactive oxygen species from the plant cells, but at higher concentrations (500 mg/mL), the TiO_2 nanofertilizer causes a severe disruption of cell membrane in the culture media.

Similarly, it was noticed that foliar application of ZnO at lower concentration (10 mg/L) increased the chlorophyll, phosphorus and total soluble leaf protein concentration on clusterbean. There was an increased acid phosphatase activity, alkaline phosphatase activity and dehydrogenase activity in all the treatments compared to control in 30 and 60 days after sowing on groundnut under pot condition.

Moreover, it was observed that under liquid medium watermelon has taken up significant amount of Fe_2O_3 nanofertilizers and translocated into different parts of the plant. The various concentrations of Fe_2O_3 nanoparticles increased the seedling germination and growth attributes and seedling growth, and enhanced physiological parameters. Changes in CAT, SOD and POD activities due to Fe_2O_3 nanoparticles were significantly higher compared to control treatments and the nanoparticle treatment (20 mg/L) had the most obvious effect on the increase of root activity, ferric reductase activity, root apoplastic iron content and plant biomass. The suitable concentration (20 mg/L) of Fe_2O_3 nanofertilizers could not only increase the seed germination and seedling growth, but also improve the various physiological functions and resistance towards the several environmental stresses on watermelon. The effect of nanofertilizers on the physiological changes of plants is shown in tabular forms (Table 37).

Table 37: Effect of nanofertilizers on the physiological changes in plants

Nano-fertilizer	Concentration	Physiological changes
TiO_2	10–2000 mg/L	Exposure of *Lemna minor* to TiO_2 nanofertilizer increased the activities of various enzymes (POD, SOD, and CAT) below concentration of 200 mg/L because of eliminating accumulated reactive oxygen species in plant cells.
Al_2O_3	10–1000 mg/L	The activity of SOD and CAT were increased with the treatments of Al_2O_3 nanofertilizer at a concentration of 200 and 500 mg/L.
ZnO	10 mg/L	Treatment of clusterbean with foliar sprays of ZnO nanofertilizer caused a significant increase in chlorophyll content (276.2 %), total soluble leaf protein content (27.1%), acid phosphatase (73.5%), alkaline phosphatase (48.7%), and phytase (72.4%) activity over control.
Fe_2O_3	20 mg/L	Increase in root activity, activity of catalase, peroxidase, superoxide dismutase, chlorophyll, malondialdehyde contents, ferric reductase activity, the root apoplastic

Nano-fertilizer	Concentration	Physiological changes
		iron content were recorded by the translocation of the significant amount of Fe_2O_3 nanofertilizer suspended in a liquid medium to various tissues of plants.
S, ZnO	500–4,000 mg/L	Application of various concentrations of both sulfur and zinc oxide nanoparticles is significantly increase the total lipids, proteins, amino acids, thiol and chlorophyll contents compared to untreated control, but no significance difference has been observed among the treatments with various concentrations for both nanoparticles.
TiO_2	200 mg/L	Application of TiO_2 nanofertilizer had a noticeable effect on chlorophyll a and b and carotenoid contents on Mentha.

9.19: Uptake and Translocation of Nanofertilizers in Plants

The various concentrations of CeO_2 nanofertilizers on the growth of cucumber, alfalfa, tomato and corn seedlings were tested. The maximum growth (140%) was recorded at a concentration of 2000 mg/L and minimum growth (50%) at a concentration of 500 mg/L compared to control treatments. The uptake, storage and differential toxicity of CeO_2 nanofertilizers in four edible plants through X-ray absorption spectroscopy concluded that the differential toxicity could be due to the physical and chemical interaction of nanoparticles with root structures and exudates. The exposed CeO_2 nanofertilizer particles as aerosol or suspension on corn leaves stated that about 50 μg of cerium per gram of leaves either absorbed or incorporated into maize leaves, but there is no evidence of translocation of cerium into the newly developed leaves. These results clearly indicated that the absorption and the translocation ability of CeO_2 nanoparticles are dependent on the plant species. Similarly, the effect of Fe_2O_3 nanofertilizer particles on the uptake, translocation and accumulation on pumpkin seedlings in hydroponic concluded that the signal for the uptake of Fe_2O_3 nanoparticles is not only confined to plant parts, but also depend on the growth medium and the uptake of nanoparticles are species dependent process because no uptake has been recorded in lima bean. The uptake of Fe_2O_3 nanofertilizer particles on the pumpkin had speculated that if the sizes of the applied nanoparticles are large then it directly affects the penetration of nanoparticles through cell walls and their transport across the plasma membranes. However, the effect of suspended Fe_2O_3 nanoparticles in liquid medium on watermelon has been directly taken up and translocated into various plant parts, which may result in the increase in seedling germination and enhanced physiological function to some degrees. The positive effects of nanoparticles increased quickly and then slowed with an increase in the concentration of the treatments.

An ultra-small TiO_2 complex with Alizarin red S nano-conjugates were directly taken up and translocated the nanomaterials into the seedlings of *A. thaliana*. The root of the experimental plants found to released mucilaginous substances, which may form a pectin hydrogel capsule around the root and playing a significant role in the inhibition or facilitation of the TiO_2 complexes with Alizarin red S. However, it was reported that TiO_2 nanoparticles is neither taken up nor translocated by the excised maize roots due to their large size. The experiment conducted on rye-grass confirmed that the adsorption and accumulation of the ZnO nanofertilizer to the root surfaces. The cross sectional transverse electron microscopy images of rye-grass root clearly showed the presence of ZnO nanoparticles in the apoplast, cytoplasm and nuclei of the endodermal cells and the vascular system. The uptake and accumulation of ZnO nanoparticles on soybean seedlings with the concentration of 500 mg/L concluded that high concentrations of nanoparticles increased the probability of agglomerate formations which may restrict the passage of nanoparticles through the cell pore and resulted into the reduce uptake and accumulation of nanoparticles inside the plant tissues.

The Al uptake in red kidney beans when exposed Al nanoparticles results that the Al concentrations in the red kidney beans were not significantly different compared to untreated control treatments. However, investigated the uptake of Ag nanoparticles in zucchini compared to bulk (control) treatments stated that the silver concentration in shoots was higher, when the plants were exposed Ag nanoparticles compared to bulk and concluded that release of silver ions from Ag nanoparticles is accounted for an increase in the concentration of silver in shoots. Similarly, treated *Brassica juncea* with Ag nanoparticles find out the real mechanism behind the formation of nanoparticles by plants and stated that there is no accumulation of Ag nanoparticles in any form in the plant tissues. However, carbon-coated Fe_2O_3 nanofertilizer particles were exposed to the leaf petioles of pumpkin plants showed the sign of trace of nanoparticles into the cells located away from the introduced area. It was found that the gold and silver ions present in a different valence state in growth media supplemented with agar, reduced and accumulated as Au and Ag nanoparticles inside alfalfa seedlings. Similarly, transformation and accumulation of silver and platinum ions into Ag and Pt nanoparticles in alfalfa and mustard seedlings were also investigated on various plants, and strongly concluded that the gold, silver and platinum nanoparticles could be accumulated into the seedlings of various tested experimental plants.

9.20: Effect of Nanofertilizers on the Cellular and Genetic Toxicity in Plants

The toxicity of multi-walled carbon nanotube suspension of rice cells found that multi-walled carbon nanotubes at low concentrations cause cell death and at high concentration cause cell mortality. However, the concentration and size of single-wall carbon nanotubes cause the toxicity in the rice and Arabidopsis protoplast cells and the breakdown of genetic material in Arabidopsis cell directly correlated with the genetic toxicity of the single walled carbon nanotubes. The cellular toxicity of CeO_2 and ZnO nanofertilizer particles on soybean seedlings detected the genetic toxicity of both nanoparticles by RAPD assay. It was elaborated that the interaction of genetic material with the zinc ions is the main reason behind the increase in the toxicity level. However, in another study, it was found that TiO_2 nanoparticles were genotoxic and cytotoxic due to the generation of superoxide radicals during lipid peroxidation. None the less, it was observed the dose dependent and duration dependent responses in the root meristem of *Allium cepa* due to decrease in mitotic indices arisen by various types of chromosomal abnormalities against Ag nanoparticles. The treatment of *A. cepa* cells with various concentrations of Ag nanoparticles caused the stickiness and breakage of chromosomes, which may finally lead to the genotoxicity of cells. Similarly, the biocompatible magnetic nanoparticles coated with perchloric acid decreased the level of nucleic acid and postulated that the magnetic nanoparticles induced the chromosomal aberrations in corn cells.

Further Studies

Abdel-Aziz, H. M. M., Hasaneen, M. N. A. and Aya, M. O. (2018). Foliar application of nano chitosan NPK fertilizer improves the yield of wheat plants grown on two different soils. Egypt. *J. Experimental Biol.* **14:** 63–72.

Alshaal, T. and El-Ramady, H. (2017). Foliar Application: from Plant Nutrition to Biofortification. *The Environment, Biodiversity and Soil Security,* Vol. 1: pp.71–83.

Amin, F. and Mohammad M, O. (2015). Effect of nano-zinc chelate and nano-biofertilizer on yield and yield components of maize (*Zea mays* L.), under water stress condition. *Ind. J. Natur. Sci.,* **5:** 4614–4620.

Delfani, H., Firouzabadi, M. B., Farrokhi. N. and Makarian, H. (2014). Some physiological responses of black-eyed pea to iron and magnesium nanofertilizers. *Comm. Soil Sci. Plant Anal.* **45:** 530–540.

Naderi, M. R. and Danesh-Sharaki, A. (2013). Nanofertilizers and their role in sustainable agriculture. *Intl. J. Agril. Crop Sci.* **5:** 2229–2232.

10

Nanonutrient use Efficiency

Improving nutrient use efficiency is a worthy goal and fundamental challenge facing the present fertilizer industry, and agriculture in general. Now, nutrient use efficiency is driven by a growing public believe that the crop nutrients are excessive in the environment. Farmer concerns about rising fertilizer prices and stagnant crop prices, exerted pressure to person concerned to improve nutrient use efficiency. Most agricultural soils in India have low native fertility and successful and sustained crop production on these soils requires regular nutrient inputs. The quantum of nutrients available for recycling via crop residues and animal manures is grossly inadequate to compensate for the amounts removed in crop production. Thus, mineral fertilizers have come to play a key role in areas with low fertility soils, where increased agricultural production is required to meet growing food demand. Chemical fertilizers as source of plant nutrients are considered as the major contributor to enhancing crop production and maintaining soil productivity.

Fertilizers have been acknowledged for 50 per cent of the yield increases in developing countries including India. Though the consumption of chemical fertilizers in India increased steadily over the years, the use efficiency of nutrients applied as fertilizers continues to remain awfully low (30–50% for N and K, 18–20% for P and 2–5% for Zn, Fe, Mo, B & Cu). When nutrient inputs are used inefficiently then both cost of cultivation and threat for biosphere pollution increase. Thus, economic and ecological highlights the compulsive need for more efficient use of nutrients in crop production. Since fertilizer nutrients are expensive and used in large quantities at national level, any increase in use efficiency will lead to a substantial cut in nutrient requirement and huge economic benefit at national level. For example, if we consider fertilizer consumption are 14.05 Mt N and 5.66 Mt P_2O_5 with nutrient use efficiency of 50 % for N and 20 % for P. The increase in the efficiency of N and P use by just 2 percentage that would lead to a saving of 5.38 lakh tonnes of N and 5.15 lakh tonnes of P, which together translate to a saving of Rs.14,920 million annually. This amply highlights the importance of enhancing nutrient use efficiency. This staggering annual economic benefit is in addition to reduced risk of environmental pollution. Since fertilizers, particularly synthetic fertilizers have a major potential to pollute soil, water and air. In recent years, many efforts were done to minimize these problems by agriculture practices and the design of the new improved fertilizers. The appearances of nanotechnology open up potential novel applications to solve these problems.

10.1: Role of Nanofertilizer

Nanostructure fertilizer exhibits novel physico-chemical properties, which determines their interaction with biological substances and process. The application of nanotechnological formulation to agricultural crop inputs is one of the proposed tools for sustainable intensifications. These applications include an increase in uptake efficiency in plants, developing DNA based nano-sensor in a polymer coated fertilizers which would release only as much fertilizer as "demanded" by plant roots. Nanofertilizer mainly delays the release of the nutrients and extends the fertilizer, and their effect period. A nano-fertilizer refers to those products that supply the plant nutrients either by encapsulating the nutrients inside nanomaterials or nano-porous materials, or coated with a thin protective polymer film. It possesses, such unique physico-chemical properties that they can fulfill plant root requirements more efficiently as compared to the conventional fertilizers. Encapsulation of fertilizer nutrients within a nanoparticle can be done in the following ways:

a) The nutrient can be encapsulated inside the nano-porous materials

b) The nutrient can be coated with a thin polymer film

c) The nutrient can be delivered as tiny particles or emulsions of nanoscale dimensions.

Nano-sized fertilizer provides nutrients more available to nano-scale plant pores, and therefore results in greater nutrient use efficiency. Nano-materials may help to improve nutrient use efficiency because of their small size (between 1 to 100 nm), more surface area and their slow rate of release, which facilitate to the plants to take up most of the nutrients without any waste. It is claimed that controlled nutrient release and increased water retention in the soil are responsible for better yield under nanofertilzer application. Nano-sized TiO_2 and Mg fertilizer promoted photosynthesis and nitrogen metabolism, carbon nano-tubes penetrate tomato seeds and affect their germination and growth rates. Nanofunctionalized carbon nanotubes enhanced root elongation. Nanofertilizers combined with nanodevices synchronize the release of fertilizer N and P with their uptake by crop, so preventing undesirable nutrient losses to soil, water and air via direct internalization by crops, which comes by avoiding the interaction of nutrients with soil, microorganisms, water, and air. Nano-clay and zeolites that are a group of naturally occurring minerals with a honey comb-like layered crystal structure, which can filled with nitrogen, potassium, phosphorous, calcium and a complete set of minor and trace nutrients. This helps to achive maximum nutrient use efficiency. Coating and bonding of nano and subnano-composites are able to regulate the release of nutrients from the fertilizers capsule and the application of a nano-composite consists of N, P, K, micronutrients, mannose and amino acids enhance the uptake and use of nutrients by grain crops. Fertilizers incorporation in to co-chleate nanotubes had improved crop yield. The nanofertilizers release the nutrients in a controlled manner in response to reaction to different signals, such as heat, moisture and etc. It is known that crops secrete carbonaceous compounds in to rizosphere under nutrient stress which can considered as environmental signals for incorporation in to novel nanofertilizers. These fertilizers can also improve the performance of fertilizers in other ways. For instance, due to its photocatalytic characteristic, nanosize titanium dioxide added in to fertilizers as bactericidal, nanosilica particles that absorbed by roots can improve the plant's resistance to stress and thus increases the crop yield. Therefore, such nano-fertilizers have a significant influence on energy, the economy and the environment. Conceptual rhizosphere management for enhancing nutrient use efficiency (NUE) is shown (Figure 78).

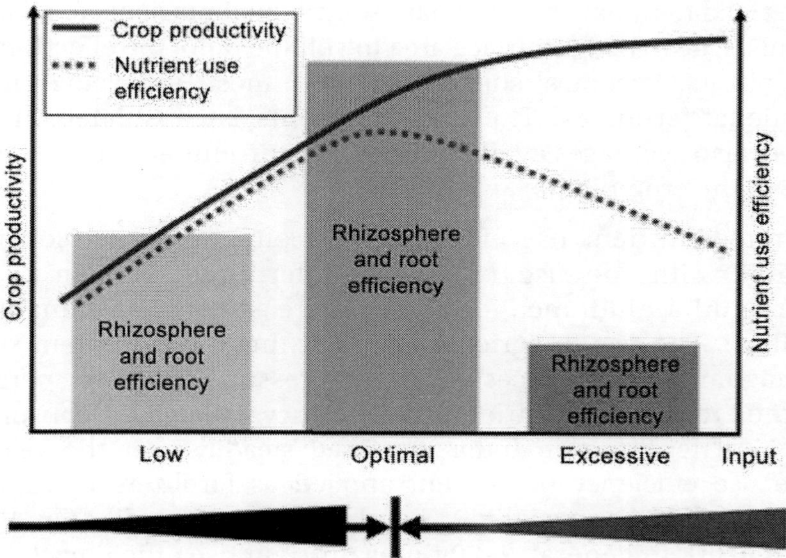

Figure 78: Conceptual model of rhizosphere management for optimizing nutrient inputs toward both high crop productivity and high nutrient use efficiency through maximizing root/rhizosphere efficiency in nutrient mobilization and acquisition. The efficiency of root/rhizosphere processes in increasing nutrient acquisition and crop growth is limited at low or excessive input, while at the optimal level of nutrient supply, there is a maximum root/rhizosphere efficiency in achieving high nutrient use efficiency and high crop productivity.

The strategies of rhizosphere management toward sustainable crop production reflect a shifting emphasis away from the conventional nutrient management with increased inputs of soluble inorganic nutrients to regulating root morphology and managing rhizosphere processes toward improving nutrient use efficiency and crop productivity.

Nutrient use efficiency is one of the most important concepts for evaluating crop production systems, which can be highly influenced by various factors, such as fertilizer management and soil, and plant-water relationships. Improving the nutrient use efficiency has been listed among todays most critical and formidable research issues. The nutrients are used with an objective of increasing the overall performance of crops and cropping systems by providing optimum nourishment and supporting sustainable agricultural production systems through contributions to soil fertility and other important soil quality components. The higher mobility of the nano-fertilizer particles lead to transport of the nano-formulated nutrients to all parts of the plants. Due to a high surface area to volume ratio, the effectiveness of nanofertilizers becomes superior to the most innovative modern conventional fertilizers. The plant nutrients encapsulated in nano-particles also increases availability of the nutrient elements and thus uptake to the crop plants.

Limited nutrient use efficiency and environmental constraints associated with the use of chemical fertilizers remain a major problem and a hindrance for achieving reasonable sustainability in agriculture. The rise in agriculture prices, huge government subsidy and stagnant crop prices exerted pressure on the personnel concerned to improve nutrient use efficiency. Though the consumption of chemical fertilizers in India increased steadily over the years, the average use efficiency of nutrient applied as fertilizers continues to remain awfully low specially P, S and micronutrients like Zn, Fe, and Cu (Table 38). In general, 2–20 times nutrient use efficiency has been increased when applied as nanoform of fertilizer particles.

Table 38: A comparison of per cent nutrient use efficiency between chemical and nanofertilizer

Nutrient use efficiency (NUE)	Chemical fertilizer (NUE %)	Nano-fertilizer (NUE %)
N	30–35	80–85
P	15–20	58–65
K	35–40	82–88
S	17–22	75–78
Fe	4–5	80–82
Zn	3–4	78–80
Cu	2–5	77–81

Nano-fertilizers possess certain specific properties, which facilitate higher nutrient use efficiency. The important properties are:

1. The nano-fertilizers have a higher surface area, which is mainly due to the very smaller size of the nanoparticles that provide more sites to facilitate the different metabolic process in the plant system. This results in the production of more photosynthesis with less consumption of nutrient elements.
2. They have high solubility in different solvents, such as water.
3. The particle size of nano-fertilizers is very small (less than

100 nm), which facilitates more penetration of nanoparticles into the plant system.

4. Nano fertilizer elements have the large surface area and particle size smaller than the pores of root and leaves of the plants. This increases penetration into the plant system from applied surfaces and thus improves uptake and nutrient use efficiency of the nano-fertilizers.
5. The reduced particle size of nano-fertilizer results in increased specific surface area and a number of particles per unit area of fertilizer, which provide more opportunity for contact of nano-fertilizers and it leads to more penetration and uptake.
6. The fertilizer elements encapsulated in nano-particles increases the availability and hence uptake of the plant nutrients to the crops.

Zeolite-based nano-fertilizers are capable of releasing the nutrients slowly to the crop plant, which increase the availability of nutrient elements to the crops throughout the growth period. This prevents loss of nutrients through denitrification, volatilization, leaching and fixation in the soil especially NO_3 and NH_4 forms of nitrogen.

The nano-fertilizers have a higher surface area and it is mainly due to very less size of particles which provide more site to facilitate different metabolic processes in the plant system result in the production of more photosynthets. Due to higher surface area and very less size they have high reactivity with other compound and they have high solubility in different solvents. Smaller particle size of nano-fertilizers facilitates more penetration of nano-particles in to the plant from applied surface, such as soil or leaves. Nano-fertilizers have large surface area and particle size less than the pore size of root and leaves of the plant which can increase penetration into the plant from applied surface and improve uptake and nutrient use efficiency of the nano-fertilizer. Reduction of particle size results in increased specific surface area and number of particles per unit area of a fertilizer that provide more opportunity to contact of nano-fertilizers which leads to more penetration and uptake of the nutrient. Fertilizers encapsulated in nano-particles will increase availability and uptake of nutrient to the crop plants. Zeolite based nanofertilizers are capable to release nutrient slowly to the crop plant which increase availability of nutrient to the crop through out the growth period which prevent loss of nutrient from denitrification, volatilization, leaching and fixation in the soil especially NO_3-N and NH_4-N. Particle size below 100 nm nano-particles can use as fertilizer for efficient nutrient management which are more eco-friendly and reduce environment pollution. Main reason for high interest in nanofertilizers is mainly their penetration capacity; size and very higher surface area which is usually differ from the same material found in bulk form. This is partially due to the fact that nano-particles show a very high surface: volume ratio. Thus, the reactive surface area is proportionally over-represented in nano-particles compared to larger particles. Particle surface area increases with decreasing particle size and the surface free energy of the particle is a function of its size.

The improvement of NUE is an essential pre-requisite for expansion of crop production in marginal lands with low nutrient availability. When nutrient inputs are used inefficiently then both cost of cultivation and threat for biosphere pollution increase. Thus, economic and ecological highlights the compulsive need for more efficient use of nutrients in crop production. Since fertilizer nutrients are expensive and used in large quantities at the national level, any increase in use efficiency will lead to a substantial cut in nutrient requirement and huge economic benefit of the country.

10.2: Nanoparticles Based Delivery Systems

It is already mentioned that the nanomaterials used as fertilizers have potential contributions in slow release of plant nutrients. The higher surface tension provided by the surface coatings of nanomaterials on fertilizer particles helps to hold the materials more strongly than the conventional material surfaces. Following are the nanoparticle-based delivery systems (Figure 79).

10.2.1: Chemical system

Fertilizer materials with sulfur nano-coating are very useful slow release fertilizers as these materials supply sulfur elements in addition to the primary nutrient elements. The fertilizer dissolution rate is reduced by the stability of the coating and thus it allows slow sustained release of sulfur coated fertilizer. In addition to sulfur coatings, encapsulation of urea and phosphatic fertilizers with nano-materials are also beneficial to meet the soil and crop demands. Some other useful nanomaterials with potential application include kaolin and polymeric biocompatible nanoparticle.

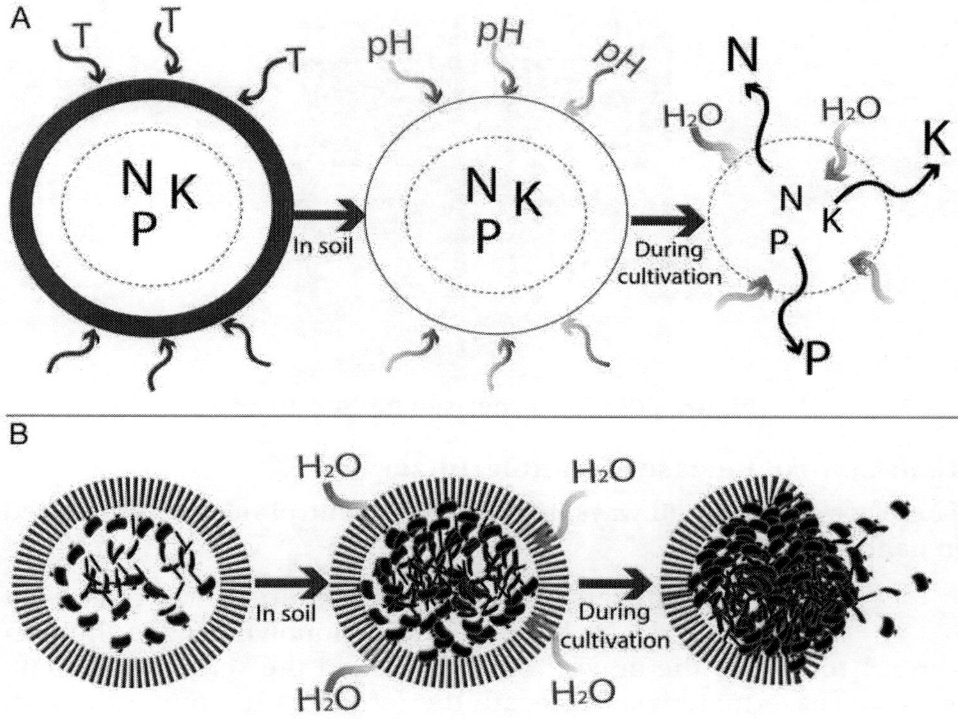

Figure 79: Schematic representation of smart delivery systems: (**A**) advanced polymeric materials, degraded under external stimulus, such as temperature, pH, and with water permeability to achieve a slow nutrient release; (**B**) microorganism encapsulation.

10.2.2: Biological system

The formulations of bio-fertilizer inocula, their storage and method of application are critical to their success in the crop field. The polymeric nanoparticles are potentially suitable for their application for coating of bio-fertilizer preparations to yield desiccation resistant formulations. Figure 80 shows biosynthesised nanofertilizers P and Zn, which can increase the P use efficiency from 15% to 61% and Zn use efficiency from 4% to 80%.

10.2.3: Field application of fertilizers

The mode of application of fertilizer influences their efficiency and environmental impact. The constraints associated with the droplet size of foliar sprays may be overcome by using nanoparticles encapsulated or nano formulated fertilizer elements that will contribute to efficient spraying and thus will reduce spray drift and splash losses.

Figure 80: Biosynthesised nanofertilizers

10.3: Control Release of Nanofertilizers

There are a number of ways, by which the control release is achieved in nano-fertilizers:

1. The nano-fertilizers are mixed with different materials, such as hydrogels, special films or other biopolymers like chitosan to reduce the uncontrolled release in the soil environment. These materials aggregate the fertilizers in complexes with mineral nanoparticles obtained from the clay in soil or other types of ceramic materials, that are used for manufacturing controlled- release blocks, pots, or film. These respond to different environmental stimuli, such as temperature or irradiance, and modify the release of the nano-fertilizers according to the plants' need.
2. Another alternative for regulating the release of nano-fertilizers to the environment is by applying as foliar spray, especially for the elements having limited bioavailability in soil, such as Fe, Cu and Ni. For this purpose, the emulsions or encapsulated organic nanoparticles can be useful.
3. Finally, another way to enable slow release of nutrient elements from the nano-fertilizers in the environment is by matching their quantities with the stage of crop growth.

The nanoparticles used as nano-fertilizers are absorbed by the plant roots efficiently due to their very small size. These are then transported through apoplastic and symplastic pathways to the xylem, cross the endodermis and then they move through the vascular bundles to the different parts of the plant. This mode of transport inside plant has been observed for mesoporous silica nanoparticles and SiO_2 nanoparticles. It has been reported that different classes of nanoparticles are transported in the plant to the inside of the cells through endocytosis or through pores or channels. Presence of an essential plant nutrient element in the substrate or soil in the form of nano-fertilizer allows better dissolution, faster absorption and assimilation by the plants as compared to traditional fertilizers and this has been demonstrated for nutrient elements, such as N, P, K, Ca, Mg, S, B, Fe, Mn, Zn, Cu and Mo . It is very much important that the nano-fertilizers be released in a correct form and at a rate suitable for plants, which minimizes losses by leaching, gasification or by competition with other organisms. Nano-fertilizers increase the availability of nutrient to the growing plant parts, which increase chlorophyll formation, rate of photosynthesis, dry matter production and thus improve overall growth of the plants. It was reported that the nano-TiO_2 treated seeds produced, such plants that showed more dry weight, higher photosynthetic rate and chlorophyll-a formation as compared to the plants produced from untreated seeds. Similarly, nano-Mg treated seeds influences more light absorption and phosynthetic rate compared to untreated seeds.

10.4: Influences of Nanofertilizers

Nutrient use efficiency (NUE) depends on the plant's ability to take up nutrients efficiently from the soil, but also depends on internal transport, storage and remobilization of nutrients. Nitrogen (N) is a fundamental element regulating plant growth and development. Plants have evolved inorganic and organic N-uptake systems to cope with heterogeneous N availability in the soil. However, NUE is dependent on root growth and root architecture. Endophytic bacteria have a direct influence on root growth and increase nutrient uptake. It has been found that nutrient in nanoform can influence the endophytic bacterial population as well as root growth and development. Under reciprocal exchange, trading carbon for nutrients, plant and bacteria establish a symbiotic association. The increase in endophytic bacteria might contribute to efficient nutrient uptake, especially organic nitrogen, through bacterial cell degradation or by externally activating nitrogen transporters for improving NUE in plants. Recent research suggests that their nano-dimensions allow their uptake through stomatal openings and the base of trichomes.

Strategies for increasing nutrient uptake by overexpressing genes encoding high-affinity nutrient transporters are likely to be mainly applicable to situations where reasonal nutrient concentrations can be maintained at the outer surface of the plasmalemma. Maintaining, such a concentration is a major problem at present, but that can be done with the application of nanofertilizers instead of conventional fertilizers. There may be scope for manipulating the expression of genes by nanofertilizers, involved in the internal mobilization of nutrients within the plants, thereby inproving the utilization.

Nanofertilizers also influences in genetic variation within and among crops, which may also affect the NUE as it is well recognized. Genotypic variability, that affects NUE and nutrient uptake, influences some processes and plant mechanisms, including differences in uptake, movement in root, shoot demand, and biomass production. Intraspecific genetic variability also affects NUE. Variations in the Ca efficiency ratio (milligrams of total plant dry weight per milligrams of Ca in total plant) of cauliflower and collard cultivars have been reported. In addition, genetic variation in Ca use efficiency of tomato strains under low Ca^{2+} stress has been demonstrated, with efficient strains producing 30% more total plant dry weight than inefficient genotypes.

Further studies

Ghormade, V., Deshpande, M.V. and Paknikar, K.M., (2011). Perspectives for nano-biotechnology enabled protection and nutrition of plants. *Biotechnol. Adv.* **29:** 792–803.

Liu, X., Feng, Z., Zhang, S., Zhang, J. and Xiao, Q. (2006). Preparation and testing of cementing nano sub nano composites of slow or controlled release of fertilizers. *Sci. Agril. J.* **39:** 1598–1604.

Singh, M. D., Chirag, G., Prakash, O. P., Mohan, M. H. Prakasha, G. and Vishwajith (2017). Nanofertilizer is a new way to increase nutrients use efficiency in crop production. *Int. J. Agril. Sci.* **9:** 3831–3833.

Tarafdar, J. C, Rathore, I. and Thomas, E. (2015). Enhancing nutrient use efficiency through nanotechnological interventions. *Indian J. Fert.* **11(12):** 46–51.

Wilson, M. A., Tran, N. H., Milev, A. S., Kannangara, G. S. K., Volk, H.,and Lu, G. H. M. (2008). Nanomaterials in soils. *Geoderma.* **146:** 291–302.

11

Nanonutrients for Gum Production and Stress Tolerance

Organic polymers can play important role in ecosystems by accumulating biologically important elements and also by retaining soil moisture after aggregating soil particles. Extracellular polymeric substances (EPS) play an important role in cell aggregation, cell adhesion, and biofilm formation that subsequently protect cells from a hostile environment. Furthermore, certain polysaccharides from microbial sources are surface active, and thus attempts have been made to use them as metal chelaters, emulsifiers and flocculants in industrial and environmental fields/domain. Such use of microbial polysaccharides has infused renewed interest in its production and characteristics. Worldwide efforts are being done in this direction. Polysaccharide is relatively complex carbohydrate consisting of multiple monosaccharides joined together by glycosidic linkages. Polysaccharide producing microorganisms are bacteria, fungi and yeast. Twenty different types of microbial polysaccharides of commercial importance are known. These polysaccharides had great application in food, cosmetic, pharmaceutical, oil industries and agriculture. The structural cofactor (component) of polysaccharides is generally Mg, Zn and Fe. It has been found a number of efficient polysaccharide producing organisms and their polysaccharide production can be enhanced dramatically with the action of Zn and Fe nanofertilizers.

As Zn is the structural component of phosphatases and phytase enzymes as well as polysaccharides, it can be well taken that application of nano-Zn may help more secretion of polysaccharides for better soil aggregation as well as phosphatases and phytase enzymes release, which may be involved in mobilizing P for plant nutrition mainly from unavailable organic sources. A list (Table 39) of some important polysaccharide producing organisms responded to enhance their polysaccharide production with the action of Zn and Fe nanofertilizers.

Table 39: Some important polysaccharide producing organisms responded to nanofertilizers

Name of the organism	NCBI Accession No.	Type of polysaccharide
Bacillus subtilis JCT 1	JN 194187	Dextron
Caulobacter vibrioides JCT 4	JQ 675300	Curdlan
Agrobacterium tumefaciens JCT 8	JQ 675304	Xanthan and Curdlan
Sporosarcina ginsengisoli JCT 12	JQ 675308	Xanthan and Dextron
Aspergillus terreus TFR 2	JN 194186	Pollulan
Aspergillus flavus TFR 7	JQ 675294	Pollulan
Stenotrophomonas maltophilia JCT 9	JQ 675305	Dextran and Pollulan

Microbial polysaccharides have immense commercial importance. They are employed in the stabilization of foods, and production of several industrial and pharmaceutical compounds. The commercial value of a polysaccharide is based on its ability to modify the flow characteristics of solutions (technically known as rheology). Polysaccharides can increase the viscosity and, are therefore useful as a thickening and gelling agents. Microbial polysaccharides are of great importance in oil industry. By conventional methods, only 50% of the oil can be extracted. And the rest is either trapped in the rock or too viscous to be pumped out.

It is now possible to recover, such oils also by a technique called microbial enhanced oil recovery (MEOR). This can be done by injecting surfactants and viscosity decreasing biological agents (*i.e.,* the microbial polysaccharides, *e.g.,* xanthan). Recently, it has been found that it has great importance in soil aggregation, moisture retention and soil carbon build up. The mechanism of poly-saccharides exudation by the microorganisms is shown in Figure 81. They are mainly secreting the polysaccharides of dextran, xanthan, pollulan and gallan. The important group of microorganisms capable of producing/secreting polysaccharides is shown in the figure 81.

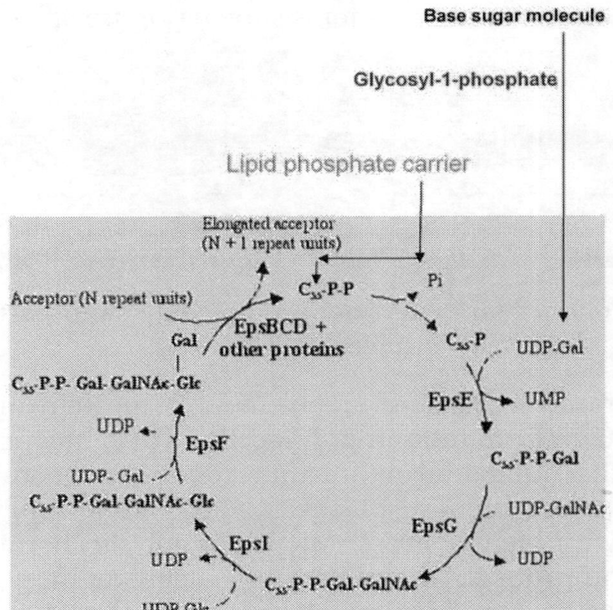

Etner-Doudoroff-pathway for elongation of polysaccharide chain

Figure 81: Pathways of Microbial Polysaccharide production

11.1: Production of microbial polysaccharides

The synthesis of polysaccharides favourably occurs in the excess supply of carbon substrate in the growth medium while limiting nitrogen supply. A carbon/nitrogen ratio of around 10:1 is considered to be favourable for optimal polysaccharide synthesis. The production process is mostly carried out by batch culture fermentation. By manipulating the nutrient supply, differential synthesis of polysaccharides can be achieved. By limiting nitrogen supply in the medium, mostly neutral polysaccharides are produced. When metal ions are limited, acidic polysaccharides are mainly synthesised. Molecular oxygen supply of around 90% saturation is ideal for good growth and polysaccharide synthesis.

The major component of the microbial extracellular polymeric substances (EPS) would be the extracelluar polysaccharides, which are associated either with the cell surface or excreted in the growth medium. It has been shown that ZnO nanofertilizers may enhance polysaccharide secretion of fungi *Aspergillus terreus* CZR1 and *Aspergillus flavus* CZR2 by 8–9 folds. The electrophoresis result revealed that EPS released by *Aspergillus terreus* CZR1 contained pullulan and some unknown polysaccharide, while the content in case of *Aspergillus flavus* CZR2 is still to be identified. Some of the microbial polysaccharide powder is shown (Figure 82).

Alcallgenes faecalis *Bacillus subtilis* *Aspergillus terreus* *Aspergillus flavus*

Figure 82: Polysaccharide powder produced by nano-particle induced microorganisms

The literature regarding involvement of protein in the heterogeneous domain formation in polysaccharides is inconclusive. Since EPS is the first fungal structure that comes into contact with the environment, its structural plasticity may enable fungi to respond better to environmental stress. Microbial polysaccharides may help in aggregation process. In simple terms, aggregation is defined as a physical process where in micro-particles collide and remain stack to each other to form clumps of particles called aggregates. These aggregates are fractal in nature, highly porous, fragile and of varying size range. Soil aggregation is very important especially in arid and semi-arid areas. The process of aggregation is influenced by particle density, collision frequency and binding properties of the EPS. Recently, increasing attention has been paid to microbial polysaccharides due to the possibility of easy and quick mass production. They can readily be prepared by microbial fermentation under various environmental conditions including climate, ecological adoption etc. Further, ZnO nanofertilizers due to their small size and huge surface energy, stabilized in the solutions, which may be bioavailable. Zn is an essential element for organisms, but it is toxic at high levels. The response of Zn nanofertilizer towards different polysaccharide producing fungi and bacteria is shown (Figure 83). With increase in the concentration of Zn nanofertilizer, the polysaccharide secretion has increased dramatically. In case of fungi, the best concentration to trigger the external polysaccharide production was varied between 40–45 mg/L concentration of Zn nanofertilizer. The enhancement was almost 9 times more in *Curvularia lunata* and almost 6 times more in *Chaetomium globosum* and *Aspergillus fumigatus* over without application of nano-Zn fertilizer treatment. In case of bacteria, the concentration to trigger the maximum polysaccharide secretion was widely varied among the polysaccharide producing bacterial species when nano-Zn fertilizer was applied. The peak of secretion of bacterial polysaccharide is varied between 4 and 36 mg/L of Zn nanofertilizer concentration. Therefore, for microbial polysaccharide production one has to be careful to apply the exact doses of nanofertilizer required for particular microorganisms for triggering the maximum polysaccharide secretion externally. The sucrose concentration also affects the microbial biomass as well as polysaccharide production. The Zn nanofertilizer application at a concentration between 4 and 20 mg/L was found to be the best for most of the bacteria to release maximum capacity of their potential of polysaccharide release externally. The increase in polysaccharide secretion is noticed up to 9 times than the original capacity of secretion before triggering by nanofertilizer.

Many studies showed phytotoxicity at a concentration of 43 to 996 mg Zn L^{-1} to various plant species. Nanofertilizers of Zn are the structural component of phosphatases, phytase and polysaccharides. They are known to stabilize the enzyme complexes in plants. One can utilized this property for increasing higher exudation of enzymes and polysaccharides which in turn would increase P mobilization and polysaccharide secretion for soil aggregation. This approach would enable in breaking of existing barriers in utilization of native P and reduce dependence on P fertilizers as well as more soil aggregation for moisture retention especially under sandy soils.

11.2: Effect on Soil

Soils of arid and semi-arid areas are generally course textured, contain low soil organic carbon and have poor structure. As a result, these soils dry rapidly and are extensively prone to wind erosion. These twin factors of wind erosion and rapid drying adversely affect the survival of microorganisms in these areas. Exo-polysaccharides produced by microorganisms can absorb moisture and can also bind soil particles together to improve soil aggregation. Thus increasing the production of more exo-polysaccharides can help in reducing the erodibility of soils, increasing soil moisture retention, survival of the microorganisms and increasing the carbon and organic matter build up in the system. Nanoparticles Zn and Fe may help to increasing more polysaccharide release by microorganisms to increasing moisture retention capacity through better aggregate formation and C as well as microbial build up in the soils. Thus increasing production of exo-polysaccharides can help in reducing erodibility of soil, increasing soil moisture retention, survivability of microorganisms and C build up into the system.

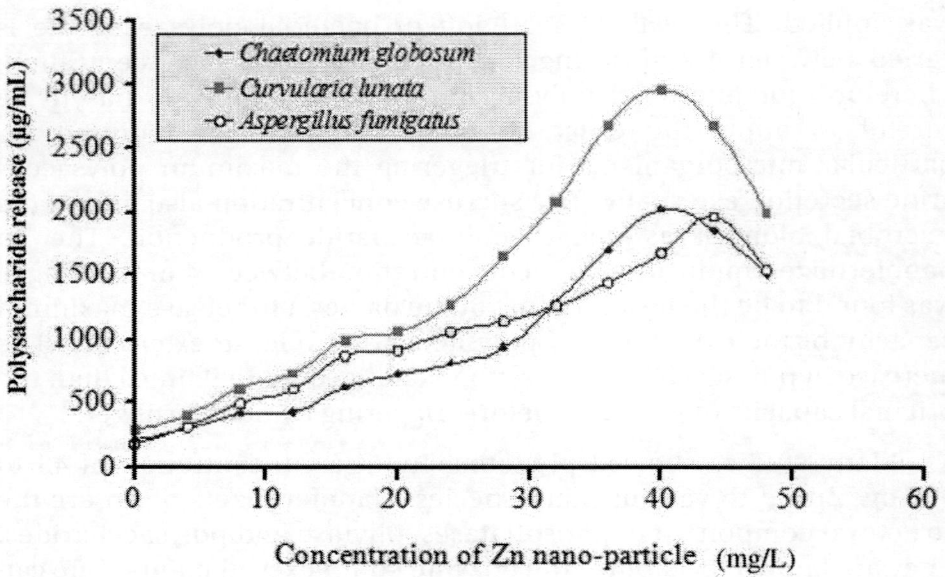

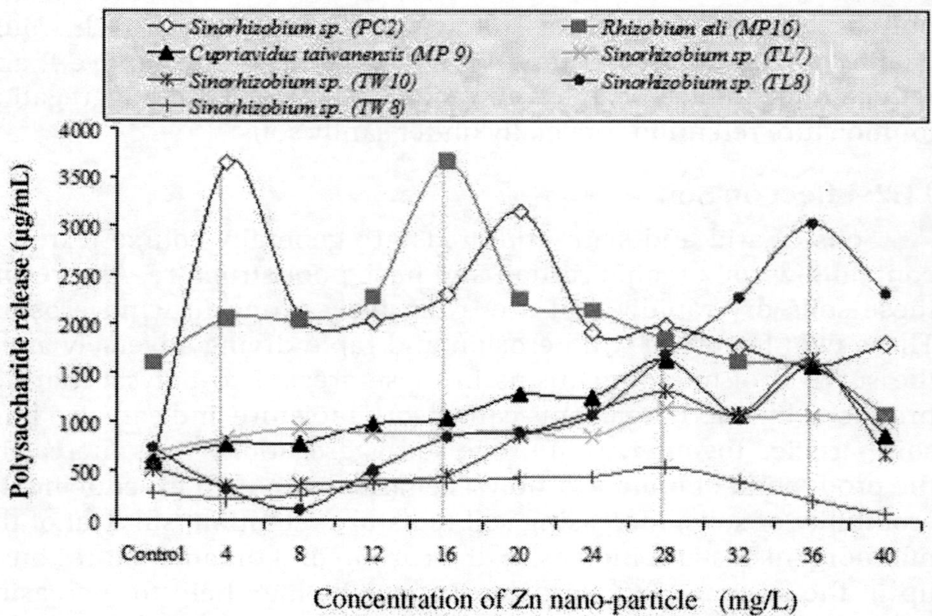

Figure 83: Role of Zn nanofertilizer in enhancement of microbial polysaccharide production

11.3: Microbial Gum

The microbial gum produced by the organisms were purified (Figure 84) by using 20% sucrose solution followed by drying the pellet at 100°C. The pellet was powdered for application in arid degraded soil to test soil aggregation and moisture retention capacity of the gum.

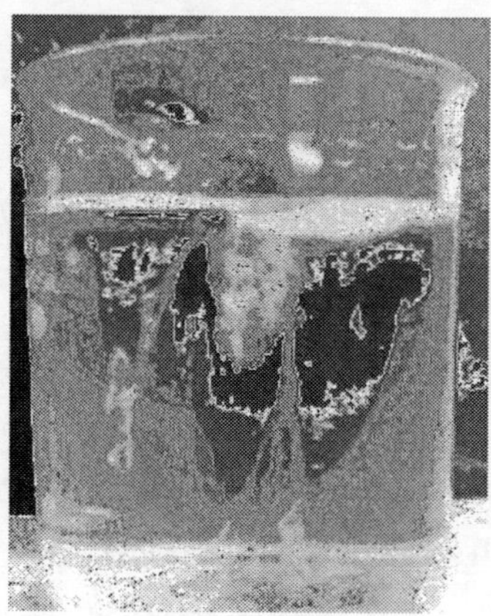

Figure 84: Purification of polysaccharides from developed microorganism

Nanoparticles Zn (10 ppm) and Fe (30 ppm) may help to increasing more polysaccharide release by microorganisms, which can be powdered (Figure 85) to increasing moisture retention capacity through better aggregate formation (Figure 86) and C as well as microbial build up in the soils.

Water is a vital component for plant survival and essentially required for transport of nutrients, therefore water deficiency leads to drought stress, which resulted into weakened vitality of plants. Nanofertilizers promise the significant effort to mitigate the drought stresses. Several recent studies have evaluated nanoparticle-mediated in different stresses. The improvement of moisture retention by nano-fertilizer induced microbial polysaccharide powder was studied at different concentration of polysaccharides ranging from 1–7%. The moisture retention capacity was measured after four weeks of application. It was found that the moisture retention improved between 10.7 and 14.5 per cent at different level of concentrations. The results (Table 40) clearly indicate that 1%, nanofertilizer induced, polysaccharide application in the soil gives economically higher moisture retention, although the maximum moisture retention was observed at 7% polysaccharide application in the arid soil (aridisol).

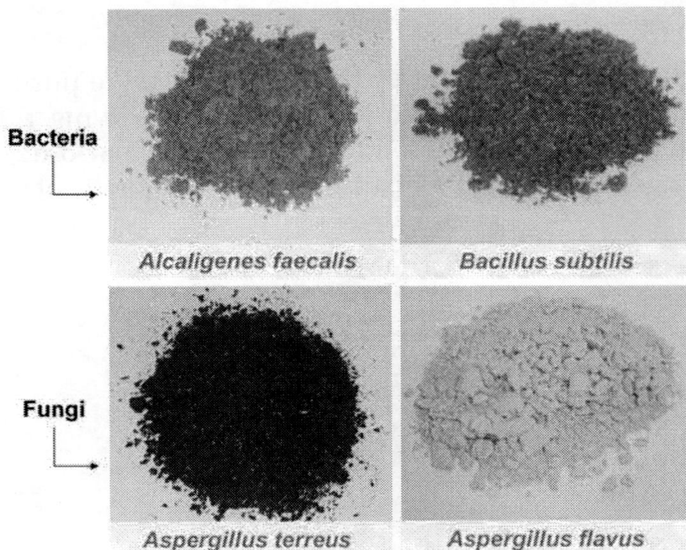

Figure 85: Nanoinduced polysaccharide powder from different microorganisms

Table 40: Improvement of moisture retention in arid soils using nanofertilizer induced polysaccharide powder

Polysaccharide applied (%)	Improvement in moisture retention (%)
1	10.7
2	12.2
3	12.5
4	12.8
5	13.6
6	14.2
7	14.5
LSD (p= 0.05)	2.2

Nanoparticle induced polysaccharide enhances the water stress tolerance via enhancing more soil aggregation, root hydraulic conductance and water uptake in plants, and showing differential abundance of proteins involved in oxidation-reduction, ROS detoxification, stress signaling, and hormonal pathways.

The carbon build up with the application of nanofertilizer induced microbial polysaccharide powder in arid soils has reflected a huge initial build up immediately after the application. But with time, there was gradual decline in carbon concentration due to the corresponding microbial build up. The enhancement of carbon build up due to microbial polysaccharide at crop harvest was noticed between 3 and 5% (Table 41). The initial organic carbon content of the experimental soil was 0.28%.

Table 41: Effect of nanofertilizer induced polysaccharide powder on soil organic C build up in Aridisol

% polysaccharide applied	% improvement of organic C in soil (days after application)				
	7	**14**	**21**	**28**	**Crop Harvest**
1	41.0	30.6	27.1	21.0	3.1
4	61.0	52.0	47.3	36.8	4.7
5	63.0	54.0	48.5	44.6	5.0

Xanthan, dextran, curdlan and pollulan are the four most important microbially produced exo-polysaccharides which play an important role in soil aggregarion. Enhancement in the soil aggregation depends on the concentration of the respective polysaccharide powder in soil.

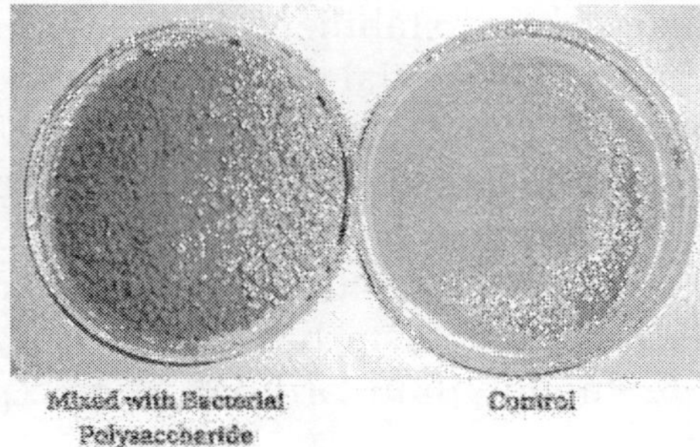

Figure 86: Improvement of soil aggregation by application of microbial polysaccharides

Table 42: Percent improvement of soil aggregation of diffrent sizes of soil aggregate

Treatment	Per cent improvement of aggregate size over control after 30 days (1% w/v)		
	1.0 mm	**0.5 mm**	**0.18 mm**
Polysaccharide from *Bacillus coagulans*	80.7	No change	No change
Polysaccharide from *Alcaligenes faecalis*	33.4	82.9	56.4

The improvent in soil aggregation depends on the size of the particle as well as the polysaccharide sources. In general, 33–81% improvement in aggregation was noticed in 1.0 mm size particles

(Table 42). The results clearly indicated (Figure 86) that microbial polysaccha-ride may help in aggregation process of soil. In simple term, aggregation is defined as a physical process wherein microparticles colloid and remain stack to each other to form clumps of particles called aggregate. Generally, one month time is required after application to form good aggregates. These aggregates are fractal in nature, highly porous, fragile and of varying size range. To form soil aggregates is very important especially under arid and semi-arid areas to get rid of wind erosion and moisture retention. The process of aggregation is influenced by particle density, collision frequency and the binding property of the nano-particle induced polysaccharides. Recently, increasing attention has been paid to microbial polysaccharides due to the possibility of quick and easy mass production. They can readily be prepared with the help of nanonutrients under various environmental conditions includeing climate, ecological adoption etc.

Further, most effective nanoparticles ZnO due to their smaller size and huge surface energy, stabilized in the solutions, which may be bioavailable. Zinc is an essential element for the organisms, but it is toxic at high concentrations. The solubility of Zn is size and pH dependent. It has been found that to enhance polysaccharide secretion from the microorganisms the safer concentration is up to 40 mg/L. The best size is below 40 nm. It was also seen that gum content and its viscosity in clusterbean seed of ZnO nanoparticle sprayed plants were higher than ordinary size ZnO particles and control (Table 43).

Table 43: Gum content and viscosity of clusterbean seeds at crop maturity

Treatment	**Gum content (%)**	**Gum viscosity (mPa)**
Control	28.86	3030
Mega ZnO	30.93	3050
Nano ZnO	31.03	3126
LSD (p=0.05)	1.5	4.8

Nanoparticles of Zn are the structural component of P mobilizing enzymes, such as phosphatases and phytase. They are known to stabilize the enzyme complexes in plants. So, this approach would enable in breaking of existing barriers in utilization of native P and reduced dependence on P fertilizers as well as more soil aggregation for moisture retention especially under sandy soils.

11.4: Factors Affecting Nanoinduced Polysaccharide Production

There are many factors which affect the nanoparticle induced polysaccharide secretion from different microorganisms. The most important ones are shown below:

11.4.1: Temperature

Temperature has a very significant effect on the polysaccharide secretion by the different microorganisms. In general, maximum polysaccharide secretion was noticed between 30 and 35˚C. Among the different bactetia *Bacillus coagulans* was found to be most efficient in nanoparticle-induced polysaccharide production. The effect of ZnO nanoparticles induced in extra secretion of polysaccharide from different bacterial cultures is presented as Table 44.

Table 44: Effect of temperature on polysaccharide (g) secretion

Temperature (˚C)	*Alcaligenes faecalis*	*Bacillus coagulans*	*Rhizobium sullae*	*Sinorizobium terange*
25	162.3	121.7	108.3	123.4
30	244.0	232.1	212.6	158.3
35	253.0	288.0	196.3	193.3
40	225.3	263.8	182.5	107.4
45	121.8	121.5	135.0	84.5
50	92.4	81.9	99.5	52.6
LSD (p=0.05)	8.3	9.1	7.7	6.9

Irrespective of organisms the induction of polysachharide production through the nanoparticles reduced considerably at below 25°C and above 40°C.

11.4.2: pH

The pH is an important criterion for production of nanoparticle induced polysaccharides. The more polysaccharide secretion was observed between pH 6 and 7.5 for bacteria and between pH 5 and 6 from fungi. The effect of pH was shown as Table 45. In general, marginal effect was observed below pH 4 and above pH 10.

Table 45: Nanoparticle induced extra polysaccharide secretion (mg/L) at different pH by bacteria

pH	*Alcaligenes faecalis*	*Bacillus coagulans*	*Rhizobium sullae*	*Sinorizobium terange*
4.0	10.4	9.1	95.7	47.3
6.0	181.1	165.7	225.7	235.1
6.8	252.8	272.6	208.4	220.6
7.5	205.1	156.7	201.0	175.2
8.0	135.5	107.7	103.6	168.8
10.0	51.3	40.2	81.3	97.7
LSD (p=0.05)	8.9	7.3	6.4	6.5

11.4.3: Incubation time period

The optimum incubation period for maximum induction of polysaccharide from polysaccharide producing organisms by nanoparticles was found between 72 and 96 hour. An appreciable induction of polysaccharide release by the microorganisms was noticed within 12 hours time. The effect of incubation time on different efficient bacterial polysaccharide secretion is presented (Table 46).

Table 46: Nanoparticle induced extra polysaccharide secretion by bacteria (mg/L) at different time of incubation

Incubation time period (h)	*Alcaligene s faecalis*	*Bacillus coagulans*	*Rhizobium sullae*	*Sinorizobium terange*
12	53.1	48.1	47.8	43.0
24	123.1	107.7	93.6	89.7
36	133.6	198.4	123.0	107.5
48	204.5	226.7	141.7	134.6
72	338.8	368.5	272.5	264.5
96	334.5	395.5	264.7	243.5
120	335.1	390.6	265.1	246.7
LSD (p=0.05)	12.1	10.9	8.2	7.6

11.4.4: Sucrose concentration

The concentration of sucrose has a great role in polysaccharide production. The effect of nanoparticles on polysaccharide secretion was more when sucrose concentration varies between 20 and 25%. It has been found that without the addition of any sucrose during nanoparticles induction, an appreciable nanoinduced polysaccharide production was also noticed (Table 47). The effect of sucrose was more on bacteria than polysaccharide releasing fungi.

Table 47: Effect of sucrose concentration on nanoparticle induced polysaccharide (mg/L) secretion

Sucrose Concentration (%)	*Alcaligenes faecalis*	*Bacillus coagulans*	*Rhizobium sullae*	*Sinorizobium terange*
0	109.7	94.6	78.2	73.6
5	123.1	107.7	93.6	89.7
10	137.4	119.4	112.5	103.8
15	193.6	174.3	174.7	164.7
20	264.7	286.4	246.8	223.6
25	263.8	268.7	247.3	221.9
30	263.6	268.9	247.1	222.7
LSD (p=0.05)	10.5	9.4	9.9	8.7

11.5: Effect of Zn Nanofertilizer on Production and Effect on Fungal Polysaccharide

The best organisms were identified to induce fungal polysaccharides production are *Chaetomium globosum* and *Curvularia lunata*. Preliminary experiment showed both the organisms released huge amount of phosphatases, phytase and polysaccharides. *Chaetomium globosum* and *Curvularia lunata* were able to secrete polysaccharides extracellularly and nano-size. ZnO enhanced extracellular polysaccharide concentration between 8 and 9 folds as compared to control where as bulk ZnO fertilizer increased polysaccharide secretion by only 10 to 376%. The effect of nano ZnO fertilizer with respect to exopolysaccharide was more pronounced in *Curvularia lunata* as compared to *Chaetomium globosum* (Table 48). Effect of nano-ZnO on intracellular polysaccharide concentration in both the fungi showed between 5 and 8 fold enhancement in concentration than control (Table 49).

Table 48: Effect of nano ZnO fertilizer on extracellular polysaccharide production by *Chaetomium globosum* and *Curvularia lunata*

Treatment	Extracellular polysaccharide concentration (μg mL^{-1})	
	Chaetomium globosum	*Curvularia lunata*
Control (without ZnO)	206.8	286.7
ZnO (mega size)	985.3 (376.3)*	315.8 (10.1)
Nano – ZnO	2016.5 (874.9)	2960.0 (932.5)
LSD (p = 0.05)	23.7	18.9

* Figures in parenthesis indicate % increase over control

Table 49: Effect of nano-ZnO fertilizer on intracellular polysaccharide concentration in *Chaetomium globosum* and *Curvularia lunata*

Treatment	Intra-polysaccharide concentration (μg mL^{-1})	
	Chaetomium globosum	*Curvularia lunata*
Control (without ZnO)	108.4	243.6
ZnO	284.9 (162.9)*	309.6 (27.1)
Nano-ZnO	986.4 (809.6)	1468.4 (502.9)
LSD (p = 0.05)	23.9	19.5

* Figures in parenthesis indicate % increase over control

In general, higher polysaccharide concentration was observed in *Curvularia lunata*. ZnO compound also increased intracellular polysaccharide concentration (27 and 163%). In absolute terms, more intracellular polysaccharide is being produced by *Curvularia lunata* under all these treatments. However, percent increase in response to either ZnO or nano-zinc oxide was more in *Chaetomium globosum*. This indicated that Zn is very important element for fungal polysaccharide production. Electrophoresis result indicates that *Chaetomium globosum* mainly secreted pullulan. The type of polysaccharide produced by *Culvuraria lunata* is dextran and pollulan.

The effect of nano-ZnO on phosphatases released by both the fungi showed up to 108% enhancement in acid phosphatase and

209% increase in alkaline phosphatase by *Chaetomium globosum* (Table 50). The phosphatases secretion by *Curvularia lunata* was enhanced by 50.8% (acid phosphatase) and 80.4% (alkaline phosphatase). Zn application as either normal oxide or nano-oxide was invariably associated with higher acid and alkaline phosphatases release. The increase was more in case of nano-zinc oxide application and that too in *Chaetomium globosum,* although the production of acid phosphatase by both the fungi was higher than the alkaline counterpart.

Table 50: Effect of nano-ZnO fertilizer on phosphatases secretion by *Chaetomium globosum* and *Curvularia lunata*

Treatment	Acid phosphatase (EU × 10^{-4})		Alkaline phosphatase (EU × 10^{-4})	
	C. globosum	*C. lunata*	*C. globosum*	*C. lunata*
Control	7.3	6.5	3.3	4.6
ZnO (1000 µg)	13.3 (82.3)*	8.2 (26.2)	8.9 (169.7)	6.8 (47.8)
Nano-ZnO (1000 µg)	15.2 (108.2)	9.8 (50.8)	10.2 (209.0)	8.3 (80.4)
LSD (p = 0.05)	0.21	0.24	0.30	0.18

EU: Enzyme unit; * Figures in parenthesis indicate % increase over control

The improvement of phytase activity due to application of the nanoparticle varied between 136.8 and 173.8% (Table 51), which was much higher than that due to ZnO application (11.9 to 89.5%). The release of phytase by both the fungi was much higher than phosphatases.

Table 51: Effect of nano-ZnO on phytase secretion by *Chaetomium globosum* and *Curvularia lunata*

Treatment	Phytase activity (EU × 10^{-2})	
	Chaetomium globosum	*Curvularia lunata*
Control	4.2	1.9
ZnO (1000 µg)	4.7 (11.9)*	3.6 (89.5)
Nano-ZnO (1000 µg)	11.5 (173.8)	4.5 (136.8)
LSD (p = 0.05)	0.19	0.27

EU: Enzyme unit;

* Figures in parenthesis indicate % increase over control

11.6: Reclamation of Salt Affected Soils

Nanofertilizer and nanoparticle induced polysaccharides have great advantages in reclaimed the salt-affected soils. The major advantages are documented under Table 52.

Table 52: Advantages from nanonutrients intervention in salt-affected soils

Processes	Advantages
Reducing salt concentration in soil solution	Enhancement Na^+ removal, soil stability and hydraulic characteristics using polymeric carriers in nano-reclaimants via clay binding processes
Improving soil drainage	Removing Na^+ from soil solution leads to improve crop growth by improving soil structure in subsurface and sub-soils using nano-Ca2+, nanoferrites as well as biofriendly nanopolymers
Replacing Na^+ by Ca^{2+}	Nano-Ca^{2+}, nano-Mg^{2+} and nano-K^+ can be used in removing Na^+ from soil because of the high selectivity and spontaneous reactions (negative ΔG_0) for Ca^{2+} over Na^+ for all clay minerals
Changing carbonate chemistry	Capping/encapsulating Na_2CO_3 with nanopolymers and nano-composites selectively may yield products which are insoluble and may be leached through preferential flows as well as formation of nanoorganic carbonates could be another possibility
Prevention of Na_2 CO_3 formation	Nanomaterials including nanocalcium carbonates, nano-Ca^{2+}, nano-Mg^{2+}, nano-K^+ and nano iron oxides may be used in preventing Na_2CO_3 formation in soils
Addition of K^+	It could be used some nanomaterials like nano-K^+ in clay minerals like illite in accelerating ion exchange reactions (Ca^{2+}, Mg^{2+} and Na^+) to reduce exchangeable sodium saturation
Solubilizing of $CaCO_3$	Calcium carbonate could possibly be solubilized by the application of nano-Ca^{2+}, nano-Mg^{2+} and nano-Fe oxides
Precipitation	Nanopolymers and nano-organic substances may be used in forming complexes or insoluble salts with harmful ions and precipitated

Processes	Advantages
Common ion effect	Nano-iron oxides, nano-Ca^{2+} and nano-Mg^{2+} may be applied to counter adverse effect of Na^{+} as a well-known phenomenon

Therefore, it could be concluded that, nanomaterials have a great impact on different agroecosystems including both positive and negative effects. Concerning the negative effects, nanomaterials may cause many toxic effects on plants, soil microbes, or aquatic organisms with elevated doses. Whereas, the positive side for nanomaterials includes using these materials in remediation both polluted soil and water, improvement of soil health in delivery nanonutrients or other substances, in alleviation the damage effects of abiotic stresses on plants, etc. So, a wise and sustainable application of nanomaterials in agroecosystems should be used.

It is reported that, nanomaterials can be used for sustainable crop production, in reducing loss of nutrients, in suppressing of diseases and then enhancing the yields. It is also documented that, nanomaterials have several benefits including enhancement of plant growth stages starting from seed germination, seedling vigor, initiation and growth of roots, and photosynthesis reaching to flowering under low concentrations. Concerning the protection of plants against the oxidative stress, nanomaterials may behave or mimic the role of antioxidative enzymes like peroxidase, superoxide dismutase and catalase. Phytotoxicity may happen under higher concentration of nanomaterials as a result of the generation of reactive oxygen species. Also, the accumulation of these reactive species may damage both of membrane of cells and proteins as well as nucleic acids. Some nanofertilizers had excellent roles in alleviating the abiotic stress, such as nano-silica, nano-silver, nano-zinc, and nano-titanium as follows:

Nano-Si plays an important role in the mitigation of salt stress. Many reports about the ability of nano-Si to counteract the negative effects of salt on plant growth rates were recorded. The use of nanosilica can be more efficient than the large particles and facilitate its uptake through plant root system. Also, nano-silica affect xylem humidity, water translocation and enhance turger pressure, thus leaf relative water content and water use efficiency will be increased. It is reported that, nano-SiO_2 particles absorbed better and faster than micro SiO_2, Na_2SiO_3, and H_4SiO_4 when applied on seeds or roots of maize plants. On the other hand, silicon deposition in the tissues helps to mitigate water stress by reducing the rate of transpiration, enhancing photosynthesis process by improving light receiving efficiency through keeping the leaf rigid and erect, preventing chlorophyll degradation and increasing tolerance insect attack. Silicon is also known to reduce uptake of Na^+ by improving K^+: Na^+ ratio and ameliorate the toxicity of other heavy metals. The application of 8 g L^{-1} of nano-SiO_2 significantly enhanced tomato seed germination potential and improved percent seed germination, average germination time, seed germination index, vigor index, seedling fresh weight and dry weight. Nano-silicon is responsible for a number of improvements which lead finally to enhancement plant growth and productivity under both salinity and drought. This includes nutrient elements homeostasis, modification of gas exchange attributes, osmotic adjustment by regulating the synthesis of compatible solutes, stimulation of antioxidant enzymes and gene expression in plants. Nano-Si fertilizer also enhances plant growth under heavy metal stress. The suggested mechanisms include reducing active heavy metal ions in growth media, chelation and stimulation of antioxidant systems in plants, complexation and co-precipitation of toxic metals with Si, compartmentation and structural alterations in plants and regulation of the expression of metal transport genes. However, these mechanisms might be associated with plant species, genotypes, metal elements, growth conditions, duration of the stress imposed and so on. Various studies described the ameliorative effect of Si application on plants under heavy metals or dangerous toxic elements, like Al, Cd, Pb, Cr or Cu. The results indicated that nano-Si is more efficient than common (bulk) Si in terms of alleviation of the toxic effects of lead (Pb) on rice growth, it prevents Pb transfer from rice roots to the shoot system of the plant and reduce Pb accumulation in grains, especially in high-Pb accumulating-cultivars and the soils with high levels of Pb pollution.

The possibility of application of silver nanoparticles as fertilizer in improvement of plant productivity was assayed. The improvement in plant growth and enhancing photosynthetic quantum efficiency and chlorophyll content was reported at lower doses. Silver nanoparticles are also used as antimicrobial agents to manage plant diseases. Application of silver nanoparticles has been found quite effective in improving resistance against salinity during germination of fennel and cumin. It is reported that exposure to Ag nanofertilizers alleviated the adverse effects of salt stress and improved the germination, root length and seedling fresh and dry weight of tomato under NaCl stress. The role of important nanomaterials under abiotic stresses is presented as Table 53.

Table 53: Role of nanomaterials in different abiotic stresses

Abiotic stresses	Role of nanomaterial
Salinity	• Keep leaf rigid and erect to improve light receiving • Prevent chlorophyll degradation
Drought	• Osmatic adjustment by regulating the synthesis of compatible solutes • Helping regulation of gene expression • Enhance xylem humidity and water translocation
Cold	• Reducing transpiration rate • Stimulation of antioxidant system
Heavy metal	• Reduction of Na uptake
Heat	• Co-precipitation
Flooding	• Complexation

Zinc oxide nanoparticles (nano-ZnO) as fertilizer, is among the important key micronutrient required for the optimum growth and development of plants which carries vital metabolic reactions within the plants to promote growth and development. Despite its role in the growth and development of plants, Zn also plays a vital role in reducing toxic heavy metals uptake by plants, thereby prevents plants from the heavy metal toxicity, such as Cd. Zinc also has an important role in plants survival under environmental stress conditions, where it plays a role in enhancing or preparing the plants to tolerate drought stress. It plays a considerable role in stomatal regulation owing to its ability to maintaining membrane integrity and retaining potassium content of the cells as well as a role in plant water relations. In an experiment on chickpea, it is indicated that Zn deficiency reduced the plant ability to adjust osmotic pressure under drought conditions. Zinc influence auxin production, where it represents a coenzyme for tryptophan synthesis, a precursor to indole acetic acid production which promotes the root development and water deficiency tolerance in plants. The supplementation of micro-element in small size (nano-size) facilitates its uptake by plant and reduces the used dose. So, the previous functions of zinc can be achieved more efficiently when using nano zinc. It is investigated that, in full irrigation treatment, the highest seed and biomass were related to bulk ZnO treatments, but under water stress conditions the highest biomass and water use efficiency was related to nano-ZnO treatment. A number of experimental results indicated that application of nano-ZnO as fertilizer increased seed yield and water use efficiency.

Titanium has significant biological effects on plants, being beneficial at low levels, but toxic at higher concentrations. Photocatalytic degradation of pesticides with TiO_2 and other catalyst has shown promise as potential water remediation method. NanoTiO_2 can improve photosynthetic apparatus and enhance a plant's ability to capture sunlight, that affects the manufacture of pigments and the transformation of light energy to active electron and chemical activity and thus increase photosynthetic efficiency as in maize, especially under drought stress. Nano-TiO_2 as fertilizer observed to promote the growth of spinach through an increase in photosynthetic rate and nitrogen metabolism in spinach. Nano-TiO_2 can enhance plant water and nitrogen use and stimulate some antioxidant enzyme activities, such as SOD, POD and CAT like in canola and wheat plants. The investigation on the effects of nano-TiO_2 and nano-SiO_2 on chemical constituents and yield characteristics of cotton plant under drought stress showed that pretreatment of cotton plants under drought stress with nano-TiO_2 or nano-SiO_2 caused increasing of pigments content, total soluble sugars, total phenolics, total soluble proteins, total free amino acids, proline content, total reducing power, total antioxidant capacity and antioxidant enzyme activities and enhancement of yield characteristics. The optimum concentration of nano-TiO_2 and nano-SiO_2 to alleviate the drought stress in cotton plants was 50 ppm and 3200 ppm, respectively. Finally, it can be concluded that foliar application of nano-TiO_2 or nano-SiO_2 could improve the drought tolerance of cotton plants. On the other hand, application of nano-TiO_2 can decrease cadmium (Cd) stress and increase Cd uptake in soybean plants. The suggested mechanism is the formation of new bonds in the plant tissue with the Cd/nanoTiO_2 particles. Therefore, it could be concluded the role of low concentrations of nanomaterials in ameliorating the plant abiotic stress through some mechanisms, such as protecting plants against different abiotic stresses by (1) activating the plant cell signals due to over production of reactive oxygen species (ROS) and/or reactive nitrogen species (RNS) and (2) stimulating plant defense system (enzymatic and non-enzymatic antioxidant activities), accumulating the osmolytes and free amino acids as well as nutrients, whereas these nanomaterials may cause toxicological effects on plants under high concentrations.

Further Studies

Lin, S., Reppert, J., Hu, Q , Hudso, J. S., Reid, M. L. and Ratnikova, T. A. (2009). Uptake, translocation, and transmission of carbon nanomaterials in rice plants. *Small* **5:** 1128–1132.

Laware, S. L. and Raskar, S. (2014). Effect of titanium dioxide nanoparticles on hydrolytic and antioxidant enzymes during seed germination in onion. *Int. J. Current Microbiol. Appl. Sci.* **3:** 749–760.

Kim, J. H., Oh, Y., Yoon, H., Hwang, I. and Chang, Y. S. (2015). Iron nanoparticle-induced activation of plasma membrane H+ - ATPase promotes stomatal opening in Arabidopsis thaliana. *Environ. Sci. Tech.* **4:** 1113–1119.

Sharif, F., Westerhoff, P. and Herckes, P. (2013). Sorption of trace organics and engineered nanomaterials on to wet land plant material. *Environ. Sci. Process Impacts* **15:** 267–274.

Tarafdar, J.C., Agrawal, A., Raliya, R., Kumar, P., Burman, U. and Kaul, R.K. (2012). ZnO nanoparticles induced synthesis of polysaccharides and phosphatases by *Aspergillus* fungi. *Adv. Sci., Eng. Med.* **4:** 1–5.

12

Nanonutrients on Crop Yield

Macronutrients (*e.g.*, nitrogen (N), phosphorus (P), potassium (K), magnesium (Mg), sulphur (S) and calcium (Ca) as well as micronutrients like iron (Fe), zinc (Zn), boron (B), copper (Cu), manganese (Mn), molybdenum (Mo) have been combined with nanomaterials for the purpose to deliver an accurate amount of nutrients to the crops and minimize their bulk requirements with extra benefits of decreasing purchasing and transportation costs. These macronutrient and micronutrient nanofertilizers comprise one or more nutrients in encapsulated form with specific nanomaterial. NPK consumption in the agriculture sector is projected to increase 265 million tons in 2020. As such, there is an urgent need to carry out research from a practical point of view to develop new fertilizers with high nutrient efficiency and being friendly to the environment to replace the conventional macronutrient fertilizers. As a nitrogen source, urea-modified zeolites, hydroxyapatite and mesoporous silica nanomaterials have been investigated as slow/control release nanofertilizers showing promising results. Biosafe nanofertilizer was developed as a source of P that is a nanostructured water-phosphorite suspension (particle size of 60–120 nm). It was the first phosphatic nanofertilizer acquired from raw phosphorite of Tatarstan's Syundyukovskoe deposits through ultrasonic material dispersion. It was observed that the morphometric indices, fresh yield and fruit yield as well as production quality of tested plant species increased several-fold. It has been found that the present generations of fertilizers (nanofertilizers) are most promising for increasing crop yield and productivity as compared to the other generations of fertilizers (Table 54).

Table 54: Percent average yield increase with different generations of fertilizer

Generation	Type of fertilizer	Average expected yield increase over control (%)
First Generation (1G)	Chemical fertilizer	12–18
Second Generation (2G)	Organic fertilizer	8–12
Third Generation (3G)	Growth stimulator	5–10
Fourth Generation (4G)	Nanofertilizer	24–32

The average expected yield was much higher under nanofertilizers as compared to chemical and organic fertilizers as well as different growth stimulators irrespecticve of crops. A comparison of the average yield increase due to application of different generations of fertilizers showed the per cent increase in yield obtained so far due to application of nanofertilizer especially, biosynthesized ones is much higher as compare to any generations of fertilizers.

12.1: Size Directs the Function of Nanofertilizers in Plants

It is axiomatic to state that the function of nanomaterials is directed by their size. It is observed through the diverse applications of nanomaterials that, size guides their success in almost all the instances. Considering their applications in plants, various studies have reported different results when nanomaterial of different sizes was applied to plants. For instance, ZnO nanofertilizer (approx. 50nm in size) were more growth-friendly with lesser inhibition of rooting from seedlings of rapeseed compared to ZnO macroparticles and Zn^{+2} ions which inhibited root growth to a higher extent. Different studies are available which have studied the size dependency, indicated that Ag nanofertilizers having 15 nm average particle size enhanced the growth of soybean compared to smaller (2 nm) or higher (up to 80 nm) particle sizes. This effect was found to be related to the synthesis of mitochondrial proteins. Proteomic analysis revealed that the abundance of voltage-dependent anion channel protein was increased upon exposure to 135 nm Al_2O_3 nanomaterials. It was concluded that varying sizes of Ag nanofertilizer have varying effect on growth parameters and physiology of plants. Similarly, in case of CeO_2 nanofertilizers, size was found to result in altered biotransformation in Lactuca plants and reduced toxicity of different CeO_2 nanoparticles. Furthermore, although it is already established that the nanoparticles are effective antimicrobial agents and play a prominent role during *in vitro* cultures, the potential of nanofertilizer particles in the getting rid of microbial contamination has also been shown to depend on their size and dimensions. Several studies have reported that size is a defining factor for the potency of nanoparticles in killing microbes during *in vitro* cultures. However, from numerous studies it has been concluded that nanofertilizer particle size less than 20 nm is most effective in the plant growth and development.

12.2: Concentration Affects the Function of Nanofertilizers in Plants

Nanoparticles can act in a concentration dependent manner, and the effects of the nanofertilizer on plant growth and seed germination is directly related to their concentrations. In some cases, higher concentrations may prove helpful for the growth, while in majority of the cases higher concentrations have deteriorating effect on plants. It was found in banana plants that the shoot regeneration frequency was enhanced with an increase in concentration of ZnO nanofertilizer. Similarly, there was a report that high concentration (up to 20 mg L^{-1}) of ZnO nanofertilizer applied through foliar spray increased the biomass of alfalfa, cucumber and tomato plants. Application of other nanoparticles, such as CeO_2 nanofertilizer (250 mg L^{-1}) and In_2O_3 nanofertilizer (500 mg L^{-1}) resulted in a significant increase in the biomass accumulation of *A. thaliana* seedlings. However, generally, lower concentrations of nanoparticles have been found to be more plant-friendly. For example, lower concentrations of ZnO nanofertilizer were shown to be more beneficial during seed germination in wheat. Lower concentrations are favourable since they do not hamper cell division and seedling growth as reported in onion. Similarly, lower concentration of ZnO nanoparticles (1mg L^{-1}) applied to *Stevia rebaudiana* cultures resulted in an elevated level of steviol glycosides. However, higher concentrations of ZnO nanoparticles (1000 mg L^{-1}) reversed the positive effects of these cases.

Higher concentrations of nanofertilizers have been shown to produce deteriorating effects on growth and development of different plants. It has been shown that higher concentration (1000 mg L^{-1}) of ZnO nanofertilizer produced yellow colouration of beans that represents harbouring of *Pseudomonas chlororaphis* in roots of beans. ZnO nanofertilizer in higher concentrations (up to 1500 mg L^{-1})

added to *in vitro* cultures of *Brassica nigra,* inhibited seed germination, root and shoot lengths significantly. Similarly, higher concentrations of Ag nanoparticles were shown to cause a reduction in both root and shoot length of *A. thaliana* as well as Au nanoparticles (100 mg L^{-1}) hampered the growth of *A. thaliana.* CeO_2 nanoparticles (250 mg L^{-1}) and In_2O_3 nanoparticles (1000mg L^{-1}) decreased seedling growth and increased ROS in *A. thaliana.* With increase in concentrations of Zn in plant system in the form of ZnO nanoparticles leads to toxic effects on the growth of plants. In cabbage application of ZnO nanofertilizer resulted in a dose dependent inhibition of seed germination. The concentrations of nanoparticles also affect secondary metabolism in plants. Higher concentrations of ZnO nanoparticles (1000 mg L^{-1}) lowered stevioside production in shoot cultures of *Stevia rebaudiana.* It was also demonstrated that the production of specific secondary metabolites is dependent upon the concentration of nanoparticles applied. The highest content of flavonoids, phenolic compounds and anthocynanins was produced in MS medium having lower concentrations of ZnO nanoparticles (25, 75 and 100 mg L^{-1}), the small size of ZnO nanoparticles, greater toxicity was caused in roots of wheat. The increase in ZnO nanoparticles increased the number of abnormal cells in garlic cultures.

The exposure time is also important during the application on nanoparticles in plants. Permanent or long-term exposure to nanoparticles has a detrimental effect on plants. For instance, prolonged exposure of alfalfa to ZnO nanoparticles inhibited the growth and accumulation of biomass in the plant. Another aspect of the detrimental role of nanoparticles is the disturbance they cause in the rhizosphere. ZnO nanoparticles, for example, kills soil microbes thus affecting the nitrogen and carbon fixing role they play in *Phoenix dactylifera.* In addition to ZnO nanoparticles, other nanoparticles, such as Ag, Au, CuO, CeO_2, and Al_2O_3 nanoparticles have also shown detrimental effect in various plant species. Ag nanoparticles have been reported to destroy cell morphology in Asian rice, and increase pollen lethality in kiwi plant. Similarly, Au nanoparticles decreased the growth and induced biochemical stress in *Brassica juncea* which is linked with increased ROS formation as a response to Au nanoparticles supplementation during growth. It also resulted in dark roots, yellow leaves and decreased biomass during *in vitro* cultures of Barley. The detrimental effects in response to exposure to CuO nanoparticles include reduced shoot and root growth of soybean, lignification of root cells in *A. thaliana,* hampered seed germination in cucumber, and decreased germination rate and biomass in *Oryza sativa.*

In addition to the above, other nanoparticles and their toxic effects include CeO_2 nanoparticles induced smaller and weaker tomato seedlings, altered nutritional value of coriander, and Al_2O_3 nanoparticles induced decreased cell viability through ROS generation in tobacco cell suspension cultures. Therefore, accurate nanofertilizer concentration is needed to be applied to the plants to get maximum benefit. It may change the phytochemical profile of the plants and help in growth, and development which ultimately contributes to crop yield. The proposed mechanism of nanofertilizer particles is shown as Figure 87.

12.3: Micronutrient Nanofertilizers

Micronutrients are those elements that are required by the plant in trace/low quantities, but are essential to maintain vital metabolic processes in plants. Micronutrients are those elements that are essential for plant growth, but are required in much smaller amounts than those of N, P and K. The micronutrients are boron (B), copper (Cu), iron (Fe), manganese (Mn), molybdenum (Mo), zinc (Zn) and chloride (Cl). Even though micronutrients are required in minute quantities, they are essential for healthy plant growth and profitable crop production. Micronutrient deficiency is widespread in many Asian countries due to the calcareous nature of soils, high pH, low organic matter, salt stress, continuous drought, high bicarbonate content in irrigation water and imbalanced application of NPK fertilizers. Some of the adverse effects of micronutrient deficiency-induced stress in plants include low crop yield and quality, imperfect plant morphological structure (such as fewer xylem vessels of small size), widespread infestation of various diseases and pests, low activation of phytosiderophores, and lower fertilizer use efficiency. The results from various experiments suggest that natural zeolites, particularly chabazite and bentonite minerals, have a high potential for Zn and Fe sorption with a high capacity for slow release fertilizers.

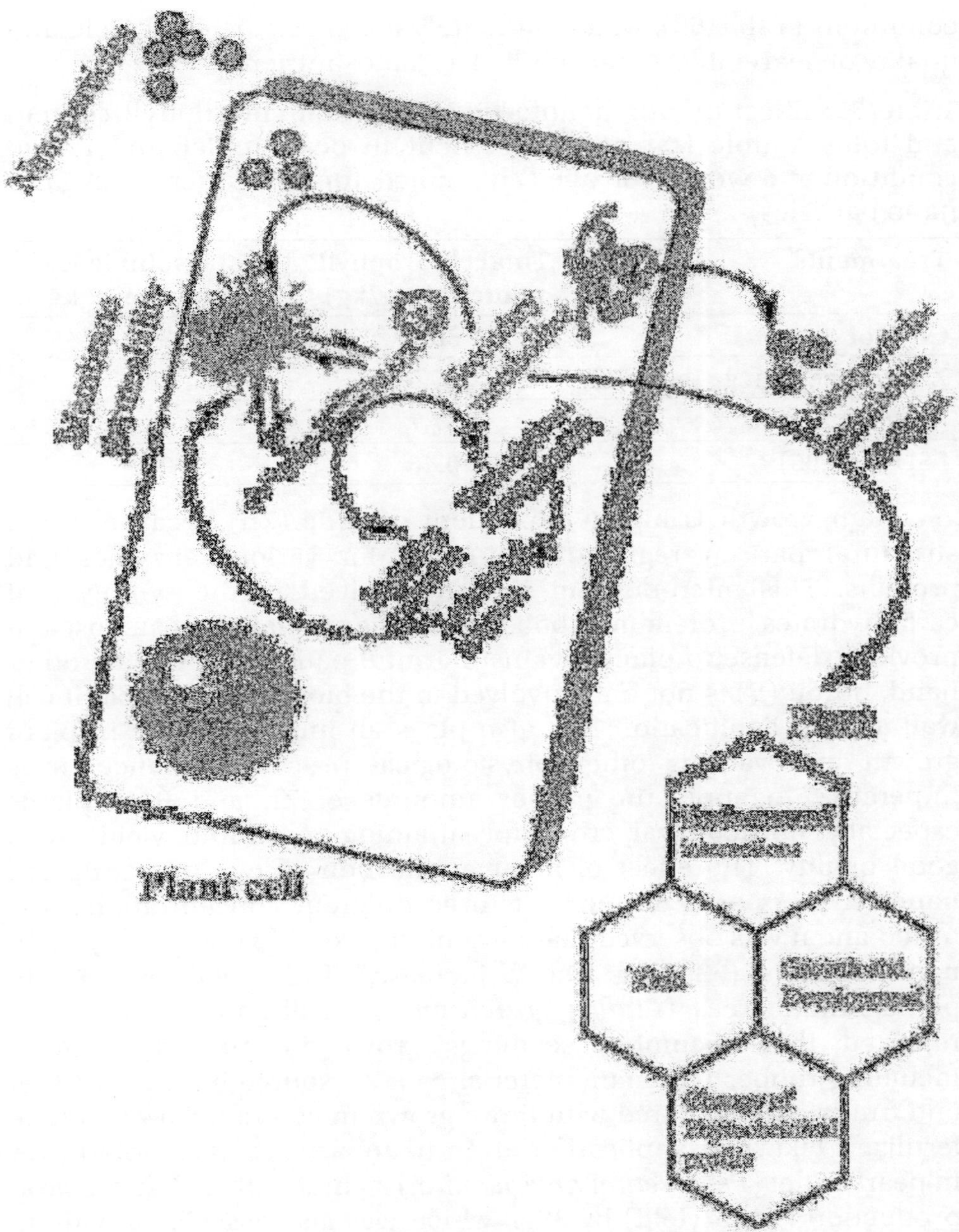

Figure 87: The proposed mechanism of action of nanofertilizer particles on growth, cell signalling and secondary metabolism in plant cell cultures

It has been found that application of nano-Zn fertilizer are much more powerful to enhance total chlorophyll and leaf protein content of many plants including pearl millet even under arid environmental conditions (Table 55), which ultimately enhances the extra yield and quality of pearl millet when applied as nano-nutrients.

Table 55: Effect of zinc nanofertilizer on total chlorophyll content and total soluble leaf protein content in pearl millet under field condition at 6 week crop age (Zn applied 10mg/L either as mega or nano particles)

Treatments	Total chlorophyll content (mg/kg)	Total soluble leaf protein (mg/ kg)
Control	30.3	37.7
Zn fertilizer (mega particle)	31.5	43.6
Nano-Zn fertilizer	37.7	52.3
LSD (p=0.05)	0.46	0.49

Plant growth is highly dependent on zinc (Zn) because it is a structural part or regulatory co-factor for various enzymes and proteins. This micronutrient is also involved in the synthesis of carbohydrates, protein metabolism, and the regulation of auxins, and provides defense to plants against harmful pathogens. On the other hand, boron (B) is not only involved in the biosynthesis of plant cell wall and its lignification, but also plays an important role in plant growth and various other physiological processes. Hence, it is imperative to apply the proper amounts of Zn and B to plants especially horticultural crops for attaining maximum yields with good quality. The effect of foliar application of two micronutrient nanofertilizers of Zn and B at three different concentrations was tested and it was observed that low amounts of B (34 mg tree^{-1}) or Zn nanofertilizers (636 mg tree^{-1}) increased fruit yield by 30% in pomegranate trees (*Punica granatum* cv. Ardestani). It was also reported that cucumber seedlings grown in nutrient solution including rubber type nanomaterial as a Zn source increased shoot and fruit yield compared with those grown in commercial Zn-sulfate fertilizer. Likewise, application of Zn nanoparticles as nanofertilizer in pearl millet (*Pennisetum americanum*) significantly enhanced crop production (grain yield) by 38%, which was also associated with an improvement of 15% in shoot length, 4% in root length, 24% in root area, 24% in chlorophyll content, 39% in total soluble leaf protein and 12% in plant dry biomass compared to the control in a period of 6 weeks. A considerable increase in yield by using Zn nanoparticles as a nutrient source in rice, maize, wheat, potato, sugarcane and sunflower was noticed. Moreover, stabilized maghemite nanoparticles applied through irrigation in solution form in soil as a nanofertilizer improved the growth rate and chlorophyll contents compared to the control (chelated iron) in *Brassica napus*. The best response of nano-ZnO on mung bean was observed at 20 mg/L and on chickpea at 1 mg/L, beyond these concentrations the seedling showed retardation in growth and development thereby suggesting the toxic effect of nano-ZnO particles. With the increasing in the concentration of nanofertilizer Zn application, the accumulation has been increased in the plants (Table 56), which can be accurately estimated through ICPMS or ICPAES. But no significant improvement in accumulation was noticed with the application of very high concentration of Zn nanofertilizer. It has been found that the tolerance capacity of nanofertilizer varies with the crops which were much higher at mung bean than chickpea. Therefore, it is suggested that concentration should be optimized before application to a particular crop.

Table 56: Concentration of zinc element in mung bean and chick pea seedling

Mung bean		**Chick pea**	
Concentration added (mg/L)	**Concentration recovered(mg/L)**	**Concentration added (mg/L)**	**Concentration recovered (mg/L)**
0	2.80	0	2.68
10	7.15	1	2.80
20	12,17	2	3.71
50	15.35	5	4.23
100	18.72	10	6.14
500	23.38	20	9.85
1000	28.28	50	13.13
2000	31,18	100	16.23

The slow release of Zn is attributed to the sparingly solubility of minerals and sequestration effect of exchange, thereby releasing trace nutrients to zeolite exchange sites where they are more readily available for uptake by plants. Experiments on foliar treatments with ion exchanged zeolite on winter wheat for three years showed zinc zeolites treatment had more increase of raw protein than that of copper-zeolite treatment. The zeolite in soil can aid in the release of some trace nutrients and in their uptake in plants. The release of cationic micronutrients has been enhanced by the presence of zeolite in neutral soil. The concentration of Cu and Mn in Sudan grass was significantly related to the zeolite/P-rock in experimental systems that used two different NH_4 saturated zeolites, two different soils and two different forms of P rock. Zinc oxide nanoparticles were shown to enter the root tissue of ryegrass and improved the germination. It has also been reported that Zinc-rich nanofertilizers (ZnO) increased the level of IAA in roots (sprouts), which in turn indicate the increase in the growth rate of plants as zinc is an essential nutrients for plants. Boron is an essential micronutrient for plants, but can be toxic to organisms when accumulated in high concentrations.

Iron (Fe) is also an important nutrient required by plants in minute quantities for maintaining proper growth and development. Although it is required in trace amounts, its deficiency or excess leads to impairment in key plant metabolic and physiological processes, thereby leading to reduced yield. Therefore, application of Fe is imperative to optimize yields in horticultural crops. In this respect, the effect of iron oxide nanoparticles and ferric ions was studied at different concentrations on the physiological and molecular changes in *Citrus maxima* plants. It was observed that iron oxide nanoparticles entered the plant roots, but their translocation from root to shoot did not occur. Among the different levels used, 20 mg/L iron oxide nanoparticles had no impact on plant growth, while 50 mg/L significantly improved the chlorophyll contents and root activity by 23% and 24%, respectively, compared to controls. In contrast, 100 mg/L negatively influenced all these characteristics, thus indicating that the effect of iron oxide nanoparticles is concentration dependent. The impact of different micronutrient nanofertilizer under non-sressful conditions are documented (Table 57).

Table 57: Impact of micronutrient nanofertilizers on some selected crops

Nano-fertilizer	Concentration used	Crop	Effect
Fe	50, 500 and 2000 mg/L	Cucumber	Dose-dependent effects on biomass and antioxidant enzymes
Fe	10, 20 mg/L	Lettuce	Reduced growth and chlorophyll contents, and increased antioxidant enzyme activities
Fe	30–60 mg/L	Garden pea	Improved seed mass and chlorophyll content

Nano-fertilizer	Concentration used	Crop	Effect
Cu	0, 100, 500 mg/L	Squash	Higher ionic Cu found in media amended with bulk Cu than with nanofertilizer Cu
Cu	130, 660 mg/Kg	Lettuce	Increased shoot/root length ratio
Cu	0, 10, 20 mg/L	Lettuce	Negative effects on nutrient content, dry biomass, water content and seedlings growth
Cu	0–1000 mg/L	Cucumber	Reduced growth and increased antioxidant enzymes
Cu	10–1000 mg/L	Grasses, Radish	DNA damage, growth inhibition
Cu	50–500 mg/L	Tomato	Improved fruit firmness and antioxidant content
Cu	0, 20, 80 mg/Kg	Cilantro	Reduced germination and shoot elongation
Cu	100, 250, 500 mg/L	Bean	Growth inhibition and nutrition imbalance
Cu	100–500 mg/L	Garden pea	Reduced plant growth and enhanced reactive oxygen species production and lipid peroxidation
Zn	1000 mg/kg	Cucumber	Root tip deformation and growth inhibition
Zn	500 mg/kg	Garden pea	Decreased chlorophyll and H_2O_2 contents
Zn	1000 mg/L	Spinach	Growth reduction
Zn	1000 mg/L	Tomato, Egg plant	Reduced fungal disease
Zn	100, 200, 500 mg/L	Chili, Pepper	Improved germination
Zn	0–400 mg/kg	Coriander	Improved pigment contents and defense responses
Zn	5,10, 20 mg/L	Onion	Inhibition of root growth
Zn	10 mg/L	Wheat, Potato	Improved root growth, dry matter and yield

Nano-fertilizer	Concentration used	Crop	Effect
B	2 mg/L	Pearl millet, Mung bean	Improved enzyme activities, root growth and crop yield
Mo	6 mg/L	Moth bean	Improved N fixation, enzyme activities and crop yield

Manganese (Mn) plays a vital role not only in metabolic and physiological processes but it also provides the plant with the ability to endure various environmental stresses by acting as a co-factor of various enzymes. It is also essential for photosynthesis, the biosynthesis of ATP, chlorophyll, fatty acids and proteins, as well as secondary metabolites, such as lignin and flavonoids. The effects of Cu, Zn, Mn, and Fe oxide nanoparticles were assessed in low concentrations (< 50 mg/L) as micronutrients, on the germination of lettuce (*Lactuca sativa*) seeds. The results showed that CuO nanoparticles were slightly more toxic than Cu ions, while the toxicity of ZnO nanoparticles was similar to that of Zn ions. However, MnO nanoparticles and FeO nanoparticles were not only less toxic than their ionic counterparts, but they also stimulated the growth of lettuce seedlings from 12% to 54%. A combination of precision deficient irrigation and application of nanofertilizers play a critical role for enhancing the efficient use of water and improving the productivity of maize production systems in Mediterranean semi-arid regions.

Molybdenum nanofertilizer particles have enhanced the rhizosphere microbial population under chickpea at a concentration between 2 and 10 mg/L. The higher microbial activity in the rhizosphere may be due to the higher root exudates as a result of increased photosynthesis. The population of different organisms (fungi, bacteria, actinomycetes, *Nitrosomonas, Nitrobacter* and *Azotobacter*) was examined at critical growth period under different treatment. The results (Table 58) indicate highest fungal population when nano-molybdenum was applied at 4 mg/L concentration. Moreover, bacteria, actinomycetes population were also noticed higher with the application of 4 mg/L concentration of nano-molybdenum. More *Nitrobacter* population was observed at 2 mg/L concentration although higher *Nitrosomonas* population at 4 mg/L concentration was observed. The results, in general, showed appreciable increase in microbial population over control in most of the cases. The overall results showed 4 mg/L concentration may be the best to keep optimum microbial population in the rhizosphere, however, it is clear from the results that 10 mg/L concentration of nanomolybdenum application resulted decline in most of the beneficial microbial population like fungi, bacteria and *Nitrobacter* as compared to control soil, which indicates their adverse effect on these groups of organisms. However, the *Azotobacter* population has been improved till 10 mg/L molybdenum application.

Table 58: Effect on microbial population under chickpea at critical growth stages with the foliar application of nano-molybdenum (nMo) fertilizer (CFU-colony forming unit)

nano-Mo fertilizer concen-tration	Fungi (CFUx10^3 /g)	Bacteria (CFUx10^5 /g)	*Nitro-somonas* (CFUx10^2 /g)	*Nitro-bacter* (CFUx10^2 /g)	Actino-mycetes (CFUx10^3/ g)
0	170	71	385	302	195
2 mg/L	180	102	350	320	170
4 mg/L	245	115	410	312	188
6 mg/L	198	104	402	265	183
10 mg/L	145	82	200	224	184
LSD (p=0.05)	35.8	21.7	41.2	18.9	23.4

Total C in shoot and root was 28.8% and 19.9% higher in pearl millet and clusterbean plants treated with nano-ZnO fertilizer. Reasons for higher C fixation in presence of nano-ZnO are not clear. But, two indirect effects of ZnO on plants may influence photosynthesis (i) high biocompatibility and fast electron transfer kinetics of nano-ZnO which may increase rate of some biochemical reactions and (ii) protection of plant membrane from the damage by the UV radiations due to lipid peroxidation. The malonialdehyde content, an indicator of lipid peroxidation, was 4.3 and 10.7% less in nano-ZnO treated pearl millet and clusterbean leaves, respectively than the corresponding controls. Application of nano-ZnO significantly decreased the C exuded from roots in both pearl millet and clusterbean resulted more biomass accumulation.

CuO nanofertilizer particles affect the growth of soybean and chickpea seedlings at different concentrations. The maximum growth was found at 100 mg/L concentration for soybean and 60 mg/L concentration for chickpea seedlings. Beyond this concentration, the growth was inhibited. The effective growth at certain optimum concentration and inhibited growth beyond this concentration may be attributed to the accumulation and uptake of nano-CuO particle by the roots. It was found that the accumulation and uptake of nano-particles was dependent on the exposure concentrations. In particular, the exposure of plants to nano-materials and the impacts of, such an exposure on plant systems could open a new direction research on nanofertilizer technology. Future studies should be directed towards understanding the mechanism uptake and translocation of nanofertilizer particles, physical and chemical properties of nano-particles in rhizosphere etc.

12.4: Macronutrient Nanofertilizers

Macro nanofertilizers (N, P, K, Ca, Mg, S) are nutrient carriers of nano-dimensions ranging from 5–40 nm and capable of holding bountiful of nutrient ions due to their high surface area and release it slowly and steadily that commensurate with crop demand.

12.4.1: Nitrogen

About 90% of Indian soils are deficient in N and thus there is a universal response of the addition of nitrogenous fertilizers. Indiscriminate use of nitrogenous fertilizers caused major detrimental impacts on the diversity and functioning of the non-agricultural ecosystems due to eutrophication of fresh water and marine water. In addition, there can be gaseous emission of N reacting with the stratospheric ozone and the emission of toxic ammonia into the atmosphere.

Nitrogen is an important component of many important structural, genetic and metabolic compounds in plant cells. It is a major component of chlorophyll, the compound by which plants use sunlight energy to produce sugars from water and carbon dioxide (*i.e.* photosynthesis). It is also a major component of amino acids, the building blocks of proteins. Some proteins act as structural units in plant cells, while others act as enzymes, making possible many of the biochemical reactions on which life is based. Nitrogen is a component of energy-transfer compounds, such as ATP (adenosine triphosphate) which allows cells to conserve and use the energy released in metabolism. Finally, nitrogen is a significant component of nucleic acids, such as DNA, the genetic material that allows cells (and eventually whole plants) to grow and reproduce. Since nitrogen exists in three general forms - organic nitrogen compounds, ammonium (NH_4^+) ions, and nitrate (NO_3^-) ions which is available for plant. Most of the nitrogen is not available for plant completely.

This is because negatively charged nitrate normally does not have higher affinity for soil particle surfaces and thus does not readily sorb on soil. To overcome the problems associated with the nitrogen leaching during fertilization, different approaches, such as polyolefin resin coated urea, neem coated urea, sulphur coated urea were taken to control the N release. However, slow-releasing fertilizers are often expensive and the release of N is slow at the time of high N requirements. N loss can also be reduced using cation exchanger as additives in fertilizer to control NH_4^+ release. The retention and timely release of needed nutrients by zeolite improves overall crop yield. Clinoptilolite zeolite (CZ), a porous mineral with high cation exchange capacity (CEC, up to 300 c mol (p+) kg^{-1}) and with great affinity for NH_4^+, has been used to reduce NH_3 emission from farm manure, and to eliminate NH_3 toxicity to plants. Amendment of clinoptilolite zeolite (CZ) to sandy soil has been reported to lower NO_3^- and NH_4^+ concentrations in the leachate and to increase moisture retention in the soil due to increased soil surface area and CEC. Urea nitrogen has been the most used N-source due to lower cost per unit of N. But N use efficiency of urea may be reduced because of losses from agricultural system by volatilization of ammonia to atmosphere. This is one of the main factors responsible for the low efficiency of urea, and may reach extreme values, close to 80% of N applied.

Ammonium retained by CZ is generally subjected to slow release through cation exchange and nitrification in soil. Clinoptilolite zeolite (CZ) has been used to reduce NH_3 emission from farm manures and as NH_4^+ loaded exchange fertilizer because of its high CEC. Application of $(NH_4)_2SO_4$ loaded into CZ was observed to minimize N leaching and to increase N utilization by crops in sandy soils compared with $(NH_4)_2SO_4$ alone. It was concluded that clinoptilolite not only improves nitrogen fertilization efficiencies, but also reduces nitrate leaching by inhibiting the nitrification of ammonium to nitrate. The zeolite has the potential for reducing ammonia volatilization by sequestering ammonium-N on exchange cites. An addition of 6.25% zeolite resulted in a 50% reduction in ammonia volatilization. An additional potential advantage is that zeolite bound ammonium is a good slow-release N source for plants. Unlike the commonly used fertilizers, the plant-growth material dramatically reduces loss of nutrients to groundwater and the environment. The urea-N losses can also be reduced using zeolites as additives in the fertilizers to control the retention and release of NH_4^+. There are reports in the literature showing that the addition of zeolite to the source of N can improve the nitrogen use efficiency and it was demonstrated the feasibility of using surfactant-modified zeolite (SMZ) using hexadecyltrimethyl ammonium as fertilizer carrier to control nitrate release, and concluded that SMZ is a good sorbent for nitrate, whereas slow release of nitrate is achievable. In laboratory, the batch experiments were conducted to investigate the sorption of nitrate from aqueous solutions using hydrothermally synthesized and surface modified zeolite nano-particles. The ability of surface modification with hexadecyltrimethyl ammonium (HDTMA) and dioctadecyldimethyl ammonium greatly enhance the sorption and slow release of nitrate. It was confirmed that mixing urea with zeolite and sago waste water had great advantage over urea alone as the mixture encouraged the formation of ammonium and available nitrate ions over ammonia. The mixture also improved retention of exchangeable ammonium and available nitrate within the soil. Recently, the nutrient release pattern of nanofertiliser formulations carrying nitrogen had shown that nano-fertilizer released nutrients upto 1200 hrs, while conventional fertilizer could support only for 300–350 hrs. The data had clearly demonstrated that zeolite as nano-fertilizer could be an ideal strategy to promote N use efficiency.

12.4.2: Phosphorus

Phosphorus (P) is an essential nutrient both as a part of several key plant structure compounds and as a elements in the catalysis of the conversion of numerous key biochemical reactions in plants. It is often the limiting nutrient in agricultural ecosystems due to its low availability in soils. Phosphorus is the most important nutrient for its function as an energy transfer and storage in plants. Adenosine di- and triphosphates (ADP and ATP) are the most important forms of phosphates that are involved in the energy transfer. It is also an important structural component of phosphor- proteins, phospholipids, sugar phosphates, coenzymes, nucleic acids, and metabolic substrates in plants. Supply of P at the early stage of crop is essential for its development of reproductive structures. Some specific growth factors that have been associated with phosphorus are: stimulated root development, increased stalk and stem strength, improved flower formation and seed production, more uniform and earlier crop maturity, increased N-fixing capacity of legumes, improvement in crop quality and increased resistance to plant diseases. After examining the solubility and cation-exchange in mixtures of rock phosphate and NH_4^+ and K-saturated clinoptilolite, it was revealed that mixtures of zeolite and phosphate rock had the potential to provide slow-release fertilization of plants in synthetic soils by dissolution and ion-exchange reactions. The zeolites (clinoptilolite), when saturated with mono-valent nutrient cations, such as NH_4^+ and K^+, it enhances the solubility of phosphate rock (PR). The efficiency of fertilizer P use by crops ranged from 18 to 20% in the year that it is applied. The remaining 78 to 80% becomes part of the soil P pool which is released to the crop over the following months and years. The release of P from fertilizer-loaded unmodified zeolite, surface modified zeolite (SMZ) and from solid KH_2PO_4 was performed using the constant flow percolation reactor. The results showed that the P supply from fertilizer loaded SMZ was available even after 1080 h of continuous percolation, whereas P from KH_2PO_4 was exhausted within 264 h. The results indicated that SMZ is a good sorbent for $PO4^{3-}$, and a slow release of P is achievable. These properties suggested that SMZ has a great potential as the fertilizer carrier for slow release of P. In general, phosphate ($H_2PO_4^-$) can be released to plants from phosphate rock composed largely of the calcium phosphate mineral apatite by mixing the rock with zeolite having an exchange ion, such as ammonium.

Phosphorus (P) is part of amino acids, nucleic acids, phospholipids, and high-energy molecules, such as adenosine triphosphate (ATP) in all living organisms. In plants, P is an essential element required for development and reproduction, and it is one of the main components of the fertilizers necessary to sustain modern agriculture. In soils, the concentration of inorganic P (available to plants) ranges from 35 to 70% of the total P. This form of P shows low diffusion and high fixation rates in soils through ligand exchange by 1:1 clay minerals, Fe and Al oxides and hydroxides, and is thus precipitated as Fe, Al, and Ca phosphates. Additionally, phosphate fertilizers are obtained from phosphoric rock, a non-renewable resource, and whose reserves are running out. Consequently, the importance of promoting the efficiency of P uptake and use in agricultural crops like rice, wheat etc. becomes evident, while supplying nanofertilizers with phosphorus represents an alternative.

The approximate reaction in soil solution is as follows: (P-rock) + (NH_4 - zeolite) = (Ca-zeolite) + (NH_4) + (H_2PO_{4-}). The zeolite takes Ca^{2+} from the phosphate rock, thereby releasing both phosphate and ammonium ions. Unlike the leaching of very soluble phosphate established equilibrium, the fertilizers (for example, super phosphate), the controlled-release phosphate is released of a specific chemical reaction in soil. As phosphate is taken up by plants or by soil fixation, the chemical reaction releases more phosphate and ammonium in the attempt to re-establish equilibrium. The rate of phosphate release is controlled by varying the ratio of P rock to zeolite. Phosphorus is also released from the rock by the lowering of soil pH as ammonium ions are converted to nitrate. It was reported that zeoponic is a plant demand driven nutrient delivery system and from this phosphate (PO_4^{3-}) and other nutrients released by controlled dissolution of synthetic apatite. N, P and K were delivered when plant needs them from zeoponics. Basically, the process is a combination of dissolution and ion exchange reactions. The absorption of nutrients from the soil solution by plant roots drives the dissolution and ion exchange reactions, pulling away nutrients as needed. The zeolite is then "recharged" by the addition of more dissolved nutrients. In effect, zeoponic systems increase nutrient retention, reduce environmental nutrient losses and reduce fertilizer requirements by establishing a replenishable and balanced nutrient supply in the plant root zone. The PO_4^{3-} release pattern of surface modified using various nanoclays and zeolite in a percolation reactor was studied. Nano-formulations have been shown to release phosphate for an extended period of 40–50 days and the conventional fertilizer let out nutrients only upto 10–12 days. The results obtained so far suggests that surface modified zeolite could be potential strategy to promote P use efficiency which hardly exceed 18–20% in conventional system.

12.4.3: Potassium

Plant processes and development cannot function properly without potassium. It is one of the three major nutrients required for quality assurance, excellent appearance and great harvest of crops. It is the key to various processes ranging from reproduction, growth, photosynthesis, protein synthesis, enzyme activation, water, and stomata regulation among others. When there is deficiency of potassium in the soil for plant absorption, the plants become prone to diseases, and the overall growth and development process of the crop is affected. Other symptoms may include appearance of spots under the leaves, curling of the leaf tips, interveinal chlorosis, etc. Potash or potassium fertilizers can be obtained from several sources either organically or inorganically. Soil potash could be improved from compost made by agricultural wastes or food by-products, such as banana peel. Wood ash obtained from burning of wood can also be a good source of potash and can be used to enrich the soil. Majority of inorganic potash used in agriculture can be obtained from potassium chloride (KCl). Other contributors are potassium sulfate, potassium nitrate, and potassium–magnesium salts. To process the fertilizer, potassium bearing ores are obtained from natural mineral deposits and crushed to reduce the sizes. Clay deposits are further removed and flotation process of separating the potassium from other compounds is carried out. Other methods, such as sizing, refining, and crystallization are also carried out to obtain the fertilizers in different stages.

Potassium is vital to many plant processes. While potassium is not a constituent of any plant structures or compounds, it plays a part in many important regulatory roles in the plant. It is essential in nearly all processes needed to sustain plant growth and reproduction. Potassium plays a vital role in photosynthesis, translocation of photosynthates, protein synthesis, control of ionic balance, regulation of plant stomata and water use, activation of plant enzymes and many other processes. It is known to activate at least sixty enzymes involved in plant growth. Plants deficient in potassium are less resistant to drought, excess water, high and low temperatures. They are also less resistant to pests, diseases and nematode attacks. Potassium is also known as the quality nutrient because of its important effects on quality factors, such as size, shape, colour, taste, shelf life, fiber quality and other quality measurements. Some natural zeolites contain considerable amounts of exchangeable K^+ that can enhance plant growth in potting media. For example, the slow release effect of K from K-zeolite has been reported. The application of chemical fertilizer at the rate 625 kg ha^{-1} mixed with zeolite 125 kg ha^{-1} indicated the largest amount of potassium in the soil because zeolite had potential to adsorb potassium from chemical fertilizer and reduce it from leakage. Natural zeolites are highly selective for K^+ rather than for sodium or divalent cations, such as calcium and magnesium, because of the location and density of the negative charge in the structure and dimensions of the interior channels. It was suggested that among all, potassium is the only element with the highest ion exchange capacity of 216 c mol (p+) Kg^{-1} and consequently, it is very easily released from the crystal zeolite structure into the soil solution, eventually increasing its total content in the soil. The slow and steady release of K from nano-zeolite may be due to the ion exchange ability of the zeolites with selected nutrient cations, zeolites can become an excellent plant growth medium for supplying plant roots with additional vital nutrient cations and anions. It was suggested that the zeolite can be "recharged" by the addition of more dissolved nutrients. Their selectivity of ion exchange on zeolite was determined in an order of K^+ > NH4 $^+$ > Na^+ > Ca^{2+} > Mg^{2+}. The potassium sorbed on zeolite increases with the increase in equilibrium K concentration. In general, the zeolites dominated by exchangeable K may be well suited for plant growth applications. It has been reported that used potassium (K^+) - loaded zeolite (K-Z) as a slow-release fertilizer and investigated the growth characteristics of hot pepper as well as the changes in the nitrogen and potassium contents of the evaluated soil. The experiment on slow release of K fertilizer showed the potassium fixation in soil and dynamic equilibrium collectively sustain the availability of potassium in soil, which can be further when use as nanosize particles as well as the release of nutrients also regulated and improved further.

12.4.4: Other macro-nutrients

Sulphur (S), calcium (Ca) and magnesium (Mg) are secondary nutrients required in relatively large amounts for good crop growth. S and Mg are needed by plants in about the same quantities as P, whereas for many plant species, Ca requirement is greater than that for P. Sulphur reactions in soil are very similar to N reaction, which are dominated by organic or microbial fractions in the soil. In contrast, Ca^{2+} and Mg^{2+} are associated with the soil clay fraction and behave similarly to K^+. Zeolite showed slow release fertilizer for calcium and magnesium too. It was reported that, zeolite improved calcium and magnesium content in soil. Moreover, zeolite is able to freely exchange nutrient ions, such as calcium and magnesium. The feasibility of using surfactant-modified zeolite (SMZ) as fertilizer additives to control sulfate release, was tested in batch and column leaching experiments which indicated that SMZ could be a good carrier for sulfate. Thus, leaching sulfate can be greatly reduced and slow release of sulfate can be achieved if SMZ is used as fertilizer additives. Secondary nutrients are less studied as the extent of deficiencies is very meager or found in isolated products.

Magnesium is an alkaline earth metal with oxidation number +2, essential to all living cells, where they play a crucial role in manipulating important biological polyphosphate compound like ATP, DNA and RNA. Hundreds of enzymes require Mg for catalytic function. Mg is also the metallic ion at the centre of photosynthetic pigment, chlorophyll, and thus a common addition to fertilizers. It is a nutrient for plants, microorganisms and animals. Magnesium oxide (MgO) is an important material for its utilization as catalysts, toxic waste remediation, and additives in refractory, paints and superconductor products. Furthermore, it is also a valuable crystal employed in the electronic and optical fields. Mg as nanofertilizer has found very effective in photosynthesis, absorption of solar radiation and leaf protein content.

12.5: Nanobiofertilizers

Biofertilizers are formulations or preparations containing one or more microorganisms that enhancing soil productivity, by fixing atmospheric nitrogen, solubilizing phosphorus or stimulating plant growth through synthesis of growth-promoting substances. Therefore, nanobiofertilizers could be defined as the integration of biofertilizers with nanostructures or nanoparticles in order to improve the growth of plants. To achieve this goal is essential to control the delivery of biofertilizers in the soil and extend the useful life of formulations.

Some of the most important aspects in nanobiofertilizers development are the interaction between nanoparticles and microorganisms, the shelf life of biofertilizers and its delivery. The interaction between gold nanoparticles and plant growth promoting rhizobacteria was shown to exert positive effects. By contrast, silver nanoparticles cannot be used with biofertilizer because it causes adverse effects on biological processes in microorganisms, like alteration of cell membrane structure and functions. On the other hand, the shelf life of biofertilizers is a limiting factor in these formulations and the use of nanomaterials can improve it. Use of nanoformulations can be helpful to enhance the stability of biofertilizers with respect to desiccation, heat, and UV inactivation. For example, polymeric nanoparticle coatings can be used to develop formulations resistant to desiccation and consequentially improve the useful life of these products.

Moreover, nanomaterials can be used to improve the delivery of biofertilizers to soil and plants. Trials using hydrophobic silica nanoparticles to the water-in-oil emulsion have shown an improvement in the delivery of the product, as well as an enhancement in shelf life by reduction of desiccation. Nevertheless, there is a fundamental problem in nanobiofertilizers production, since the nanoscale constructs are usually smaller than cells. In this regard, macroscopic filters made of radially aligned carbon nanotube walls, which can absorb *Escherichia coli*, could be used as a promising technology to collect other microorganisms from fermentation processes and deliver them to the plants. Therefore, nanobiofertilizers can solve some limitations of biofertilizers, but this technology still requires further research and development.

There is tremendous improvement in soil microbial populations with the introduction of biosynthesized nanofertilizers (Table 59). The bacterial population was found to be encouraged more due to different nanofertilizer applications at recommended doses.

Table 59: Percent improvement in microbial population in the crop rhizospheres with the recommended bio-nanonutrient doses

Bio-nanonutrient	Fungus	Bacteria	Actinobacteria
N	15.6	32.9	29.0
P	23.7	43.8	35.9
K	12.2	21.1	18.2
Mg	10.1	37.2	18.0
Fe	31.5	44.2	17.6
Mo	6.7	11.2	19.0
Zn	47.8	39.9	31.0

Doses applied: N-80ppm, P-40ppm, K-40ppm, Mg- 20ppm, Fe-30ppm, Mo-6ppm, Zn-10ppm

Similarly, considerable improvement of root length, root area, dry biomass and nodulation (in case of leguminous crops) was observed under 14 different crops tested. The average improvement of different root parameters is listed below (Table 60).

Table 60: Effect of Bionanofertilizers on different root parameters

Bio-nanofertilizer	Concentration sprayed	Root parameters (% improvement over control)
N	80 ppm	Length : 2– 12% Area : 4–20% Dry biomass : 2–6% Nodulation (legumes): 5–58%
P	40 ppm	Length : 25–32% Area : 18–21% Dry biomass : 7–10% Nodulation (legumes): 17–68%
Mg	20 ppm	Length : 2–8% Area : 5–15% Dry biomass : 1–5% Nodulation (legumes): 4–32%
Fe	30 ppm	Length : 3–12%

Bio-nanofertilizer	Concentration sprayed	Root parameters (% improvement over control)
		Area : 6–18% Dry biomass : 3–7% Nodulation (legumes): 8–51%
Zn	10 ppm	Length : 1–5% Area : 3–12% Dry biomass :2–6% Nodulation (legumes): 7–45%

Nanofertilizer has found greater effect so far tested on different crops, both under research and farmers field. The result available so far with different biosynthesized nanofertilizer was summarized (Table 61). The results reflected more effect on cauliflower with the application on nano P and nano-Zn fertilizer.

Table 61: Yield increase (%) due to nanofertilizer application under different crops (modified after Tarafdar *et al.*, 2018)

Crops	Nanofertilizer applied	Research field	Farmers field
Capsicum	P, Zn	19–24	—
Castor	P, Zn	—	24–37
Clusterbean	P, Zn, Fe, Mg	17–36	21–37
Cauliflower	P, Zn	38–44	47–54
Maize	P	23–32	25–38
Moth bean	P, Zn, Fe	17–31	18–27
Mung bean	P, Zn, Fe, Mg	17–42	23–38
Pearl millet	P, Zn. Fe, Mg	17–48	18–43
Potato	P, K, Zn	24–32	22–37
Rice	P	22–28	25–33
Tomato	P, Zn	18–23	25–29
Wheat	P, K, Zn	21–32	27–32

Doses applied: P-40 mg/L, Fe-30 mg/L, K-40 mg/L, Mg-20 mg/L, Zn-10 mg/L

The changes in physio-chemical properties of the pearl millet growing soils with the foliar application of biosynthesized nitrogen nanoparticles at different concentrations (0 to 100 mg/L) resulted in no significant changes in organic carbon percentage (Table 62) as compared to control. In general, marginal decline in soil available phosphorus at N-20 and N-100 concentration, otherwise there was no significant difference of available P status with the application of nano-nitrogen at different concentration used. There were absolutely no changes in electrical conductivity of soil with the foliar application of nano-nitrogen. Similar results on soil pH were also observed. The results clearly suggest that foliar application of biosynthesized nano-nitrogen has absolutely no adverse effect on soil physio-chemical properties even with the application of 100 mg/L concentration.

Table 62: Changes in physio-chemical soil properties of pearl millet under different doses of nano-N application at critical growth stage

Treatment	pH	EC (dSm^{-1})	Organic C (%)	Olsen P (mg/ kg)
Control	8.76	0.12	0.32	11.05
N-20 (mg/L)	8.96	0.11	0.34	10.21
N-50 (mg/L)	8.87	0.11	0.34	11.36
N-80 (mg/L)	8.92	0.14	0.35	11.64
N-100 (mg/L)	9.01	0.11	0.33	10.29
LSD (p=0.05)	0.12	0.03	0.04	0.63

A significant increase in beneficial enzymes (acid and alkaline phosphatases, esterase, and dehydrogenase) and microbial population (fungi, bacteria, actinomycetes, *Nitrosomonas, Nitrobacter, Azotobacter*) was noticed with the application of optimum dose (80 mg/L of N nanofertilizer) at critical growth period (6 weeks) in pearl millet. Application of 80 mg/L concentration of nano-nitrogen on two weeks old pearl millet plant resulted in an appreciable improvement in root area (539%), root length (159%), root perimeter (46%), number of tips (14%), average root diameter (76%) and total biomass (157%). Several studies indicated the possibility of biosynthesized nano-N for efficient use as nano-fertilizer. The rhizosphere soil analysis data on Pearl millet at critical growth stage (six weeks old) were reflected more beneficial enzyme activities (alkaline phosphatase, acid phosphatase, esterase and dehydrogenase) at 80 mg/L concentration. The more dehydrogenase (4.5%) and esterase (40%) activities also indicate that more microbial build up in the rhizosphere with the application at optimum concentration.

Nanobiofertilizers offers a new set of tools to manipulate the genes using nanoparticles, nanofibres and nanocapsules. Properly functionalized nanomaterials serve as a platform to transport large number of genes as well as chemicals that trigger gene expression in plants.

12.6: Effect of Nanocapsules on Plant Growth

The kaolinite nanocomposite prepared by the methods of organic material intercalation under certain temperature and pressure can be used as coating material of slow and controlled release fertilizer because of its strong adsorption and thickness to macronutrients and organic carbon. The addition of these nano composites fertilizers benefits the soil, and raises the utilized efficiency of fertilizers because of its excellent characteristics. The physical adsorption and chemical combination occurred between nutrient elements and nanocomposites due to surface reaction and small size reaction of nanocomposites. They formed the efficient multifunctional fertilizer, which heightened the adsorption of nutrient elements by plants, lowered the leaching in soil, and the fixation of particle in the soil.

The nano-subnano composites significantly affected and controlled the structure and penetrability of the soil, increased the organic matter granule of the soil, improved fertilizer storage and water holding capability in the soil, promoted the action of microorganisms, regulated the ratio of C/N, enhanced the fertility of the soil and so on. Increased yields of various crops have been claimed for fertilizers that are incorporated into co-chleate nano-tubes. The use of nitrogen by urea hydrolysis has been controlled through the insertion of urease enzymes into nanoporous silica,

Non-functionalized carbon nano-tubes affected root length more than functionalized nano-tubes. Carbon nano-tubes were found to penetrate tomato seeds and affect their germination and growth rates. The germination was found to be dramatically higher for seeds that germinated on medium containing CNTs compared to control. It has been found that CNTs are able to penetrate thick seed coat and support water uptake inside seeds, a process which can affect the seed germinations and the growth of seedlings. The development of functional nanoscale films and devices has the potential to produce significant gains in the nutrient use efficiency and crop production. The capability of single-walled carbon nano-tubes (SWNTs) to penetrate the cell wall and cell membrane of intact plant cells has been demonstrated in many crops. Cellular uptake of SWNT/fluorescein isothiocynate and SWNT/DNA conjugates, demonstrating that SWNTs also hold great promise as nano-transporters for walled plant cells. Moreover, the results suggested that SWNTs could deliver different cargoes into different plant cell organelles.

In general, on an average the seeds incubated in CNTs for two days possessed a moisture content of 58% as compared to the 39% for the control. This is due to the new pores generated during penetration of seed coat by CNTs which added better water permeation. Another reason for more moisture was that the efficient gating of water channels by the CNTs in the seed coat. It was observed that CNTs induced increased biomass, rapid germination, and growth rates will get better yield under short duration. Furthermore, nano-silica particles absorbed by roots have been shown to form films at the cell walls, which can enhance the plant resistence to stress and leads to improved yields. For this, zeolites can be used as the carrier for the nanofertilizer which releases nutrients in slow and steady state.

12.7: Nano-composites

Nanocomposites have been developed in order to supply wide range of nutrients in desirable properties. These compounds are capable of regulating the inputs depending on the conditions of soil or requirements of crops. Zinc-aluminium layered double-hydroxide nano-composites have been used to the controlled release of nutrients that regulate plant growth.

In soil niches, nano-materials are porous and hydrated, and as, such they control moisture retention, permeability, solute transport, and availability of plant nutrients in soil. These nano-materials also control exchange reactions of dissolved inorganic and organic species between the soil solution and colloidal surfaces. The physico-chemical properties of nano-composites surfaces provide much of reactivity to soil biological and abiotic processes. The organic agents and clay formed nano-composites through hydrogen bond combination. In general, the sizes of nano-composites are ranging from 10 to 20 nm. It was reported that nano-composites can effectively be used to improve the yield and quality of the crops especially, in cereal crops like wheat and maize.

12.8: Clay-based Nanofabricated Material

Clay nanofabricated material could be used in controlling release of nitrogen (zeolites, hydroxyapatites), nutrient release and microflora control in soil rhizospheres, ion transport in soil-plant system, controlling emission of dust and aerosols from agricultural soil, zeoponics, and precision water farming. Some clay based nanofabricated fertilizers for agricultural application are presented (Table 63).

Table 63: Clay-based product and its application

Clay-based product	Application
Nanocides	• Pesticides encapsulated with nanomaterial for greater release • Nanoemulsion for greater efficiency
Buckyball fertilizer	Ammonia from buckyballs
Nanoparticles	Adehesion-specific nanoparticles for removal of *Campylobacter jejuni* from poultry
Food packaging	Airtight plastic packaging with silicate nanoparticles
Use of agricultural waste	Nanofibres from cotton waste for improved strength of cotton
Nanosensors	• Contamination of packaged food • Pathogen detection
Precision agriculture	Nanosensors linked to GPS for real-time monitoring of soil
Livestock and fisheries	Nanoveterinary medicine (nanoparticles, buckyballs, dendrimers, nano-capsules for drug delivery, nanovaccine, smart herbs, nanocheck for cleaning fish ponds, iron nanoparticles feed for fish)

Transmission Electron micrographs showing adsorbed phosphorus onto kaolin in nano-form. Phosphorus is spread uniformly as white globular deposition onto kaolin surface (Figure 88).

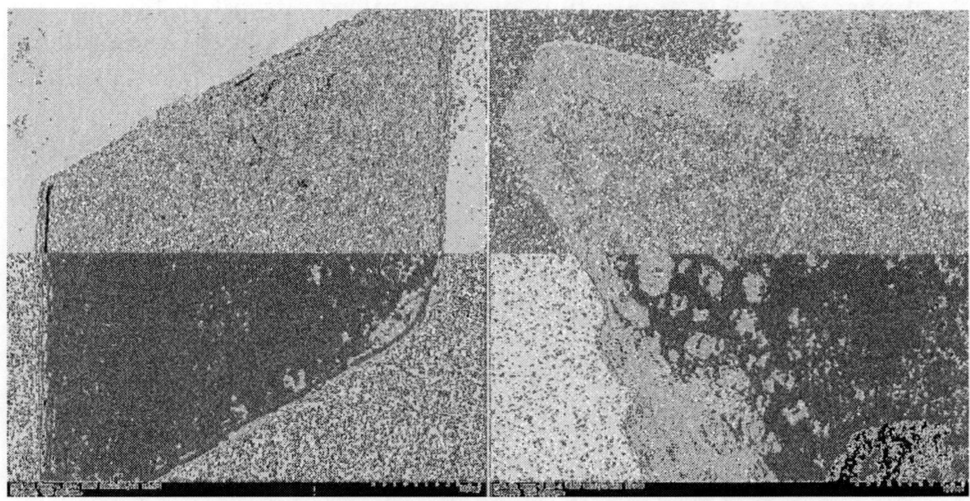

Figure 88: TEM micrograph of nanofabricated P-kaolin

With nano-fertilizers emerging as alternative to conventional fertilizers, build ups of nutrient in soils and thereby eutrophication and drinking water contamination may be eliminated. In fact, nanofertilizer technology has opened up new opportunities to improve nutrient use efficiency and minimize the cost of environmental protection. It has helped to divulge that plant roots and microorganisms can directly lift nutrient ions from solid phase of minerals (that includes so called susceptible (*i.e.* easily weatherable) as well as non-susceptible minerals. Nanofertilizer application opening new avenues to improve nutrient use efficiency and to reduce nutrient builds up in soils and thereby reducing its load in surface water bodies and checking contamination in drinking water.

Further studies

Babaei, K., Sharifi, R. S., Pirzad, A. and Khalilzadeh, R. (2017). Effects of bio-fertilizer and nano Zn-Fe oxide on physiological traits, antioxidant enzymes activity and yield of wheat (*Triticum aestivum* L.) under salinity stress. *J. Plant Interac.* **12:** 381–389.

Davarpanah, S., Tehranifara, A., Davarynejad, G., Abadía, J. and Khorasani, R. (2016). Effects of foliar applications of zinc and boron nanofertilizers on pomegranate (*Punica granatum* cv. Ardestani) fruit yield and quality. *Scientia Hort.* **210:** 1–8.

Janmohammadi, M., Amanzadeh, T., Sabaghnia, N. and Dashti, S. (2016). Impact of foliar application of nano micronutrient fertilizers and titanium dioxide nanoparticles on the growth and yield components of barley under supplemental irrigation. *Acta Agricultur. Slovenica.* **107:** 265–276.

Nair, R., Varghese, S. H., Nair, B. G., Maekawa, T., Yoshida, Y. and Kumar D. S. (2010). Nanoparticulate material delivery to plants. *Plant Sci.* **179:** 154–163.

Selva Preetha, P. and Balakrishnan, N. (2017). A Review of nano fertilizers and their use and functions in soil. *Int. J. Curr. Microbiol. Appl. Sci.* **6:** 3117–3133.

13

Safety Assessment of Nanonutrients

Nanofertilizers definitely have incredible potential to revolutionize many aspects of human life; the benefits may come with some price. One of the major questions faced by the world before accepting the nanofertilizers is, whether the unknown risk of nanoparticles involving their environmental and health impact prevail over their potential benefits. The risk associated with the application of nanofertilizers is yet to be fully evaluated before implementing this technology. The new branch of research on Nano-toxicology has been developed which is responsible in assessing toxicological potential as well as promoting safe design and use of nanoparticles or nanofertilizers. A systematic and thorough quantitative analysis regarding the potential of health impacts, environmental clearance, and safe disposal of nanoparticles can lead to improvements in designing further application of nanofertilizers.

Although no direct human disease has been linked to nanoparticles/ nanofertilizers so far, early experimental studies indicate that nanoparticles could initiate adverse biological responses that can lead to toxicological outcomes. Nanofertilizers enable nanoparticles to enter in the food chain allowing their distribution in every organism related to the food chain. Therefore, we have to be very careful about the doses of the application, so that it does not exceed the required for plants and environmental damage. As all substances, from arsenic to table salt, are toxic to plants, animals, or humans at some exposure level, this would not limit their use in various applications which are designed keeping in mind the critical exposure concentration. So, likewise we may fix the doses of particular nanofertilizers for application to different plants, so that it does not have any ill effect to the plants, animals, humans as well as soil microbes and environment. The risk assessment of nanofertilizers has to be performed on a case-by-case basis.

Therefore more research is needed on the aspect. The ultra-small sizes that make the nanoparticles of immense usefulness, unfortunately the same characteristic is a basic cause of several adverse effects and may also represent significant hazards to environment, animals, human beings and plants when use non-judiciously.

Nanoparticles pesticides, fertilizers or in other formulations, using as air borne may deposit on leaves and floral parts of plants. They may plug stomata and create a fine physical and toxic barrier layer on stigma preventing pollen tube penetration. They may also enter the vascular tissue and impair translocation of water, minerals and photosynthates. Animals may inhale nanoparticles deep into lungs resulting into various ill effects and disorders. The particles may enter the bloodstream.

Nanopesticides may reduce environmental contamination through the reduction in pesticide application rates, but they may also create new kinds of contamination of soils and waterways due to enhanced transport, longer persistence and higher toxicity.

Airborne nanoparticles, represents some specific hazards for human health, they may enter the body through the respiratory system. Due to entry of nanoparticles in to lungs and blood stream, there is possibility of pro-inflammatory effects on the basis of their ability to cause inflammation; protein fibrillation (caused by specific types of carbon nanotubes); induction of genotoxicity, and side effects in terms of biodistribution. The particles may also translocate to other organs, especially to lungs and kidneys and may cause plugging of veinlets, Bowmans capsule of nephron etc. The nature of chemicals loaded on the nanoparticles would further aggravate the associated adverse effects.

13.1: Nanofertilizers used as Recommended Doses

Many experiments were conducted to find out the health hazards of nanofoods with the application of recommended doses of nanonutrients. The results showed that nanofertilizers with recommended doses have no adverse effect on per cent seed germination and soluble seed protein content (Table 64).

Table 64: Effect of seed germination and soluble seed protein content with the application of recommended doses of nanofertilizer under important arid crops (NS-non-significant)

Treatment	Seed germination (%)				Soluble protein (mg/kg) seed			
	Cluster-bean	Moth bean	Mung bean	Pearl millet	Cluster-bean	Moth bean	Mung bean	Pearl millet
Control	90	82	89	80	105.6	98.5	78.5	70.0
Mega Zn	93	82	92	80	105.0	99.2	79.1	74.3
Nano Zn	92	84	91	82	108.1	106.9	79.8	76.1
Mega Fe	89	83	90	81	106.3	108.5	78.6	75.0
Nano Fe	97	84	93	84	115.4	105.3	78.9	75.4
Mega P	94	82	90	81	105.3	100.4	79.0	78.2
Nano P	93	85	93	85	110.2	103.6	81.1	77.9
LSD (p=0.05)	NS	NS	NS	NS	NS	NS	NS	NS

In general, microbial population increased significantly with the application of recommended doses of different nanofertilizers. There was no adverse effect on body weight, grain consumption rate and blood pH of mice with the feeding of nanofertilizer sprayed plants as compared to control (without nanofertilizer) plants.

Pre-clinical safety evaluation of pearl millet and mung bean grown with recommended doses of different nanofertilizers reported:

- No pre-terminal deaths in any groups investigated
- No abnormal clinical signs, behavioural activity etc. were observed in animals which received nanofertilizer sprayed test materials
- No significant effect on feed intake, body weight gain was observed between the control and nano-fertilizer treated materials intake
- There were no changes in gross necropsy and any organ weights.

In a nutshell, the plants grown under recommended doses of nanofertilizer treatment did not induce any adverse effect in rats even after feeding more than two and half times of limit dose.

The health effects of nanofertilizer derived food were evaluated in female Wistar rats using repeated sub-acute toxicity study. The nanoparticles were sprayed during the vegetative cycle of wheat plant. Later on, these matured wheat plants were harvested and selected as test substance for 90 days sub-acute toxicity study. Based on the body weight means, Wistar rats (Female) were randomly assigned to various experimental groups including control in groups of 6 animals each [Group A-Normal diet; Group B-Normal diet supplemented with control (unsprayed) wheat plant; Group C-Normal diet supplemented with magnetite nanoparticle sprayed wheat plants; Group D- Normal diet supplemented with 100 mg kg^{-1} magnetite nanoparticles; Group E- Normal diet supplemented with ferrous sulphate (100 mg kg^{-1}) sprayed wheat plant]. The test substance (33%, w/w) was properly mixed in the normal animal chow and the rats were allowed free access to both food and water. The animals were observed twice daily; body weight, food and water consumption were measured once weekly. After 90 days, the animals were anaesthetized and sacrificed by cervical dislocation followed by examination for gross and histopathological changes.

Histopathology analysis was performed at National Institute of Pathology (NIOP), New Delhi. The liver, kidneys and spleen of control (group A) and test groups (group B, C, D, E) were fixed in 10% neutral buffered formalin for 120 h and then transferred finally to 70% ethanol through 30% and 50% ethanol gradients. The tissues were processed using routine histological techniques. After paraffin embedding, 3 μm sections were cut and stained with hematoxylin and eosin (H&E) for histopathologic evaluation. The H&E staining provides a comprehensive picture of the microanatomy of organs and tissues. Hematoxylin precisely stains nuclear components, including heterochromatin and nucleoli while eosin stains cytoplasmic components including cytoplasmic granules, extracellular components including collagen and elastic fibres, muscle fibres and red blood cells. Histopathology analysis of liver, kidney and spleen tissues revealed that oral exposure of test substances produced no significant adverse effects as evidenced by the normal tissue architecture observed in the exposed animals at post-instillation time period of 90 days in comparison to the normal diet exposed controls. Mild inflammation resultant of acute biological response was observed at many sites within liver and kidney shown by the noticeable abundance of lymphocytes, however these histological alterations cannot be pronounced as an indication of cell injury due to test substance, as similar results were witnessed in control groups as well. Overall analysis of all the samples leads to the conclusion that the gross architecture was intact with no noticeable necrosis or fibrosis within the analyzed tissue (Table 65).

Table 65: Summary of histopathological analysis for estimating toxicological effect of test substance (-ve indicates no toxicity observed)

Tissue/Groups	Group A	Group B	Group C	Group D	Group E
Liver	- ve	- ve	- ve	- ve	- ve
Kidney	- ve	- ve	- ve	- ve	- ve
Spleen	- ve	- ve	- ve	- ve	- ve

A complete bioinformatics study was conducted with the application of recommended doses of nanofertilizers to cereals (pearl millet) and legumes (mung bean) at critical growth stage (6 weeks old) of the plants. GO (Gene Ontology) sequence distributions, helps in specifying all the annotated nodes comprising of GO functional groups. Unigenes associated with similar functions are assigned to same GO functional group. The GO sequence distributions were analyzed (Table 66) for all the three GO domains *i.e.,* biological processes, molecular functions and cellular component. The novel findings are presented as Figure 89.

Table 66: Gene ontology distribution for unigenes in pearl millet (6 weeks old plant)

Treatment	Biological processes	Molecular functions	Cellular component
Control	9222	12044	6223
Nano-P fertilizer	10902	14435	9415

Gene ontology analysis helps us in specifying all the annotated nodes comprising of GO functional groups, such as Cellular components, Biological process and Molecular function.

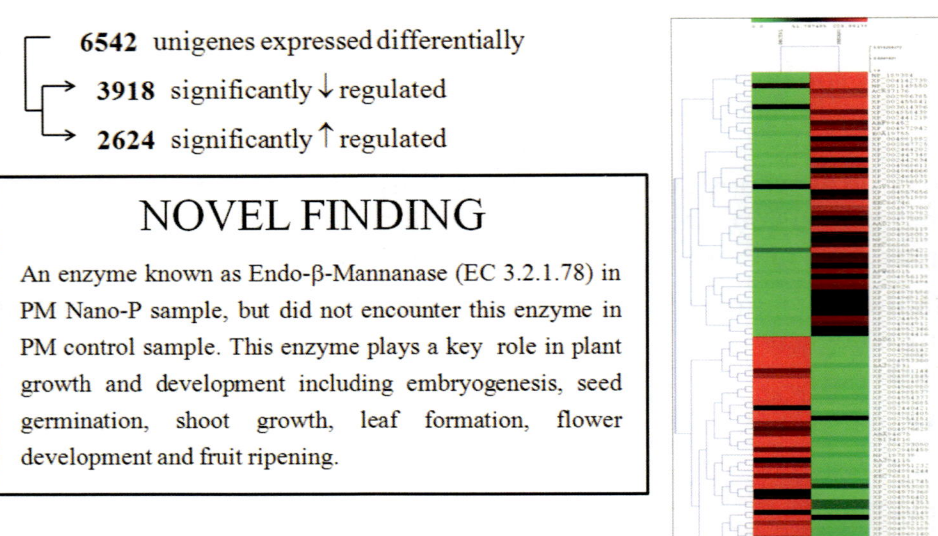

Figure 89: Differential expression of unigenes in PM (Pearl millet)-control and PM nano-P fertilizer sample

Table 67 represents the GO distribution for CDS for mung bean control and mung bean nano samples. It was observed that mung bean nano samples were having predominantly more biological process terms (8395) and molecular function terms (9862) as compared to mung bean control sample. Figure 90 represents a comperative accounting of GO terms distribution between mung bean control and mung bean nano samples. The results showed 2488 unigenes were exclusively present in mung bean samples helping in different metabolic activities.

Table 67: GO distribution for CDS

Treatment	Biological processes	Molecular functions	Cellular component
Mung bean- Control	3926	4660	2277
Mung bean- Nano	8395	9862	4481

Due to these risks, the use of nanotechnology in fertilizer and pesticide formulations has to be met very cautiously, and warrants mandatory need to critically analyze and examine the risks involved with the nano-formulations. Recommended doses should be fixed for each and every product of safe limit before universal application.

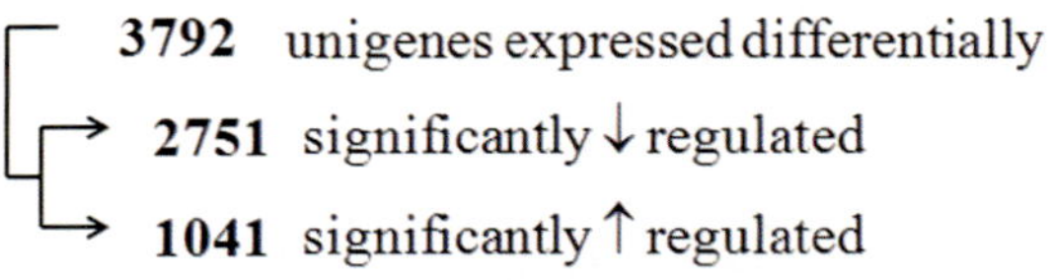

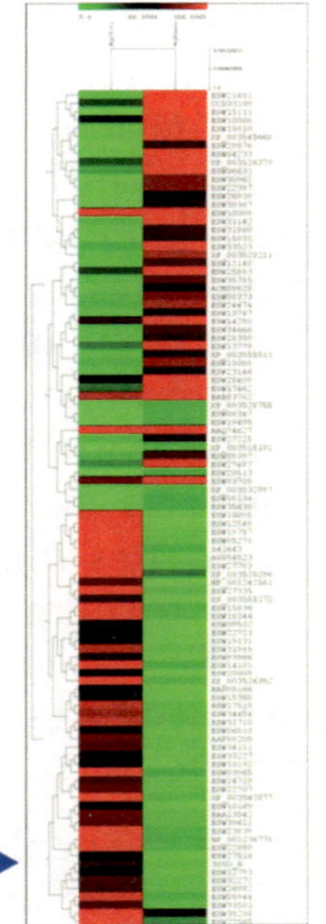

NOVEL FINDING

Encountered 2488 unigenes which are exclusively present in mung bean – nano sample but they are absent in mung bean-control samples. They are helping in carbohydrate metabolism (605), lipid metabolism (237), nucleotide metabolism (323), aminoacid metabolism (314) and biosynthesis of secondary metabolism (110) etc.

Heat map of top 100 differentially expressed genes →

Figure 90: Differential expression of unigenes in mung bean control and mung bean treated with nanofertilizer samples

Nanomaterials may exhibit certain physicochemical properties, biokinetic behaviour, biological interactions, and/or toxicological effects that are different from conventional or bulk form of the same ingredients. This guidance therefore highlights specific aspects that should be considered when testing and reporting data for nanomaterials. It points out the type of data/information that must be provided by the applicant to the commission in support of the safety of the nanofertilizers intended for use in fertilizers. It also needs to be emphasised that the guidance provided in this document is based on the currently available knowledge. As the field of nanofertilizer safety assessment is still evolving, future revisions will be carried out as necessary when new scientific knowledge becomes available.

13.2: Safety Considerations Relating to Nanofertilizers

It has emerged from numerous studies that some materials manufactured at the nano-scale show significant deviations in physicochemical properties, interaction with biological systems, and/or toxicological effects, compared to conventional equivalents. For example, nanoparticles in the lower size < 60 nm range may penetrate biological membrane barriers very well that normally prevent the entry of (larger) particulate materials into cells and tissues. It is therefore possible that, once internalised in the form of nanofertilizer particles, some insoluble or poorly soluble materials may be able to reach those parts of the body that could not have been reached by larger sized particles. As particle size at the nanoscale may be accompanied by certain specific changes in physicochemical properties, a detailed characterisation of the nanomaterial intended for use in plants as well as cosmotic products becomes crucially important. Characterisation is not only important for proper identification of the nanofertilizers in terms of chemical composition, but also in relation to other particle-related characteristics and properties that need considering in safety assessment (for example, physicochemical characterisation). The main safety concerns relating to the use of nanomaterial as fertilizer and in cosmetics can be achieved from the question whether the use of, such products could lead to:

- Local and systemic exposure of the consumer to nanoparticles
- Local and/or systemic harmful effects and,
- Health risk to the consumer as a result of the exposure.

A number of studies and reports investigating possible regulatory gaps have concluded that the current risk assessment paradigm used for conventional bulk materials should also be applicable to nanomaterials. The current hazard identification/dose-response characterisation, which is based on structured toxicological evaluation of conventional chemicals, should also identify/characterise toxic effects of nanomaterials, provided that nano-related aspects have been duly considered during testing. The conventional risk assessment approach for chemicals considers both hazard and exposure – where the absence of one means no risk to the consumer. Thus, safety assessment of nanofertilizer ingredients may, in the first instance, be driven by exposure considerations, with attention to any distinctive material characteristics at the nano-scale. This will inevitably require detailed characterisation of nanofertilizers and determination of the likelihood and extent of systemic exposure resulting from potential translocation of nanomaterials across dermal, respiratory, or gastrointestinal barriers depending on the possible route(s) of exposure. In addition, local effects will need to be considered, irrespective of the fact whether or not the use of a nanofertilizer product containing nanomaterials can lead to systemic exposure. Even in the absence of systemic availability as an nanomaterial and when no local effects are being observed, it needs to be assessed whether the chemical substance may be translocated and could cause systemic effects. Then, the safety of the nanofertilizer needs to be assessed according to its chemical nature by following the SCCS Notes of Guidance (SCCS/1602/18 or more recent version).

Physico-chemical characterisation of nanofertilizers should provide unambiguous identification of the nanomaterials that are used in the final fertilizer product and for which test data have been provided. If these are not the same material, justification should be provided for the scientific basis for considering them 'similar'. A description of the manufacturing process should be provided, along with data on batch-to-batch variation. Where there is a significant variation between batches produced by one manufacturer, or by different manufacturers, it is important that detailed specifications of the nanomaterial intended for use in a fertilizer product are provided by the applicant with indication of the range for each physico-chemical parameter. Due to potential changes in physico-chemical characteristics, the status of nanomaterial in the final fertilizer product should be determined at different stages. Determination of the stability of the core nanomaterial as well as surface moieties is important. It is recommended that, as a minimum, characterisation of nanomaterial intended for use in a fertilizer product should be determined at three stages:

- As manufactured (pristine state) to identify the basic nanofertilizer nutrient,
- After addition to the final fertilizer formulation to identify how consumers are exposed and,
- As used for toxicological investigations.

If characterisation of nanofertilizer is not feasible at any of these stages, it should be justified and documented.

13.3: Specific Considerations Relating to Testing of Nanofertilizer

13.3.1: Solubility/Dispersion

When testing insoluble or partially-soluble nanofertilizer particles, it must be kept in view that they will be present in a dosing or test medium as a nano-dispersion rather than as a solution. Therefore, special attention should be paid to the agglomeration/aggregation behaviour, and the insoluble/ partially-soluble nature of nanofertilizer. Possibilities for disagglomeration of nanoparticles under different testing and physiological conditions should also be considered. During toxicological evaluations, some properties of nanofertilizer may change due to interaction with the surrounding media. Thus, a focus of investigations should be on ascertaining that the tested nanofertilizer particles are in exact form/composition as intended for use in a fertilizer formulation, and as the formulation is delivered to the end-user. Where toxicological data on different nanomaterials, or a different form of the same nanomaterial, is presented in the dossier, justification must also be provided to indicate that the two are justifiably comparable. Special care is also needed in regard to the applied doses, because the concentration of a nanofertilizer may decrease during a test due to sedimentation, binding with other moieties in the test medium, or adhesion to glass/plastic ware. It is therefore important to ascertain the stability and uniformity of the nanofertilizer dispersion in a test medium to ensure that the applied concentration/dose is maintained for the intended period during the test. Possible interaction of the nanofertilizer with other components of a test medium/formulation will also need determining.

13.3.2: Surface interactions

The interactions of a nanofertilizer with the surrounding media and biological systems largely take place through its surface. The surface characteristics of particles are determined by the nature of the entities present on the surface due to the inherent (bio)chemical composition of the material itself, or because of other moieties that may have adhered or attached to the surface due to van-der-Waals forces or electrostatic interactions, or may have been deliberately applied as a coating. It is well known that due to high surface energies nanofertilizer particles tend to stick together to form larger agglomerates and aggregates, and may adsorb or bind various moieties on the surface, including proteins. A nanofertilizer with different surface characteristics (*e.g.,* hydrophilic versus hydrophobic surface) may have profoundly different properties and may interact differently with biological fluids, cell membranes and other biological entities. In view of the potential agglomeration/aggregation of particles, it is essential that attention is paid to the process used for dispersing nanofertilizer particles in preparations used in toxicological testing. It has been shown that composition of protein corona is highly dependent on the initial mixing steps involved. Due to the potential to bind other moieties on surface, and penetrate cellular membrane barriers, nanofertilizer particles may transport other substances into the test systems (the so-called 'Trojan Horse' effect), which may lead to altered (increased or decreased) activity/toxicity. For example, nanofertilizer particles may bind and carry certain immunogens/antigens to the immune cells and impart or trigger an immunological effect. Such a transport of certain components of the test systems by nanofertilizer particles may also lead to artefacts and false indications of harmful effects. This can be avoided by a thorough characterisation of the nanomaterials, and the use of appropriate controls within the testing scheme. Selection of controls should also consider possible interaction of the nanofertilizer particles with the read out system of the assay as it has been demonstrated for various nanomaterials for tetrazolium salts or other dye-based cytotoxicity assays. In case of a doubt over the validity of the outcome of an assay, the use of an additional independent analytical method may provide more information.

Nanofertilizers in the assay can itself have an influence on a read out system, especially if the read out is based on spectroscopy. Similarly, the composition of the culture medium in a test system may influence the outcome of the assay. Special attention should be paid to agglomeration/aggregation behaviour, and the insoluble/partially-soluble nature of nanofertilizer particles. Possibilities for de-agglomeration of nanoparticles under different testing and physiological conditions should also be considered. As properties of nanofertilizer particles may change during toxicological evaluations due to interaction with the surrounding media, investigations should also focus on whether the tested nanofertilizer are in exact form/composition as intended for use in a fertilizer formulation delivered to the end-user. The applicant should also consider any changes in the applied doses of nanofertilizer due to sedimentation, binding/adhesion with test medium or glass/plastic ware to ensure that the applied concentration/dose is maintained during the test. The so-called 'Trojan Horse' effect and possible interaction of the nanofertilizers with the readout system of the assay should also be considered. Furthermore, proper controls should be in place to evaluate possible interference especially when colourimetric methods are used as readout system.

13.3.3: General criteria for different coatings on nanofertilizer particles

Particle surfaces of reactive (*e.g.,* photocatalytic) nanofertilizer particles are generally modified, coated, or doped with other materials to 'quench' the reactivity before use in fertilizer products. Surface modification of nanomaterials, however, may also bring about profound changes in the physico-chemical properties (*e.g.,* hydrophobicity/hydrophilicity). A significant alteration in the properties and biokinetic behaviour may also alter their toxicity due to the potential penetration and accumulation of particles in organs that are not expected to be the target of an unmodified or a differently coated form of the same nanofertilizers. It is therefore important that not only the nanoparticles, and the materials used for surface modification, are assessed individually, but that they are also assessed for safety together when in the form of a surface-modified/coated nanofertilizer. In particular, a major change in hydrophobicity of the nanoparticle surfaces may affect dermal absorption. This raises the question whether a nanofertilizer with several different surface modifications/coatings will need to be tested each time. The SCCS Opinion (SCCS/1580/16) has considered the use of different coatings on a nanomaterial in the context of titanium dioxide (nano-form). In brief, where a coating material applied to an nanofertilizer surface has not already been evaluated for, such an application, it will need to be demonstrated to the SCCS to be safe and not affect the particle properties related to behaviour and/or effects. In this regard, a full dataset on the physico-chemical properties, biokinetic behaviour and toxicological effects of the nanofertilizer particles with each new surface modification/coating would be preferable. However, as a minimum, in addition to safety data on the core nanofertilizer particle, the SCCS will require the following information/data on each material used for surface modification/coating of the nanofertilizer to indicate that it is safe for use in the intended fertilizer product; *e.g.,* it is an approved fertilizer ingredient, or not a banned or restricted substance. Data on physico-chemical properties of the surface-modified/coated nanomaterial to show that they have not significantly changed compared to the same material when uncoated or with a different surface modification/coating that has already been assessed as safe by the SCCS. However, when a coating is applied for a specific purpose (*e.g.,* reduction of photo catalytic activity), the effect of the coating on the intended activity should also be demonstrated.

The coated nanofertilizer to indicate that:

- The surface modification/coating is stable in final formulation,
- The surface modification/coating does not lead to any significant absorption of the nanoparticles through the exposure route(s) anticipated for the intended use,
- The photocatalytic activity of the surface modified/coated nanofertilizer particles is relatively low (*i.e.,* not more than 10% compared to the non-coated equivalent),
- When testing a combined use of different coating materials, a combination of the individual concentrations that represents 'worst case' in terms of hydrophobicity should be used and justification why a certain combination should be considered as worst case should be given,

The SCCS would consider a nanomaterial that has been surface modified or coated with a new substance 'similar' to an already assessed surface variant of the same nanofertilizer particles if both compare well in terms of the above criteria. However, a full toxicological dataset would be required for safety evaluation if the new material used for surface modification/coating is not already known to be safe, or brings about a significant change in the physico-chemical properties, dermal absorption, and/or (photo) catalytic activity of a nanofertilizer. Where a coating material is applied to an nanofertilizer surface, it will need to be demonstrated to the SCCS to be safe and not to affect the properties relating to particle behaviour and/or effects with exception of the intended modification/purpose. As a minimum, data/information should indicate that:

- Each material used for surface modification/coating is safe for use in the intended fertilizer product.
- Data on physico chemical properties of the surface modified/ coated nanofertilizer to show that they have not significantly changed compared to uncoated form of the same material (or with a different surface modification/coating that has already been assessed safe by the SCCS).
- Data on dermal penetration, stability of the surface modification/ coating and (photo) catalytic activity of the nanofertilizer material (for use in products intended for application on skin exposed to sunlight). Where more than one coating material is applied, data should be provided on a

combination of the individual concentrations which represents 'worst case' in terms of hydrophobicity.

13.3.4: Nano-carriers and nano-encapsulated materials

Encapsulation and other forms of formulation have been increasingly used to develop nanosized carriers or delivery systems for (bioactive) substances. It is imperative that safety assessment not only considers safety of the individual components (*e.g.,* the encapsulating material and the encapsulated contents), but also safety of all the components when put together in the form of a nano-sized entity. This is because nano-sizing of substances may impart certain changes in their properties, behaviour and effects compared to corresponding conventional forms and data on safety of individual components in conventional forms may not be sufficient for safety assessment when they are in the form a nano-encapsulated entity. The applicants should therefore provide a clear description of the nano-encapsulated entity in terms of chemical composition, purity, concentration, as well as physico-chemical properties, stability, and dermal penetration of both the components and the nano-encapsulated entity. Safety assessment of, such applications will also require consideration of the potential toxicological effects and exposure estimates under foreseeable use conditions for each component, as well as the nano-encapsulated entity as a whole. The intended function and uses of the nano-encapsulated forms should also be clearly described. For encapsulated nanomaterials, a clear description of the intended function and uses, chemical composition, purity, concentration, as well as physico-chemical properties, stability, and dermal penetration of the components of the nano-encapsulated entity should be provided. Safety assessment should consider toxicological and exposure aspects under foreseeable use conditions for each component, as well as the nano-encapsulated entity as a whole.

13.3.5: Immunotoxicity

Nanofertilizer particles absorbed into the body through different routes of exposure may lead to immunological effects. Research shows that some nano-materials can stimulate and/or suppress the immune responses and that their interaction with the immune system is largely determined by their size, shape, composition, surface properties, protein binding and administration routes. Such effects may result from induction of reactive oxygen species, apoptosis, cell cycle inhibition, complement activation, enhanced secretion of cytokines and chemokines, interaction through toll-like receptors, inflammatory responses, induction of autophagy, reduced viability of the major cell types involved in the innate and adaptive immune system. Due to the potential for binding other substances on the surface, nanofertilizer particle need special attention because they may carry other substances including proteins to the immune cells and thus act as a 'Trojan horse'. This has also been exploited in the form of nanofertilizer particles carriers of various immunogens in the development of vaccines. For example, nano-silica has been proposed as a vaccination platform for allergen-specific immunotherapy. In addition, particles including nanofertilizer can modify immune responses and are known to exacerbate allergic responses in the lung.

The sensitisation potential of the nanoparticles used in fertilizer is evaluated as part of safety assessment. It is also important to ascertain that any systemically or locally available nanomaterial in fertilizer products will not exert an adverse effect on the immune system. The toxicological studies should investigate potential immunological effects where data indicate either systemic availability of the nanoparticles through the likely route(s) of exposure, or a potential for local contact of nanoparticles with the immune cells. This topic has been recently reviewed by EFSA and different methods for determining immunogenicity/allergenicity and immunotoxicity have been proposed. Currently, there are no regulatory documents specifically dedicated to evaluate immunotoxicity of nanofertilizers. Their immunotoxicity assessment is performed based on existing guidelines for conventional substances or medicinal products. Research groups involved in developing nanofertilizer material already use a wide range of *in vitro* assays to screen for essential aspects of the immunosafety profile that are not included in the current regulatory guidance. For example, a number of international projects have produced guidelines for testing strategies and test methods, including *in vitro* assays, for nanomaterial safety evaluation (*e.g.,* the FP7 EU projects NANOMMUNE and MARINA). A number of *in vitro* immune assays have provided results with a good or fair correlation to *in vivo* assay outcomes. Good correlation was indicated for the *in vitro* assays of hemolysis, complement activation, opsonization and phagocytosis, and cytokine secretion assays. Other assays can also be regarded as broadly predictive of the functional alterations of the immune system, including the Colony Forming Unit-Granulocyte Macrophage assay, the leukocyte proliferation test (immune modulatory assays), platelet aggregation, leukocyte procoagulant activity, and various plasma coagulation tests (thrombogenicity assays). Detailed protocols of many of these assays have been published recently. For systemically or locally available nanomaterial in a fertilizer product, it is important to ascertain that they will not exert an adverse immunological effect. This is particularly important for nanofertilizers that are composed of, or contain on the surface, peptides/proteins or other immunogenic/allergenic substance(s). A number of *in vitro* immunoassays can provide results with a good or fair correlation to *in vivo* assay outcomes.

13.3.6: Genotoxicity

Mutagenicity refers to the induction of permanent transmissible changes in the amount or structure of the genetic material of cells or organisms. These changes may involve a single gene or gene segment, a block of genes or chromosomes. The term clastogenicity is used for agents giving rise to structural chromosome aberrations. A clastogen causes breaks in chromosomes that result in the loss or rearrangement of chromosome segments. Aneugenicity (aneuploidy induction) refers to the effects of agents that give rise to a change (gain or loss) in chromosome number in cells, resulting in cells that do not have an exact multiple of the haploid number (2006/1907/EC). Nanomaterials may induce genotoxic damage by a) primary (direct or indirect) or b) secondary mechanisms: i) directly by interaction with DNA, by disturbing the process of mitosis, or by producing Reactive Oxygen Species (ROS) directly or after alterations of mitochondrial functions or ii) by secondary mechanisms as result of oxidative DNA attack during nanomaterial elicited chronic inflammation caused by activation/ recruitment of immune cells, such as macrophages and/or neutrophils.

For *in vitro* genotoxicity assessment, both chromosomal damage (clastogenicity and aneugenicity) and gene mutations should be evaluated. The widely used bacterial reverse mutation (Ames) test is not considered appropriate for nanomaterial mutagenicity assessment, due to limited or no uptake of nanomaterial by the bacteria. The bacterial cell wall hinders uptake and thus nanoparticle internalisation is unlikely to occur to the same extent as observed in mammalian cells, hence sensitivity of the assay is questionable. It is therefore suggested that for nanomaterials the following *in vitro* genotoxicity tests be conducted: Mammalian cell chromosome aberration/clastogenicity tests (either *in vitro* chromosome aberration test or micronucleus test). The micronucleus test can be performed using either the mononucleate or cytokinesis blocked protocols. However, if the cytokinesis blocked micronucleus mammalian cell chromosome aberration/clastogenicity tests (either *in vitro* chromosome aberration test or micronucleus test), the micronucleus test can be performed using either the mononucleate or cytokinesis blocked protocols. However, if the cytokinesis blocked micronucleus assay is to be applied then the blocking agent (cytochalasin B) addition must be post-treatment (after the nanomaterial exposure period). Alternatively, a delayed-co-treatment protocol is also acceptable if a sufficient nanomaterial exposure period has been allowed to enable uptake into the test system cells. Coexposure to both cytochalasin B and the test nanomaterial for the duration of the experiment should be avoided due to possible interference of nanomaterials with cytochalasin B in terms of the cellular uptake of the nanomaterial. An *in vitro* mammalian cell gene mutation test (*e.g.,* Hypoxanthine-guanine Phospho Ribosyl Transferase (Hprt), Thymidine Kinase (Tk) or Xanthine-guanine Phospho Ribosyl Transferase gene (Xprt) tests may be an indicator test. Other indicator tests, such as the comet assay may be included for further weight of evidence. The Comet assay modified with repair enzymes is useful for detection of DNA oxidation damage induced by nanomaterial. The Comet assay is especially suitable in a high throughput version, to cope with large numbers of nanomaterial samples, concentrations, and incubation times.

Another useful indicator test is the cell transformation assay (CTA). For *in vitro* genotoxicity studies, it is necessary to demonstrate uptake of the nanoparticles in the cell and preferably the nucleus to demonstrate exposure of cellular target structures (*e.g.,* DNA). If, such exposure cannot be demonstrated, a negative outcome of, such assay might be meaningless, as the target exposure will not be known. In addition, the amount taken up by the cells may be considered for expression of the possible dose response relationship. Properties of nanofertilizers, such as adsorption capacity, optical properties, hydrophobicity, chemical composition, surface charge and surface properties, catalytic activities as well as agglomeration can result in interference with standard toxicity test. Agglomeration of nanofertilizer particles affects their bioavailability to the cell and thus might lead to false positive/negative results. Several cytotoxicity, oxidative stress and genotoxicity assays, such as the comet assay and the micronucleus test, have been investigated for the possibilities of, such interferences and suggestions made for a modification of the micronucleus assay to ensure correct genotoxicity assessment.

For *in vitro* genotoxicity assessment, both chromosomal damage (clastogenicity and aneugenicity) and gene mutations should be evaluated. The widely used bacterial reverse mutation (Ames) test is not considered appropriate for nanomaterial mutagenicity assessment and an *in vitro* mammalian cell gene mutation test should instead be carried out. Other indicator tests should also be considered, such as Comet assay modified with repair enzymes, and the cell transformation assay (CTA). It is imperative that assessment of cellular and preferably nuclear uptake is also carried out to demonstrate target exposure during the *in vitro* genotoxicity studies.

13.3.7: Carcinogenicity

Substances are defined as carcinogenic if they induce tumours (benign or malignant) or increase their incidence, malignancy or shorten the time of tumour occurrence when they are inhaled, ingested, dermally applied or injected. A carcinogenicity study is in general only submitted when already available; *e.g.,* when carried out before the animal testing ban or when generated for compliance under a different legislative framework (due to animal testing ban). The decision on the carcinogenic potential of mutagenic or genotoxic substances may thus be made on the outcome of *in vitro* mutagenicity tests. A positive *in vitro* result in mutagenicity tests is also seen as indicative of the carcinogenic potential of the substance. There are several ongoing initiatives to develop *in vitro* tests for the indication of carcinogenicity. New *in vitro* approaches, such as cell transformation assays (CTAs) or toxicogenomic approaches may also be useful for the identification of genotoxic as well as non-genotoxic carcinogenic nanomaterials. The latter in combination with transcriptomics provide mechanistic information at the molecular level. Additionally, novel toxicity end points, such as epigenetic toxicity will have to be considered in the future. Epigenetics refers to heritable changes in gene expression that occur without alterations in DNA sequence. A growing body of evidence indicates that environmentally-induced epigenetic alterations play a role in the onset of several human diseases, including cancer, mental disorders, obesity, and other severe conditions. Several studies have shown that epigenetic toxicity can be induced by nanomaterials and can occur at sub-cytotoxic and sub-genotoxic concentrations. So far only the *in vitro* CTAs that can detect both genotoxic and non-genotoxic carcinogens have been validated. CTAs are *in vitro* tests measuring the conversion from normal to transformed phenotype of mammalian cells (primary Syrian hamster embryo, *i.e.,* (SHE), or stable cell lines, such as mouse BALB/c-3T3 or C3H/10T1/2 cells when exposed to test compounds. A guidance document on the *in vitro* SHE CTA was adopted in 2015 by the OECD. The OECD Guidance Document on *In vitro* Cell Transformation Assay Based on the Bhas 42 Cell Line was adopted in 2016. The CTAs have been used to test nanomaterials (and larger particles and fibres). SHE and BALB/c 3T3 CTAs have the potential to detect non-genotoxic as well as genotoxic carcinogens.

The most frequently used end point is morphological transformation. Morphologically transformed cells are characterised by the loss of density-dependent regulation of growth and the formation of colonies with criss crossed cells or foci of piled-up cells that are not observed in untreated control cultures. CTAs are promising tests for predicting nanomaterial-induced cell transformation as one of the crucial carcinogenicity endpoints. An international Working Group of experts convened by the International Agency for Research on Cancer (IARC) has identified 10 key characteristics of established human carcinogens that should be taken into account. That opinion, originally prepared to facilitate a systematic and uniform approach to organising the available mechanistic data relevant to carcinogens in a standard form, could also be applied for assessment of many cases of nanomaterials. High-through put assay systems, such as the US Environmental Protection Agency's (EPA) Toxicity Forecaster (ToxCast) program, which can provide *in vitro* mechanistic data on many of the key characteristics, may be useful in the overall weight of evidence assessment. However, as the ToxCast database is on conventional chemicals it may be of limited use for nanomaterials. Identifying non-genotoxic carcinogens is a challenge in the absence of recourse to animal testing. Since non-genotoxic compounds can exert carcinogenicity through different mechanisms, it is advisable that a battery of tests (as exemplified above) should be used to exclude the non-genotoxic carcinogenicity potential of the nanomaterial. Although a carcinogenicity study is only submitted when already available due to the EU ban on *in vivo* testing of cosmetic ingredients/products, a positive *in vitro* result in mutagenicity testing should be seen as indicative of the carcinogenic potential of the substance. *In vitro* approaches, such as cell transformation assays or toxicogenomic approaches may also be useful for the identification of genotoxic as well as non-genotoxic carcinogen nanomaterial.

13.3.8: Developmental and reproductive toxicity of nanomaterials

The database on developmental and reproductive toxicity of nanomaterials following skin exposure or exposure by inhalation is very limited. Indeed, it is only recently that attention has been directed towards the potential reproductive toxicity of nanomaterials. Some nanomaterials have been shown to pass through the blood-testis barrier, placental barrier, and epithelial barrier, which protect reproductive tissues, and then accumulate in reproductive organs. However, only in a few studies an effect on foetuses was noted after particle inhalation. The literature also provides some limited evidence that some nanomaterials (such as anatase TiO_2 particles) after oral exposure may affect foetal development of the male reproductive system. It has been shown that accumulation of nanomaterial in reproductive organs (testis, epididymis) may cause damage to those organs by destroying sertoli cells, leydig cells, and germ cells, causing reproductive organ dysfunction that adversely affects sperm quality, quantity, morphology, and motility. Some observations on the female reproductive system have also been made, such as a reduction in the number of mature oocytes and disruption of primary and secondary follicular development. However, reproductive function in female off-springs has hardly been studied and cannot be commented upon.

In addition, nanomaterials (such as anatase TiO_2) can disrupt the levels of secreted hormones, causing changes in sexual behaviour. Neurodevelopmental consequences of nano-TiO_2 exposure were suggested by a study in which pregnant Wistar rats were treated by oral gavage with anatase TiO_2 particles (primary size of 10 nm) at 100 mg/kg body weight. The molecular mechanisms involved in nanomaterial toxicity to the reproductive system are not clearly understood yet, but possible mechanisms include oxidative stress, apoptosis, inflammation, genotoxicity or endocrine activities. Previous studies have shown that nanoparticles can increase inflammation, oxidative stress, and apoptosis and induce ROS, causing damage at the molecular and genetic levels which results in cytotoxicity. It is also plausible that nanoparticle may translocate from the respiratory tract to the placenta and foetus. In addition, adverse effects may occur secondarily to maternal inflammatory responses. Effects of nanoparticles used in fertilizer products should be considered for potential reproductive effects including mechanisms and ED mediated mode of action. Due to the animal testing ban, NAMs should be evaluated for potential reproductive toxicity using a weight of evidence approach. Further research and development is required in this area, in particular with regard to the value of *in vitro* testing by the embryonic stem cell test, the micromass embryo toxicity assay, and the whole rat embryo culture. The database on developmental and reproductive toxicity of nanofertilizers following skin exposure or exposure by inhalation is currently very limited. Nanomaterials used in fertilizer products should be considered for potential reproductive effects, including endocrine mediated mode of action. Due to the animal test ban, a weight of evidence should be derived from NAMs for potential reproductive toxicity of nanofertilizers.

13.4: Replacement of *in vivo* Testing by *in vitro* Testing

Among the available alternatives, *in vitro* and *ex vivo* assays, and *in silico* modelling approaches take a prominent place. Generally, these methods aim to reduce, refine, or replace the use of experimental animals. However, there is no stand-alone *in vitro* or *ex vivo* test at present that can replace a standardised *in vivo* method for toxicological assessment of nanomaterials. A tiered approach based on non-testing and *in vitro* methods has therefore been proposed for the prediction of realistic biological outcomes when used in a weight of evidence (WoE) approach. The proposed approach involves thorough physico-chemical characterisation of nanomaterials, *in vitro* screening tests including '-omics', the use of non-testing approaches (in silico models, read across) and the use of OECD and EURL ECVAM validated/ approved *in vitro* methods. A model for tiered nanotoxicity screening has been proposed for risk assessment of nanofertilizers. Validated replacements methods are methods that have passed the various steps of the modular validation process established at EURL-ECVAM, and are considered by its Scientific Advisory Committee (ESAC) to comply with the process. Equally so, methods considered by EURL-ECVAM of having the status equivalent to validation, or alternative methods accepted by OECD, are in the EU recognised as validated methods. At present, various *in vitro* guidelines are being adapted and validated to accommodate nanofertilizers at OECD level. In addition, the international organization for standardization (ISO TC 229) has been developing guidelines for nanofertilizers.

As discussed before, the suitability of *in vitro* methods for nanomaterials can be affected by specific nano-related properties due to *e.g.,* aggregation/agglomeration and subsequent sedimentation, floating and other changes. Furthermore, as already stated, it is well known that nanomaterials might interfere with commonly used assays by influencing readout parameters, such as absorbance or fluorescence. As a consequence, the outcome of an *in vitro* assay for nanofertilizers is often difficult to interpret. Work is ongoing to develop suitable protocols, for dispersion, analysis of cellular doses and quality criteria for nanofertilizer particles. Recently, OECD published a state of the art report on 'alternative testing strategies in risk assessment of manufactured nanomaterials and concluded: while stand-alone alternative testing methods may contribute to basic mechanistic or toxicity knowledge, they will not be sufficient for use in quantitative risk assessment; rather, a battery of alternative testing methods will likely be used in a Weight of-Evidence (WoE) approach. Strategically, incorporating multiple alternative testing methods into alternative testing strategies will allow for an understanding of human and environmental behaviour and toxicity of nanofertilizer particles across endpoints, receptors and material groups. Among other issues, the report highlighted the following: Research must ensure that alternative tests are representative of *in vivo* eukaryotic conditions; for example, the OECD recently concluded that the commonly used Ames test, a bacterial mutagenicity assay, may not be suitable for detecting potential human genotoxicity induced by manufactured nanomaterials because of the lack of endocytosis and limited nanomaterial diffusion across the bacterial cell wall. *In vitro*, models are becoming increasingly sophisticated and better at simulating human-relevant conditions (*e.g.,* 3D cell co-cultures and (micro) fluidic models). A lack of availability of quality data that can address the issues related to categorisation and grouping of nanofertilizer based on their physico-chemical properties, mode of action or relevant exposure also hinders the development of *in silico* method.

Existing data can be harnessed to develop an adverse outcome pathway (AOP), which starts from a molecular initiating event (MIE), which links to key events (KEs) at different levels of biological organisation (*e.g.,* cellular or organ response), eventually leading to an adverse outcome at an organism or population level. It has become clear that direct correlations between the physico-chemical properties of a single nanofertilizer and *in vivo* outcomes are not possible; AOPs instead focus on groupings based on both the chemical activity and the consequent biological processes. In the absence of *in vivo* and other information on a fertilizer ingredient in nano-form, the following elements should be considered in safety evaluation: chemical structure, physico-chemical properties, non-testing information (read across, *in silico* modelling, PBPK modelling) and information from *in vitro* and other alternative methods. Currently, there is no stand-alone *in vitro* or *ex vivo* test to replace a standardised *in vivo* method for toxicological assessment of nanomaterials. A tiered approach based on non-testing and *in vitro* methods has therefore been proposed that involves physico-chemical characterisation, *in vitro* screening tests including '-omics', and the use of non-testing approaches (*in silico* models, read across). The current state-of-the-art on 'alternative testing strategies, in risk assessment of manufactured nanomaterials is published by the OECD. For fertilizer purposes, only data from validated replacement methods are accepted. However, in the absence of alternative methods that have been specifically validated for nanofertilizers, the SCCS also takes into consideration, such methods that are not yet formally validated, but can be demonstrated to be scientifically valid.

The application of nanostructures or nanoparticles as agrochemicals (fertilizers or pesticides) is systematically being explored, before nanofertilizers could be used in agriculture or farming for a general farm practice. The properties of many nanoparticles are considered to be of potential risk to human health, *viz.*, size, shape, solubility, crystal phase, type of material, and exposure and dosage concentrations. However, expert opinions indicate that food products containing nanoparticles available in the market are probably safe to eat, but this is an area that needs to be more actively investigated. To address the safety concern, detail studies are required to know the impact of nanoparticles within the human body once exposed through nanofood. Researchers have to assess and develop proper assessment strategies to assess the impact of nanoparticles and nanofertilizers on biotic and abiotic components of ecosystem. Among the various issues, the accumulation of nanomaterials in environment, edible part of plants might be the important issues before use in agriculture.

Nanoparticles acquire unique properties due to their small size to large surface area ratio. It thus supports in development of novel products and processes as well as enhances the performance of existing ones across several disciplines. Nanotechnology has recently been introduced for improvement of agricultural systems through higher crop yield and better crop protection in order to meet the changing needs and domains of providing food to the growing population of the world. The innovative nano-intervention in agriculture and food sector could generate low-cost, high-efficacy solutions in terms of products and processes, especially suitable for developing countries. However, the unique properties can also lead to nanoparticle-related toxicity in humans and environment.

The guidelines for evaluation of nanoproducts in agriculture and food are more challenging than the existing procedures for assessment of fertilizers or safety evaluation of pesticides or toxicity evaluation in food. The activity, efficacy and impact of nanofertilizer materials depend upon interaction of their physico-chemical parameters with diverse environmental factors and, therefore, require a multidisciplinary approach for development of new alternative strategies and methods for evaluation. It is imperative to modify the existing policies and also develop certain new standard guidelines for evaluation of novel products on the basis of current scientific understanding. The multidisciplinary nature of nanotechnology and its rapidly increasing scope for development of commercially viable applications pose a huge challenge to regulatory bodies across the globe. Nanotechnology involves an amalgamation of knowledge from various disciplines of science, including chemistry, materials science, physics, biology, engineering and medicine. Such an interdisciplinary nature makes nanoscience an important domain to facilitate enhanced scientific and technological prospects and development of novel applications. Moreover, nanotechnology and nanoproducts are dealt by different ministries and different departments, and thus interdepartmental and inter-ministerial convergence is also required. These guidelines have the aim to ensure not only the quality and efficacy to encourage the commercialization of nanotechnology-based innovations, but also safety of novel products by emphasizing high benefit to low risk ratio compared to bulk counterparts. There are no unanimously acceptable international guidelines for nano-agriproducts (NAPs). A few provisions that are in place globally for nanomaterials include REACH, EPA, AVMPA, OECD, and FAO/WHO with certain specific guidelines for quality, safety and efficacy. However, new innovations with alteration of functionality of nanosystems make it difficult to apply a universal set of evaluation parameters for different nanoproducts with different applications. Many a time the case-by-case basis evaluation is needed.

13.5: Biodegradable Nanofertilizer

Biodegradable nanofertilizers are used frequently as nanocarrier systems and other agri-inputs due to their unique and useful properties. A few examples of biodegradable nanomaterials are alginate, polyhydroxybutyrate, carrageenan, dextran, silk protein, micelles and emulsions (based on biodegradable surfactants/emulsifiers), PEG, albumin, PLA, PLGA, chitosan, gelatin, polycaprolactone, poly (alkyl cyanoacrylates) and nanoparticles of bioactives and nanoclay. Nonbiodegradable nanomaterials are also used in NAPs (more commonly used in controlled and slow released fertilizers). Some examples of non-biodegradable nanomaterials include metal oxides, metal nanoparticles, nanocarbon allotropes, synthetic polymers (for seed and fertilizer coatings), quantum dots, boron and carbon nitrides.

13.5.1: According to chemical nature of nanofertilizers

Nanofertilizers could also be categorized based on their chemical nature. They broadly fall under organic or inorganic categories. Besides their chemical nature, their properties at nanoscale also differ widely depending upon their method of synthesis and interaction with other atoms. Organic nanofertilizers are composed of organic compounds, such as lipids, proteins and carbohydrates. They are primarily used in agriculture due to their low toxicity. Examples of organic nanomaterials used in agriculture include synthetic nano-biochar, liposome, albumin, polymer–protein or polymer conjugates. The precursor materials used for synthesis of organic materials are generally considered to be non-toxic and biodegradable.

Inorganic nanofertilizers owing to their high stability, simple synthesis methods using bottom-up approaches, and a wide range of tunable physico-chemical properties, such as shape, size, surface charge, surface area, crystallinity and composition, are a versatile choice for agri-inputs compared to organic nanomaterials. The properties, such as optical (absorption and fluorescence), electrical (conductivity and surface charge), magnetic and thermal can be easily tailored for a specific application requirement.

Composite nanofertilizers are the materials that contain mixture of several different categories of materials. According to nano-form of the ingredient nanocarriers loaded with active ingredient. A nanocarrier is a soft and hard nanomaterial used as a carrier system for targeted agri-input nanomaterials. Common examples include polymer conjugates, polymeric nanoparticles, carbon-based materials (carbon nanotubes, graphene, carbon nanoparticles, nano-biochar), lipid-based carriers (liposomes, micelles), dendrimers, metallic nanoparticles, nanozeolites, metal oxide and so on. These also have the advantage of controlled and slow released delivery of agri-inputs.

Active molecules/compounds could be converted into nano-forms, thereby increasing their potential for improved stability and efficacy. Nanomaterials that are synthesized using bio-agents that are include metallic nanoparticles, bimetallic nanoparticles, metal oxide nanoparticles, quantum dots, nano-clusters and reduced graphene. Nanomaterials that are synthesized using synthetic chemicals as reducing, oxidizing and template are: metallic nanoparticles, bi-metallic nanoparticles, metal oxide nanoparticles, quantum dots, nanoclusters, reduced graphene and molybdenum disulphide. Nanomaterials that are synthesized using physical processes, such as ball milling, laser ablation, temperature and microwave assisted, ultrasonication, glow discharge, plasma, pulsed laser deposition and UV assisted. Examples include carbon nanotubes, graphene, metallic nanoparticles, bi-metallic nanoparticles, metal oxide nanoparticles, quantum dots, nanoclusters, reduced graphene, molybdenum disulphide and nitride. The nanofertilizers and their transformed waste disposal may have an adverse impact on the ecosystem. Therefore, the known and supposed adverse impacts on ecosystem should also be taken into consideration. The claim should be made on the basis of parameters that must include safety, efficacy, application modes and frequency, improved crop yield and productivity or any other benefit over conventional products. Addressing any issue arising out of a significantly different mode of action and assimilation than that of the conventional products. Addressing the issue of specific adverse effect/property associated with the conventional products, if any, such as soil and plant toxic effects, stability testing.

13.5.2: Storage stability and safety

The following four principal types of storage stability studies as per OECD TG 318, FAO/WHO, AVMPA may be adopted: (a) Accelerated storage stability tests, (b) Ambient storage stability tests, (c) Low temperature storage stability tests, (d) Testing for reactivity towards container materials. The following test parameters (whichever is applicable) may be considered for each product. Relevant scientific argument should be provided to explain why to exclude any one of the following test parameters: (a) Selection of containers, (b) Shelf-life statement, (c) Batch (laboratory-, pilot- or production-scale) and size of products, (d) Duration of storage stability, (e) Validation of analytical methods, (f) Technical characteristics (colour, odour, acidity or alkalinity and pH, wettability, suspensibility, dispersion stability, dilution stability, particle size distribution, emulsifiability, re-emulsifiability, emulsion stability, viscosity, flowability, crystalline state, release kinetics and leakage) and (g) microbial stability.

Safety of manufactured nanofertilizer depends upon the product type, application and exposure to humans and environment. The suitable *in vitro* methods for hazard assessment and effective regulation should be adopted from the listed items. Each of the *in vitro* assays mentioned is based on existing OECD Testing Guidelines (TGs) for application to testing manufactured nanomaterials (OECD, 2019): Dermal exposure/toxicity: OECD TG 428 (*in vitro*); OECD TG 427 *in vitro* kin corrosion: OECD TG 431 Eye irritation: OECD TG 437 Genotoxicity: OECD TG 471, 473, 476, 482 and 487 Inhalation exposure (toxicity): OECD TG 403 Cytotoxicity assays used in the OECD testing programme: ATP CellTiter-Glo, neutral red uptake, LDH release, MTT, XTT, cell impedance, trypan blue, BrdU, Alamar Blue, WST-1, live/dead cell counting, colony forming efficiency, genotoxicity assays used in the OECD Testing Programme: Comet assay and DNA double-strands breaks Ecotoxicity: (A) Aquatic test: OECD TG 201 (freshwater algae, cyanobacteria and growth inhibition test), OECD TG 202 (*Daphnia* sp. acute immobilization test), OECD TG 211 (*Daphnia magna* reproduction test), (B) Soil and sediment test: OECD TG 222 (Earthworm reproduction test), (C) Effect on soil microbiota OECD guidelines: OECD Method No. 216 and 217.

Residue analysis of used nano active ingredients, additives, and nanocarrier materials needs to be performed during the life cycle of the product. Data on nanomaterial residue, presence of any nanosized degradation products in food/feed, excipients or surface coating used on food contact material need to be declared by the manufacturer during product registration. The report must mention the following details: Method for determination (detection and quantitation limits) of residues from the used active ingredient, additive and nanocarrier, quantities of generated residues and summary of anticipated risks of generated residues. Nanopesticides (chemical and biological sources) and growth regulators are required to be registered under the existing Insecticides Act, 1968 and Rules 1972 through nodal agencies, namely, Central Insecticides Board and Registration Committee. Likewise major fertilizers (chemical and biological sources) and micronutrients are required to be registered under the existing FCO, 1985 through the nodal agency Department of Fertilizers, Government of India and agriculture department of state governments. The food and feed products developed using nanotechnology interventions are required to be registered as per the guidelines by FSSAI. The nanotechnology intervention used in these products must be registered under the existing regulations. The additional data sets required for registration of nano-agriproducts based on the active ingredient and nanocarrier formulated products are as follows: A brief description of the product, Intended use, Category, for example; Nanofertilizers (*e.g.,* major, secondary and micronutrient), Nanopesticides (insecticide, fungicide, nematicides, acaricide and rodenticide). Justification for developing nanoproducts (claims), Detailed Information on the ingredients, Information on nanomaterials used (active ingredient/nanocarrier), method of production and composition, Nanomaterials property characterization, Hydrodynamic particle size and distribution (polydispersion index), Surface charge (using zeta potential) , Crystallinity (XRD), Transmission electron microscopy (for shape, size and actual average particle size), Aspect ratio (only for 1D and 2D nanomaterials using TEM, SEM and FE-SEM), Hydrophilicity/lipophilicity using contact angle measurement, pH using pH meter, Viscosity (in case of liquid formulation using viscometer), Electrical conductivity (in case of liquid formulation using conductivity meter), Organic (HPLC and GC data); inorganic (XRF and ICP-MS data), FTIR spectrum, X-ray diffraction chromatogram.

The other important data required are: Stability data (as per OECD 318 TG), Impurities detail, Quality control checks parameters and test protocols. Sampling procedure and preparation for specific analysis Testing protocol/s as well as Certificate of analysis, Preliminary toxicity analysis data (confirmatory toxicity analysis would be performed by CIB), Cytotoxicity: ATP Cell Titer-Glo, neutral red uptake, LDH release, MTT, XTT, cell impedance, trypan blue, BrdU, Alamar Blue, WST-1, live/dead cell counting, colony forming efficiency, Genotoxicity: OECD TG 471, 473, 476, 482 and 487, Comparative field efficacy data and residue report (as per section 9(3) of the Insecticides Act, 1968; for herbicide, data to be generated for two seasons and three locations and for others, data to be generated for one season and four locations). Occupational hazards also included as exposure and fate assessment.

13.5.3: Nanofood and nanofeed

For nanofood and nanofeed products, it was suggested for inclusion in the following types of toxicity testings for nanofertilizer material.

13.5.3.1: *In vitro* studies

They help to understand biological responses of nanomaterials and underlying mechanism for toxicity screening. However, suitability of test system and possible structural and functional changes arising from interaction of nanomaterials with culture medium should be considered.

13.5.3.2: *In vitro* digestion studies

Physio-chemical and mechanical conditions of the human gastrointestinal tract can be simulated to understand dissolution and degradation of nanomaterial during digestion. This leads to limited or no further studies for hazard analyses. There are many *in vitro* digestion models available, notably dynamic gastrointestinal digestion system (present in Indian Institute of Food Processing Technology (IIFPT) under MoFPI, Dynamic Gastric Model Institute of Food Research (Norwich, UK), Model Stomach system, Human Gastric simulator (UC Davis, Food Science and Technology), TIM-1 (Netherlands), SHIME (ProDigest and Ghent University, Belgium) and Dynamic *in-vitro* human stomach, China. They help to understand the digestibility and release behaviour of ingested food components and thus fate of added nanomaterials.

13.5.3.3: *In vitro* genotoxicity testing

Use of bacterial reverse mutation assay cannot be considered for detection of genotoxicity of nanomaterials due to the fact that bacterial cells do not phagocytose particles like mammalian cells and nanomaterials cannot penetrate bacterial cell wall. Studies, such as OECD TG 476 for induction of gene mutations in mammalian cells (preferably the mouse lymphoma TK assay with colony sizing) and OECD test guideline 487 for an *in vitro* micronucleus assay should be considered for evaluating nanomaterial in food.

13.5.3.4: Other *in vitro* studies

This includes various *in vitro* models to assess the effects of nanomaterials on permeability/integrity of the gastrointestinal barrier, inflammatory responses to assess gut maintenance, immune cells and immune responses etc. Cells, like differentiated CaCo-2 cells, primary human oesophageal epithelial cells and M-cells (modified enterocytes present throughout the epithelial lining) are used to simulate the *in vivo* conditions.

13.5.4: *In vivo* studies

In vivo studies are essential to identify ADME profile, adverse responses and dose dependent toxicity. Forms of administration of nanomaterials (*e.g.,* adding to feed, water or by gavage) for *in vivo* studies also influence the toxicity profiling. For example, nanomaterials that interact with food and form complex matrices, simulant cannot be used and it should be homogeneously blended in food.

13.5.4.1: ADME studies

Absorption, distribution, metabolism and excretion (ADME) studies are essential for toxicity evaluation of nanomaterials. Appropriate measuring systems should be adopted to detect nanomaterials in organs, tissues or biological fluids. Labelling with radioactive isotopes, fluorescent dyes and comprehensive mass balance studies are to deal with nanomaterial polydispersity, and toxicokinetic changes upon repeated administration should be considered while designing ADME studies. Simple ICP-MS cannot determine the presence of nanomaterials.

13.5.4.2: *In vivo* repeated-dose 90–day oral toxicity study

Repeated-dose 90–day oral toxicity study in rodents as per the OECD TG 408 is required to access orally ingested nanomaterials. Emphasis on assessment of cardiovascular and inflammatory parameters, endocrine-related endpoints and oestrous cycles is required during oral toxicity studies.

13.5.4.3: *In vivo* genotoxicity testing

If genotoxicity is observed in any of the *in vitro* studies, or when it is impossible to conduct *in vitro* studies for selected nanomaterials, any of the following *in vivo* tests may be adopted: *in vivo* micronucleus test (OECD TG 474), *in vivo* comet assay and transgenic rodent gene mutation assay.

13.5.4.4: Other *in vivo* toxicity tests

If there is evidence of toxic effects and accumulation of nanomaterials (or degradation of products/metabolites) in organs and tissues, chronic toxicity by following OECD TG 453 may be appropriate in order to reveal progressive toxic effects or delayed toxicity and developmental toxicity and to identify a BMDL or a NOAEL. OECD test guidelines 414, 415 and 416 may be adopted for study design of reproduction and developmental studies.

13.5.5: Uncertainty analysis

Analysing possibility of uncertainty, some of the possible reasons for uncertainty in assessing nanofertilizers are as follows:

- Non-availability of standard methods for physico-chemical characterisation of various nanofertilizer structures and associated properties
- Sample preparation procedures and calibration of the analytical equipment dictate characterisation
- Accuracy
- Differences in the physical principles applied by various measurement techniques
- Aggregation/agglomeration behaviour of nanofertilizers and other factors, such as dilutions or dispersions
- Vary with their interaction by various environmental factors

Reduced information can be provided when no exposure to nanomaterials is confirmed by data indicating no migration from food contact materials or when complete degradation/dissolution is demonstrated with no absorption of nanofertilizer nanomaterials.

13.6: Other Safety Issues

Although nanofertilizer has incredible potential to revolutionize many aspects of human life, the benefits may come with some price. One of the major questions faced by the world before accepting nanofertilizer is whether the unknown risks of nanoparticles involving their environmental and health impact prevail over their potential benefits. The risks associated with the application of nanoparticles are yet to be evaluated before fully implementing this technology. This consideration has developed "nanotoxicology," which is responsible for assessing toxicological potential as well as promoting safe design and use of nanoparticles. A systematic and thorough quantitative analysis regarding the potential health impacts, environmental clearance, and safe disposal of nanoparticles can lead to improvements in designing further applications of nanofertilizer. Although no direct human disease has been linked to nanoparticles so far, early experimental studies indicate that nanoparticles could initiate adverse biological responses that can lead to toxicological outcomes. Nanoparticles which constitute a part of ultrafine particulate matter can enter in the human/animal system through oral, respiratory, or intradermal routes. Currently, there is a common assumption that the small size of nanoparticles allows them to easily enter tissues, cells, and organelles and interact with functional biomolecular structures (*i.e.,* DNA, ribosomes) since the actual physical size of an engineered nanostructure is similar to many biological molecules (*e.g.,* antibodies, proteins) and structures (*e.g.,* viruses).

A corollary is that the entry of the nanoparticles into vital biological systems could cause damage, which could subsequently cause harm to human health. However, one of the most disgustful scenarios is the lack of concrete technical data on toxicological aspect of nanoparticles giving opportunity to both nanotechnology proponents and opponents to make contradictory, unscientific, and sweeping conclusions about the safety of nanoparticles including nanofertilizers. This atmosphere of uncertainty is precisely the feature of nanofertilizer that causes cynics the greatest concern. Therefore, there is a need for proper physico-chemical characterisation and determination of appropriate exposure protocols and reliable methods for assessing nanoparticles outcome in the environment, their internalization, and their kinetics in living organisms. Once these issues are addressed, optimal experimental conditions could be established in order to identify if a particular nanoparticle poses a threat to human health. Multidisciplinary research between materials scientists, environmentalists, and life scientists should overcome these limitations in identifying the true hazards of nanofertilizers. Sadly, the risk assessments of nanofertilizer are partly subjective and likely to be highly politicized. Disastrously, no single scenario for describing risks and controls can be universally applied to conclude the outcome due to the heterogeneous and developmental nature of nanofertilizers. Also, an absence of standardized methodologies and guidelines makes it difficult to compare the safety/toxicity assessments from different research groups. The ethical issues will be specific only for the knowledge base at a given time and for a specified product and its exposure scenario. Moreover, maintaining utmost specificity regarding design of experiments, alternative assessments are needed to take into consideration ethical, social, and political values that relate policies, such as those involving nanofertilizers.

Before interpreting toxicological data, it is thus essential to calculate and determine the expected concentrations of nanoparticles that may be exposed to the biological system or present in the ecosystem. The use of nanofertilizer in agriculture is significantly important as it directly affects humans. Nano-fertilizers enable nanoparticles to enter in the food chain allowing their distribution in every organism related to the food chain. As all substances, from arsenic to table salt, are toxic to plants, animals, or humans at some exposure level, this would not limit their use in various applications which are designed keeping in mind the critical exposure concentration. As discussed in most of the studies regarding the use of nanoparticles for promoting growth of plants with a focus on using lower concentrations of nanoparticles, it can be argued that it will pose insignificant health and environmental damage. Many countries have identified the potential of nanofertilizers in the food and agriculture sectors and are investing significantly in its applications to food production. However, owing to our limited knowledge of the human health effects of these applications, these countries recognize the need for early consideration of the food safety implications of nanofertilizers.

As suggested by the scientific committee of the European Food Security Authority (EFSA), "the risk assessment paradigm (hazard identification, hazard characterization, exposure assessment and risk characterization) is applicable for nanoparticles (EFSA Scientific Committee 2011). However, risk assessment of these nanoparticles in the food and feed area should consider the specific properties of the subject nanoparticles in addition to those common to the equivalent non-nanoforms." It is most likely that different types of nanoparticles vary as to their toxicological properties. The available data on oral exposure to specific nanonutrient and any consequent toxicity are extremely limited; the majority of the available information on toxicity of nanoparticles is from *in vitro* studies or *in vivo* studies using other routes of exposure. The risk assessment of nanoparticles has to be performed on a case-by-case basis. Various parameters may be included in deciding the risk associated with the use of any particular nanoparticle in food and feed. These include physico-chemical characterization of nanoparticles, its stability in the food and feed, exposure scenario of the nanoparticles from food and feed, and toxico kinetics (absorption, distribution, metabolism/

biotransformation, excretion/elimination) within the human and animal systems. The Nanotechnology Regulatory Science Research Plan of the US Food and Drug Administration (FDA) lays out a framework and implementation plan to provide coordinated leadership on regulatory science activities and issues related to FDA-regulated products that either contain nanoparticles or otherwise involve the application of nanofertilizer to address key scientific gaps in knowledge, methods, or tools needed to make regulatory assessments of these products.

Further studies

Bouwmeester, H., Poortman, J., Peters, R. J., Wijma, E., Kramer, E., Makama, S., Puspitaninganindita, K., Marvin, H. J., Peijnenburg, A. A. and Hendriksen P.J. (2011). Characterization of translocation of silver nanoparticles and effects on whole-genome gene expression using an *in vitro* intestinal epithelium coculture model. *ACS Nano.* **24;5(5):** 4091–4103.

Chaudhry, Q., Bouwmeester, H. and Hertel, R. F. (2010). The Current Risk Assessment Paradigm in Relation to Regulation of Nanotechnologies. In: G.A. Hodge, D.M. Bowman and A.D. Maynard (eds). International Handbook on Regulating Nanotechnologies. Cheltenham: Edward Elgar, 124–143.

Dekkers, S. and Oomen, A. G. (2016). Towards a nanospecific approach for risk assessment. *Reg. Tox. Pharmacol.* **80:** 46–59.

EFSA (2011). Guidance on the risk assessment of the application of nanoscience and nanotechnologies in the food and feed chain, *EFSA Journal* **9(5):** 2140, 36 pp. https://www.efsa.europa.eu/en/efsajournal/pub/2140

Giannakou, C., Park, M. V., de Jong, W. H., van Loveren, H., Vandebriel, R. J. and Geertsma, R. E. (2016). A comparison of immunotoxic effects of nanomedicinal products with regulatory immunotoxicity testing requirements. *Int. J. Nanomed.* **11:** 2935–2952.

14

Future Prospects of Nanofertilizers

Nanofertilizer application in agriculture may serve as an opportunity to achieve sustainability towards global food production. There is a tremendous food production pressure on the sector as nutritional deficiencies in human populations are mainly because of using less nutritious food. Important benefits of nanofertilizers over conventional chemical fertilizers rely on their nutrient deficiency system. They regulate the availability of nutrient in crops through slow/control release mechanism. Such a slow delivery of nutrient is associated with the covering or cementing of nutrients with the nanomaterials. By taking advantage of this slow nutrient delivery, growers can increase their crop growth because of consistently long term delivery of nutrients to plants. For example, nutrient can be released over 40–50 days in a slow release fashion rather than 4–10 days by the conventional fertilizers. In conventional nutrient management systems, more than half of the applied fertilizer is lost in leaching or become unavailable for the plants because of the excessive availability hindering the roots to uptake or sometimes causing toxic effects on the plant. Furthermore, nanofertilizers reduce the need for transportation and application costs.

Another advantage of using small quantities is that the soil does not get loaded with salts that usually are prone to over application using conventional fertilizer on a short or long term basis. Nanofertilizer also can be synthesised according the nutrient requirement of intended crops. In this regard biosensor can be attached to a new innovative fertilizer that controls the delivery of nutrients according to soil nutrient status, growth period of a crop or environmental conditions.

14.1: Bionanofertilizer

Nanofertilizer also can be used as bionanoformulations. Bionanofertilizers are formulations containing one or more microorganisms after blending of required nanoparticles with plant growth promoting microorganisms. There are various utilities of different bio-nano formulations (see Figure 6, chapter 2). It could be defined as the integration of biofertilizer with biosynthesised nanoparticles in order to improve the better growth of the plants. This goal can be achieved to control the delivery of biofertilizer in the soil and extent the useful life of formulations. The formulations or preparations containing one or more microorganisms enhancing soil productivity, by fixing atmospheric nitrogen, solubilizing phosphorus or stimulating plant growth through synthesis of growth-promoting substances. Therefore, nanobio-fertilizer could be defined as the integration of biofertilizers with nano-structures or nanoparticles in order to improve the growth of plants. To achieve this goal is essential to control the delivery of biofertilizers in the soil and extend the useful life of formulations. Use of nanoformulations can be helpful to enhance the stability of biofertilizers with respect to desiccation, heat, and UV inactivation. For example, polymeric nanoparticle coatings can be used to develop formulations resistant to desiccation and consequentialy improve the useful life of these products. The important aspect in bionanofertilizer development is the interaction between nanoparticles and microorganisms, the shelf life of biofertilizer and its delivery. Major reaction involved in the biosynthesis of nanoparticles is oxidation/reduction. The microbial enzymes or the plant photochemical is responsible for reduction of the salts/metal compounds to their respective nanoparticles. The advantage of bio-nanoformulation is to increase the solubility of the poorly active ingredient by the release of active ingredient in a controlled manner and protection against premature degradation. Agrochemical companies also utilizing the advantage of bionanoparticles after development of nano-emulsions and nano-capsules contain active ingredients. Such advanced functionality of bio-nano particles may provide the sustainable solution in agricultural sector. It can also solve some limitations of biofertilizer, but this technology needs further refinement. Biosynthesised nanoparticles can be encapsulated that provide several benefits, such as ease of handling, enhanced stability, protection against oxidation, retention of volatile ingredients, taste making, moisture triggered controlled release, pH triggered controlled release, consecutive delivery of multiple active ingredients, change in flavour character of food and enhanced bioavailability and efficacy. In the biological system nanocapsules carry the food supplements and help the bioavaibility of the substance

14.2: Nanofertilizer for Seed Treatment

Nanofertilizer can be used as seed treatment, resulted to increase moisture content of seeds as well as protection against soil borne diseases (see Figure 5, Chapter 2). Carbon nanotube and nano-silver are found to be very effective. Pre-treating seeds of bitter almond rootstock with 50% of nano-fertilizer under saline conditions markedly increased germination measurements, stem length and diameter, elongation of secondary and primary roots/plant, and number of secondary roots/plant relative to urea and ammonium sulfate fertilizers, thus throughout its vital role in increasing the moisture content of seed which in turn increased germination percentage and rate. Additionally, all the amount of nano-fertilizer was covered by plants via slow and continuous release of fertilizer, therefore producing stronger seedlings under saline conditions. It has also been found that nano-fertilizer as nanourea modified with hydroxyapatite nanoparticles as a source for N, P, and Ca nutrients under saline condition.

In the recent years nanofertilizers are freely available in the market, but particularly the agricultural fertilizers are still not shaped by the major chemical companies. The available nanofertilizers may contain nano zinc, phosphorus, silica, iron and titanium dioxide, different core shell gold nanorods, QDs etc. It should endorse controlled release and improve its quality. Studies of the uptake, biological fate and toxicity of several metal oxide nanoparticles available in the market at present are needed for agricultural use. Moreover, the quantification of nanoparticles is necessary to reduce the toxic effect of it.

14.3: Nanofertilizer in Precision Farming

Precision farming has been a long-desired goal to applying input as per demand of the crop that maximize output (*i.e.,* crop yields), while minimizing input (*i.e.,* fertilizers, pesticides, herbicides etc.). Precision farming makes use of computers, global satellite positioning systems (GPS), geological information systems (GIS) and remote sensing devices to measure highly localized environmental conditions thus determining whether crops are growing at maximum efficiency or precisely identifying the nature and location of problems of crops and edaphological environment. Precision farming can also help to recycle agricultural waste and thus keep environmental pollution at minimum extent. Even though, tiny sensors and monitoring systems enabled by nanotechnology will have a large impact on future precision farming methodologies.

One of the major roles for nanotechnology-enabled devices will be the increased use of autonomous sensors linked into a global positioning system (GPS) for real-time analysis. These nano-sensors could be distributed throughout the field where they can monitor soil conditions and crop growth. The union of these two technologies in sensors will create equipment of increased sensitivity, allowing an earlier response to environmental changes, For example: (a) Nano-sensors utilizing carbon nano tubes or nano-cantilevers has capability to trap and measure small molecules. (b) Nano-particles or nano-surfaces can be engineered to trigger an electrical or chemical signal in the presence of a contaminant, such as bacteria and other pests or pathogens. (c) Other nano-sensors perform by triggering an enzymatic reaction or by using nano-engineered branching molecules called dendrites as probes to bind to target chemical sand proteins.

Ultimately precision farming, with the help of smart sensors, will allow enhanced productivity in agriculture by providing accurate information, thus helping farmers to make better decisions.

14.4: Nanosensors

Nanofertilizers are also being developed to increase soil fertility and crop production. Nano-sensors could also monitor crop health and magnetic nanoparticles could facilitate removal of soil contaminants (Hg, Pb and Zn). "Lab on a chip" technology also could have significant impacts on developing countries. Researchers are trying to come up with sensors for detection of pathogens up to very minute fractions by exploiting the novel properties of nanomaterials. Biosensors are being increasingly used in industrial and food processing, environmental monitoring, food storage and health care. The nanosensors used as seed storage is shown as Figure 91.

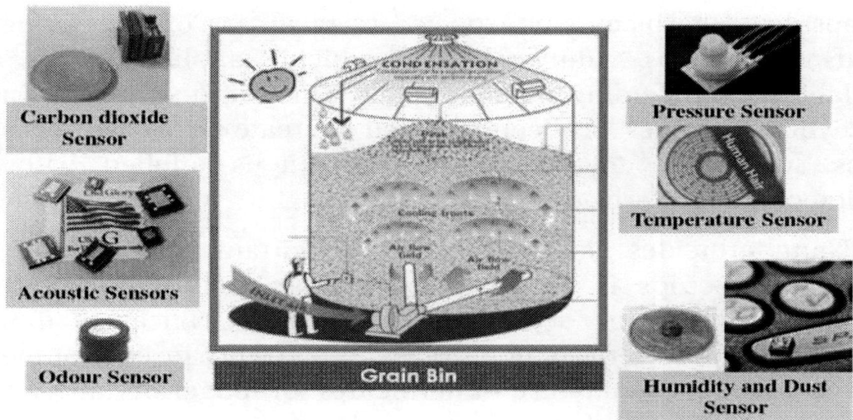

Figure 91: Different nano-sensors used as grain storage

Biosensors, incorporated with new technologies in molecular biology, microfluides and nanomaterials, have diverse application in agricultural production, food processing, and environmental monitoring for in-field, on-line and real-time detection of pesticides, pathogens, toxic materials, proteins, antibiotics, odours creating bacteria, microbes etc in soil, air, water, food, plants and animals. The biosensors are capable of forecasting the possible occurrence of soil diseases. The basic principle of soil diagnosis with the biosensor is to estimate the relative activity of favourable and unfavourable microbes in the soil determined on the basis of differential oxygen consumption in their respiration. It is feasible, therefore, to predict before hand, whether or not there will be an outbreak of soil disease in the tested soil. The biosensor does offer an innovative technique of diogonesting soil conditions based on numerical data. Some research groups are working on developing biosensors based upon the surface plasmon resonance (SPR) exhibited by metal nanoparticles.

14.5: Nanoherbicides

Multi-species approach with single herbicide or repeated use of herbicide with similar mode of action in the cropped environment resulted in poor control and herbicide resistance. Developing a target-specific herbicide molecule encapsulated with nanoparticles is aimed at a specific receptor in the roots of target weeds. On entering into roots system, it translocated to parts that inhibit glycolysis or other pathways of food reserve in the root system. This will make the specific weed plant to starve for food and gets killed. In rain fed areas, application of herbicides with insufficient soil moisture may lead to loss as photo decomposition. The controlled release action of encapsulated herbicides is expected to take care of the competing weeds with crops. Adjuvants for herbicide application are also available that including nanomaterials. One nano-surfactant based on soybean micelles has been reported to create glyphosate-resistant crops susceptible to glyphosate when it is applied with the 'nanotechnology derived surfactant'.

Nanoherbicides can play a very important role in removing weeds from crops in an eco-friendly way, without leaving any harmful residues in soil and environment. Encapsulation of herbicide in polymeric nanoparticles also results in environmental safety. Disproportionate use of herbicides for longer period of times leaves residues in soil, which cause damage to succeeding crops. Continuous use of same herbicide for constant period of time cause weeds resistance against same herbicide. Effectiveness of nano-zerovalent iron (nano ZVI) has been assessed to dechlorinate herbicide atrazine (2-chloro-4-ethylamino-6-isopropylamino-1, 3, 5-triazine) from atrazine-contaminated water and soil. Target specific nanoparticles loaded with herbicide have been developed for delivery in roots of weeds. These molecules enter into the roots system of the weeds, translocate to cells and inhibit metabolic pathways, such as glycolysis. This ultimately leads to death of plants. A list of nanoparticles used as herbicides against weeds in different vegetable crops is given in Table 68.

Table 68: Nanoparticles used as herbicides in commercial vegetable crops

Nanofertilizer particles	**Used against herbs/weeds**
Silver nanoparticles-chitosan encapsulated paraquate	*Eichhornia crassipes*
Ag, Cu, Fe, Zn, Mn	*Allium cepa* (L.)
Cu	*Cucurbita pepo*
CuO	*Raphanus sativus, Lolium perenne* and *Lolium rigidum*
CuO and ZnO	*Fagopyrum esculentum*
Cu	*Elodea densa*
CuO and ZnO	*Cucumis sativus*

14.6: Nanopesticides

The utilization of nanofertilizer particles in plant protection and production of food is under-explored area in the future. It is well recognized that insect pests are the major ones in the agricultural fields and also in its products, thus nanoparticles may play a key role in the control of insect pests and host pathogens. The recent development of a nano-encapsulated pesticide formulation has slow releasing action with improved solubility, specificity, permeability and stability. Formulation of nano-encapsulated pesticides led to reduce the dosage of pesticides, improve pesticide efficacy and human beings exposure to them which is eco-friendly for crop protection. So, the development of non-toxic and promising pesticide delivery systems for increasing crop productivity per unit time basis, while reducing the negative environmental impacts to ecosystem. Recently, few chemical companies openly promote nanoscale pesticides for sale as "microencapsulated pesticides." (Karate ZEON, Subdue MAXX, Ospray's Chyella, Penncap-M). Syngenta markets some products, such as the Primo MAXX, Banner MAXX, Subdue MAXX, etc. Despite they are known as micro-emulsions. However, they are really nanoscale emulsions and this technique is commonly used for formulations of organic nanoparticles containing active agrochemicals or substances. A range of formulation types have been suggested including emulsions (*e.g.,* nanoemulsions), nanocapsules (*e.g.,* with polymers) and nanoclays. These products can be used to enhance the use efficacy of existing pesticide active ingredients or to improve sustainability. A list of nanofertilizers are tested and their action against plant pathogens are given below (Table 69).

Table 69: Nanofertilizer against plant pathogen

Nanofertilizers	Antimicrobial action against plant pathogen
Ag	*Bacillus cereus, Escherichia coli, Bacillus subtilis, Streptococcus thermophilus, Fusarium oxysporium, Aspergillus fumigates* and *Corynebacterium* sp.
Cu-chitosan	*Alternaria alternata, Macrophomina phaseolina* and *Rhizoctonia solani*
Ag and TiO_2	*Lactobacillus strains*
S	*Aspergillus niger* and *Fusarium oxysporum*
Au	*Klebsiella pneumonia*
Zn	*Aspergillus niger* and *Fusarium oxysporum*
Cu	*Aspergillus flavus, Aspergillus niger, Candida albicans, Staphylococcus aureus, Escherichia coli, Micrococcus luteus, Klebsiella pneumoniae, Pseudomonas aeruginosa, Phoma destructiva, Curvularia lunata, Alternaria alternate, Fusarium oxysporum, Fusarium equiseti,* and *Fusarium* sp.

Application of nanofertilizer particles in pests and plant diseases management as pesticides, fungicides and herbicides is the fastest and cheapest way to control pests and diseases. Indiscriminate use of pesticides has caused many problems, such as: adverse effects on human health and on pollinating insects and domestic animals, and entering this material directly or indirectly in ecosystems. Intelligent use of chemicals on the nano-scale can be an appropriate solution for this problem. These materials are used as carriers in nano-scale has self-regulation, this means that the medication on the necessary amount only be delivered into plant tissue. Nanoparticles for liberation of active ingredients or drug molecules will be at its helm in near future for therapy of all pathological sufferings of crop plants. There are myriad of nano-materials including polymeric nanoparticles, iron oxide nanoparticles and gold nanoparticles which can be easily synthesised and exploited as pesticide, fertilizer or drug delivery piggybacks. It has been found that small sised (35 nm) sulfur nanofertilizer are very effective in prevention of the fungal growth and can be useful for the protection of important crops, such as tomato, potato, apple, grape etc., from different diseases, mainly for organic farming because of antimicrobial property of sulfur. Nano-based viral diagnostics, including multiplexed diagnostic kit development, have taken momentum in order to detect the exact strain of virus or other pathogens and stage of application of some therapeutic to stop the disease. The insecticidal activity of Ag and Zn nanofertilizer particles has demonstrated against the *A. nerii*. A representation of nanofertilizer application has shown in figure 92, which clearly mentions its role in insect-pest management and herbicide control besides nutrient supply and use as sensors.

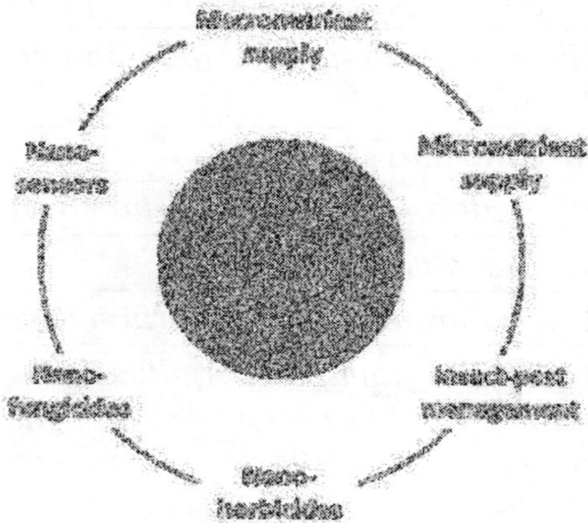

Figure 92: Schematic representation of application of nanofertilizer in crop production and management

Nanofertilizer can supply all the required nutrients for the plant growth and development. They can help in plant protection agaist insect-pest as well as precision farming in crop production.

14.7: Nanofertilizer for Water Management

Nanomaterials have unique size-dependent properties, such as high surface area and reactivity, that make them ideal for treating nontraditional water sources, and these properties can enable the development of novel nanofertilizer-based solutions for more efficient utilization of drinking water, nontraditional water sources, and wastewater treatment processes. Indeed, there are many nanotechnology-enabled approaches that could be applied during key steps in the water treatment process. For example, membranes can be designed with nanoscale pores that remove specific pollutants, while allowing water molecules and important nutrients to pass through, and the antimicrobial properties of silver nanoparticles can be utilized for point-of-use water disinfection. Given the broad application space for nanoparticle-enabled water treatment, the agencies participating in the Water NSI have identified the following key objectives where interagency coordination can lead to nanotechnology-enabled improvements in water purification technologies to increase water availability. The important objectives to increase the water availability are:

- Use nanotechnology to double the throughput and halve the cost of filtration and membrane separation systems within 5 years.
- Demonstrate nanotechnology-enabled alternatives to reverse osmosis for desalination within 5 years. Within 10 years facilitate the transfer of these technologies from demonstration to market.
- Develop nanoscale catalysts for use in water treatment that can completely replace precious metal-based catalysts within 10 years.

Nanofertilizer can help in reducing agricultural run-off as well as water management. Nano coated pesticides can reduce environmental and health risks (Figure 93).

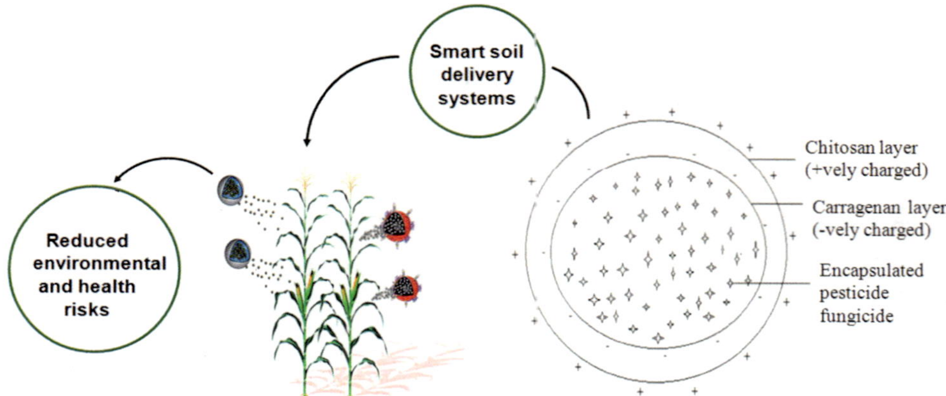

Figure 93: Utilization of nanofertilizer particles in plant protection and water management

Filter and membrane-based water purification and disinfection technologies are critically important for a wide variety of applications, including food processing, wastewater treatment, medical applications, such as artificial kidneys, and environmental protection. The incorporation of nanoscale features and nanomaterials into membranes and filters offers several compelling opportunities, such as high selectivity and antifouling functionality. For instance, in the case of membranes, the feature sizes of nanostructured surfaces allow fine-tuning of flux, selectivity, and strength to optimize performance. The use of nanofilter for water purification is shown in Figure 94. They need very low operating energy.

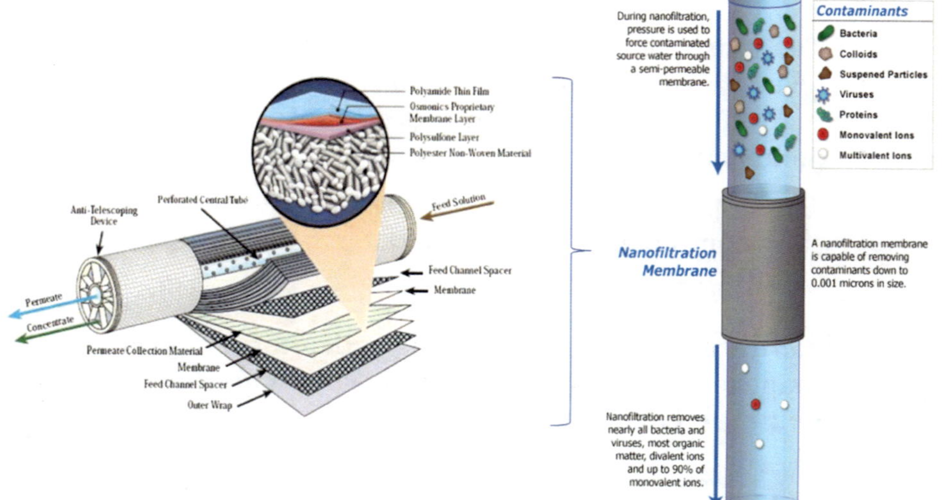

Figure 94: Nanofiltration for water treatment

More advanced developments in atomically precise membranes could separate greenhouse gases from the air and pure water from saltwater. Alternatively, the introduction of carbon nanomaterials, such as nanotubes and graphene, offers a unique combination of robustness, precise control, and ease of functionalization to create desirable membrane characteristics. Both of these approaches also show promise in reducing fouling—perhaps the key challenge in the cost-effective deployment of advanced membrane systems. In the case of filters, nanoscale morphology provides a high surface area for adsorbing pollutant molecules. Further functionalization or decoration of surface groups or inclusion of metal clusters as adsorption sites can enhance the adsorption capacity and selectivity. For example, photocatalytically active silver-titanium dioxide nanofibres could be used in a filter with chromate-absorbing iron oxide nanofibres to treat water contaminated by agrochemicals and other impurities that are common in rural areas. However, the ability to regenerate nanoscale sorbents for reuse remains a challenge. Additional improvements in nanomaterial-enabled filters may be available through broadening the range of materials used for sorbents, the types of pollutants they can treat, and the types of membranes whose capabilities they can enhance. Further, long-term technology improvements may result in systems that require less pressure and are more resistant to biofouling and scaling.

Over 96% of the water on earth is sea water, and desalination offers an opportunity to convert previously unusable saline water to fresh water. Reverse osmosis (RO), often in combination with nanofiltration pretreatment, has emerged as the predominant technology used in desalination operations in the United States. However, sea water that is treated with current commercial RO systems is significantly higher in cost than water from traditional sources. Moreover, alternatives to RO may offer opportunities for cost, energy, and carbon emissions reductions, as well as smaller infrastructure requirements. Technologies, such as forward osmosis, multi-effect distillation, multistage flash distillation, membrane distillation, freeze separation, and capacitive deionization potentially can be used in commercial desalination of both brackish water and sea water, but these technologies will require research advances to achieve "pipe parity" or achieve costs equal to those of current processes for delivering fresh water.

14.8: Nanofertilizer for Climate Resilient

Lightweight nano-composite materials can reduce emissions in vehicles by reducing their weight, in turn, decreasing fuel consumption can have an immediate and significant global impact. It is estimated that a 10% reduction in weight of the vehicle corresponds to a 10% reduction in fuel consumption, leading to a proportionate fall in emissions. In recognition of the above, there is growing interest worldwide in exploring means of achieving weight reduction in automobiles through use of novel materials. For example, use of lighter, stronger, and stiffer nano-composite materials is considered to have the potential to significantly reduce vehicle weight.

Nanoparticle coatings are a good short-term way of reducing emissions and maximizing clean energy production. For example, nano-coatings can be applied to aircraft, which can make aircraft's smoother, reducing drag and also protect the materials from the special conditions of the environment where they are used (instead of the conventional bulk metals, such as steel). Since the amount of CO_2 emitted by an aircraft engine is directly related to the amount of fuel burned, CO_2 can be reduced by making the airplane lighter. Hydrophobic nano-coatings can also improve the energy produced from solar panels.

Nanoparticle is already applied to improve fuel efficiency by incorporation of nanocatalysts. Enercat, a third generation nanocatalyst developed by Energenics, uses the oxygen storing cerium oxide nanoparticles to promote complete fuel combustion, which helps in reducing fuel consumption. Recently, the company has demonstrated fuel savings of 8%–10% on a mixed fleet of diesel vehicles in Italy. Reducing friction and improving wear resistance in engine and drive train components is of vital importance in the automotive sector. Applied nanosurfaces, reducing friction can lower the fuel consumption by about 2% and result in cutting down CO_2 emissions by 500 million tons per year from trucks and other heavy vehicles in Sweden.

Nanomaterials like silica made tyres will in the future be capable of attaining the best energy rating, the green category. Cars equipped with category A tyres consume approximately 7.5% less fuel than those with tyres of the minimum standard (category G). Residential and commercial buildings contribute to 11% of total greenhouse gas emissions. Space heating and cooling of residential buildings account for 40% of the total residential energy use. Nanostructured materials, such as aerogels, have the potential to greatly reduce heat transfer through building elements and assist in reducing heating loads placed on air-conditioning/heating systems. Aerogel is a nanoporous super-insulating material with extremely low density; silica aerogel is the lightest solid material known with excellent thermal insulating properties, high temperature stability, very low dielectric constant and high surface area. The harmful gas absorption by carbon nanotubes made from TiO_2 nanoparticles is shown in Figure 95.

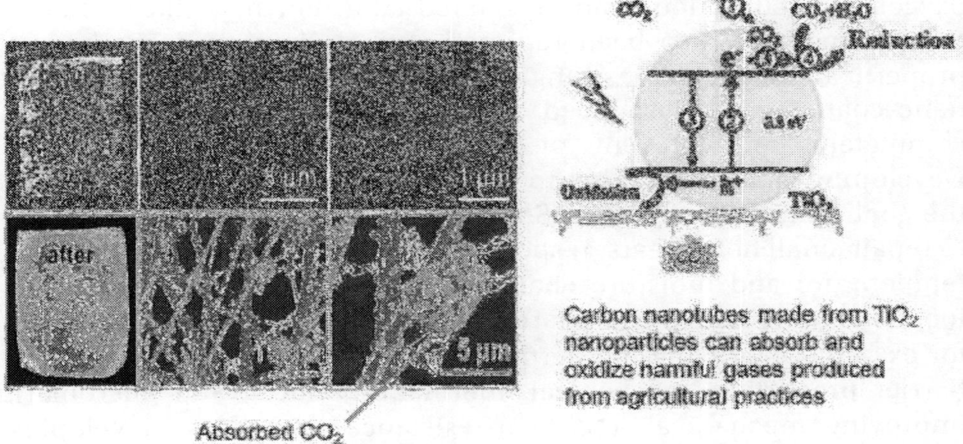

Figure 95: Harmful gas absorption by nanoparticles

Nanoparticles can improve renewables in various ways: experts are discovering means to apply nanoparticles to photovoltaics, which would produce solar panels with double or triple the output by 2025; wind turbines stand to be improved from high-performance nanomaterials like graphene, a nano-engineered one-atom thick layer of mineral graphite that is 100 times stronger than steel. Nanoparticles will enable light and stiff wind blades that spin at lower wind speeds than regular blades; nanoparticles also could play a major role in the next generation of batteries. For example, coating the surface of an electrode with nanoparticles increases the surface area, thereby allowing more current to flow between the electrode and the chemicals inside the battery. Such techniques could increase the efficiency of electric and hybrid vehicles by significantly reducing the weight of the batteries. Nanoparticle is positioned to create significant change across several domains, especially in energy where it may bring large and possibly sudden performance gains to renewable sources and Smart Grids. Nanomaterial enhancements may also increase battery power by orders of magnitude, allowing intermittent sources, such as solar and wind to provide a larger share of overall electricity supply without sacrificing stability.

14.9: Applications of Nanonutrient in Food Industry

Oxygen is a problematic factor in food packaging, as it can cause food spoilage and discolouration. Nanoparticles have been developed a new plastic that preventing the penetration of oxygen as a barrier. In other words, the oxygen for entry into package takes longer route due to nanoparticles intervention, and hence with the long route for oxygen molecules, food can be spoiled later. Polymer-silicate nano-composites have also been reported to have improved gas barrier properties, mechanical strength, and thermal stability. Recently, nano-coatings are produced for fruit that covering the fruits completely, and prevent of fruit weight loss and shrinkage. Developing smart packaging to optimize product shelf-life has been the goal of many companies. Such packaging systems would be able to repair small holes/tears, respond to environmental conditions (*e.g.*, temperature and moisture changes), and alert the customer if the food is contaminated. Nanoparticles can provide solutions for these, for example modifying the permeation behaviour of foils, increasing barrier properties (mechanical, thermal, chemical, and microbial), improving mechanical and heat-resistance properties, developing active antimicrobial and antifungal surfaces, and sensing as well as signalling microbiological and biochemical changes. In particular, silver nanoparticles have been shown to be a promising antimicrobial material. The most effective of controlling the fungus by nanoparticles is in < 24hrs. In addition, the different concentrations of silver nanoparticles controlled *A. flavus*. Therefore, the concentration of nanoparticles is effective for controlling the fungi into food stuffs package.

Nanofertilizer has a role in symbiotic exchange between soil and the plant system. When the same is applied slowly and efficiently, all the required nutrients are taken up by the plant and restores the required and efficient energy in it for which the yield increases drastically. When fertilizers are in the encapsulated form this can be achieved. Nano-porous zeolite usually help in slow release of the fertilizer to the plant, this way of doing makes the plant to grab entire amount of nutrients from the fertilizer supplied rather than the minimal uptake. Since it has larger surface area, many molecules can fit into it and get released whenever the plant requires.

Carbon nano-tubes are in cylindrical shape and mostly made up of carbon molecule with the difference in number of wall construction. These carbon molecules are held in position with strong van der Waals force making its application in packaging of particles, filtration, energy storing device, monitoring environment and many more. The reason for this numerous application is its flexibility alone. These multi-walled carbon nano-tubes have played a beneficial role in increasing the growth rate, water intake, uptake of essential nutrients from the soil. With the incorporation of varied ranges of carbon tubes, different yields were noticed along with this when an external Fe supplement given, the Ca ion made balance and maintained the yield. To other crops like maize, wheat, peanut and garlic this multi-walled tube with a concentration of 50 $\mu g\ mL^{-1}$ has given good results in an increase in the length of root and shoot, seed germination time, enhancement in growth and had well developed root system. Carbon nano-tubes help plants to retain their water content and it also increases the production rate drastically with very less amount of nano-material of 50 μg/mL.

Nano Aptamers are those which work with the principle of target like, just specific binding with high affinity, they are single stranded nucleic acid, they fit for the target in all the way forming three dimensional with strict bonding can be produced *in vitro*. Aptamer selection have certain criteria based on which it is selected for different process it includes the size range between nano or pico molar ranges. SELEX, a kind of sensors gives more specific and effective detecting plant diseases, crop resistance and yield production. Initial detection can help irradiate the problems well before. In order to know the proper regulation happening in the system there is no need to distress the cell instead a sensor to monitor cell to cell signalling with the help of photoluminescence target specific device, *i.e.,* insulin binding aptamer was deviced to monitor extinction of lights from cells to get the signal. In cases dealing with herbicide and pesticide (Atrazine and Malachite green) respectively, an efficient aptamer sensor was devised to monitor the toxicity level in food which is kept for consuming with the luminescent assay technique. Nano-particles can help by preventing surfaces becoming spoiled by rain, bacteria and environmental corrosion. Painted surfaces can remain perfect for up to 21 years and more using environmentally friendly nano solutions. Nanomaterials could even be used to control the release of the fertilizer, such that the nutrients are only taken up by the plant, and not lost to unintended targets like soil, water, or microorganisms.

Important benefits of nanofertilizers over conventional chemical fertilizers are:

- Nanofertilizer regulates the delivery of nutrients in crops through slow/control release mechanisms. Such a slow delivery of nutrients is associated with the covering or cementing of nutrients with nanomaterials. By taking advantage of this *i.e.,* slow nutrient delivery, growers can increase their crop growth because of consistently long-term delivery of nutrients to plants. For example, nutrients can be released over 40–50 days in a slow release fashion rather than the 4–10 days by the conventional fertilizers.
- Nanofertilizers required in small amount which reduce the cost of transportation and field application.
- An additional major advantage is over accumulation of salt in soil can be minimized as it required in small amount.
- Nanofertilizers can be synthesised according to the nutrient requirements of planned crops. In this regard, biosensors can be attached to a new innovative fertilizer that controls the delivery of the nutrients according to soil nutrient status, growth period of a crop or environmental conditions.
- The miniature size, high specific surface area and high reactivity of nanofertilizers increase the bioavailability of nutrients.
- Providing balanced nutrition, nanofertilizers facilitate the crop plants to fight various biotic and abiotic stresses.

It is reported that use of nanofertlizers and nanomaterials enhanced the growth and yield in several crops relative to plant treated with conventional fertilizers. However, the extensive use of nanofertilizers in agriculture may have some important limitations, which must be considered and it is crucial to determine the toxicity/biocompatibility of nanofertilizers.

Further studies

Allahvaisi, S. (2016). Effects of silver nanoparticles (AgNPs) and polymers on stored pests for improving the industry of packaging foodstuffs. *J. Entomol. Zool. Studies.* **4:** 633–640.

Chinnamuthu, C. R. and Boopathi, P. M. (2009). Nanotechnology and agroecosystem. *Madras Agril. J.* **96:** 17–31.

Khota, L. R, Sankarana, S., Majaa, J. M., Ehsania, R. and Schuster, E.W. (2012). Applications of nanomaterials in agricultural production and crop protection: a review. *Crop Prot.* **35:** 64–70.

Prasad, R., Bhattacharyya, A. and Nguyen, Q. D. (2017). Nanotechnology in sustainable agriculture: recent developments, challenges, and perspectives. *Frontiers Microbiol.* **8:** 1014–1035.

Shaimaa, H. A. E. and Mostafa, M. A. M. (2015). Applications of nanotechnology in agriculture: An Overview. *Egypt. J. Soil Sci.* **55:** 1–14.

15

Environment and Social Safeguard Aspect on Nanofertilizer Application

The ultra-small sizes that make the nanofertilizers of immense usefulness, unfortunately, the same characteristics cause several adverse effects. They may represent significant hazards to the environment, animals, human beings and plants when used non-judiciously. The possible hazards are:

a) Nanoparticles as fertilizers or pesticides, when airborne, may deposit on plant parts. They may plug the stomata and create a fine physical and toxic layer on stigma preventing pollen tube penetration. They may also enter the vascular tissue and impair translocation of water, minerals and photosynthate.

b) Animals may inhale nanofertilizers resulting in various ill effects and disorders. The particles may also enter the blood stream.

c) Nano-pesticides may reduce environmental contamination through the reduction in pesticide application rates, but they may also create new kinds of contamination of soil and waterways due to enhanced transport, longer persistence and higher toxicity.

d) Airborne nanofertilizer particles present some specific hazards for human health. They may enter the body through the respiratory system. Due to the entry of nanoparticles in lungs and blood-stream, there is a possibility of inflammation, protein fibrillation and induction of genotoxicity.

Due to these risks, the use of nanotechnology in fertilizer and pesticide formulations has to be addressed very cautiously, and this warrants mandatory need to critically analyze and examine the risks involved with the nano-formulations.

For nanofertilizer application, it is advisable to wear nitrile gloves and mask. Nanomaterial phytotoxicity is also an issue since different plant responds differently to various nanomaterials in a dose dependent manner. Therefore, before marketing we should know the critical doses of each nanomaterial for the crops applied. As nanomaterials are very reactive, because of their minute size with enhanced surface area, reactivity and variability of these materials are also a concern. This raises safety concerns of farm workers who may become exposed to xenobiotics during their application. These include not only those exposed to nanofertilizer manufacturing, but also nanofertilizer application in the field.

During manufacture of nanoparticles, use of B2 biosafety cabinet is internationally recommended. While handling of nanoparticles one should wear mask and nitrile gloves as a precautionary measure. Nanoparticle in gaseous phase should be experimented in an unsealed environment to study its exposure. The laboratory used for nanoparticle studies should be extremely clean (Figure 96). The dust particle concentration should not exceed 10^4 /m^2. Maintenance and up-gradation of machinery used in processing and distribution of nanofertilizers should be taken care of. Cleaning up of spilled chemicals and waste materials containing nanoparticles should be a primary task, as it exposes to the environment, these waste products can enter the soil and kill the necessary microbes or invade the aquatic life causing them to be extinct.

Considering the anticipated benefits, there is consequently a need to explore the feasibility and sustainability of these new smart fertilizers. Indeed a considerable concern about their transport, toxicity and bioavailability as well as unintended environmental impact upon exposure to biological systems.

15.1: Environmental Concern

It has been found that 15–33% loss of nanofertilizer to the environment is expected during spraying to the plants. The loss is more when normal sprayer is used than nebulizer (Table 70).

Other important issue is to their accumulation in plants and their food parts. The accumulation of nanoparticles depends on many factors, but mainly on plant species, tissue/organ that will be used directly as food or for food processing, and nanoparticles type and size. Due to the variability of interactions between nanoparticles and plants, nanomaterials used in nanofertilizers can accumulate in plants. In some cases, they can cause toxicity problems, not only

Figure 96: Laboratory safety protocols needed for nanofertilizer production (courtesy Washington University in St. Louis, USA)

Table 70: Loss of nanofertilizer during application in different sprayer used for spray

Nanoparticles (during spray)	Aerosol sprayer (nebulizer)	Normal spray
% Recover	85.3	67.5
% loss	14.7	32.5

to plants, but also to humans. Therefore it is necessary to work on the strategies that cope with their accumulation and potential risks for human health and the environment, while adopting the advantage of using nanoparticles in crops. Standardization of the correct dose and type of nanofertilizer recommended for each application and crop species is very important so that any potential toxicity to the environment, crops and food is minimized. Information on the nanoparticles bioavailability in edible parts is urgently needed to use nanofertilizer safely and globally.

There are numerous mechanisms by which different nanofertilizer particles induce their toxic effects, and these include cell proliferation, necrosis, apoptosis, DNA damage and oxidative stress, among others. Interestingly, however, these same mechanisms have also been shown to be caused by environmental toxicants, such as metal ions, pesticides, and other industrial chemicals. Thus, in investigating the minimum concentrations of nanofertilizers that can cause adverse effects, particularly in the actual environment that contain other toxicants. The risk assessment should involve aspects, such as additivity, synergistic, potentiation and antagonistic effects. This kind of information is hugely beneficial in terms of coming up with the regulatory framework and policies aimed at protecting human health concerning to nanofertilizer application in agriculture. This is because, besides humans being exposed to nanoparticles through nanotechnologically grown foods, they (humans) are also exposed to various other industrial chemicals through ingestion, inhalation and dermal contact. It is of no doubt, however, that the current assessment of the risks posed by nanofertilizers has several inherent limitations and uncertainties. The degree of uncertainty to a large extent is dependent on the application to which the nanofertilizers will be subjected.

Nanoparticles can easily enter into the biological systems which is not possible for the larger particles. Nanoparticles may cross cell membrane which depends on the dimension of nanoparticle. There is a possibility to inhale the nanoparticles by farm workers while spraying, which can reach the blood and may reach other target sites, such as brain, liver or heart. These nanoparticles might affect the regulatory mechanisms of enzymes and other proteins. There is a chance of nanoparticles entering into the food chains which ultimately affects the human health system. Hence, there is an urgent need to develop human resources with an understanding of the complexities of the agricultural production system to serve nanofertilizer applications in agriculture successfully.

There should be a planned program which can be implemented in case the soil, atmosphere or aquatic life is exposed to the toxicity of the nanofertilizers. Various chemicals should be kept ready to counter the harmful effects of nanofertilizer particles, if needed. Moreover, the workers should also be given proper education and training in the handling of nanofertilizer particle. For example, workers should be given respirators beside nitrile gloves and mask while spraying nano-fertilizers, so that the airborne toxic substances from the nanoparticles cannot affect them. There should be an organization responsible for surveying and keeping check on all the ongoing research and industrial activities related to nanofertilizer. In this way, the government can always keep a track on its upcoming applications as well as can ensure that the nanofertilizer production is not exceeding a designated toxicity level.

In agriculture, however, the current risk assessment data do not seem to be sufficient for both industry and consumers to make informed choices about the use of this technology. This insufficiency of data leads to some significant uncertainties that relate to consumer and environmental safety, which is critical in the regulatory framework, and a necessary ingredient in giving public confidence in the products. In addressing the current state of uncertainties, there are many of critical questions that need to be answered.

- Is current toxicity testing protocols sufficient to provide necessary information on delayed toxicity of nanofertilizers?
- Which dose metric best describes the toxicology of nanofertilizers, particularly through those that gain entry into humans through ingestion?
- Are there currently some validated techniques and methods that can detect the presence of nanoparticles in the food matrix?
- Is there sufficient regulatory framework to ensure safety of nanofertilizers related to their use in agriculture?
- Is the nanofertilizer toxicity data from cell lines sufficient to inform regulatory framework?
- Are there some guidelines on the generation of nanofertilizer risk assessment data to ensure comparability of, such data?
- Are risk assessment protocols used for both aquatic and terrestrial organisms sufficient to provide credible information for the exposure of humans through ingestion?
- What impacts will these nanosystems have on beneficial soil microorganisms?
- And finally, to what extent do these nanofertilizer accumulate and biotransform in plants?

These and several other questions will be dealt with the immediate effect. The question of whether the current toxicity testing protocols are sufficient to provide necessary information on delayed toxicity of nanofertilizers is one that speaks to the adequacy/inadequacy of the design of the risk assessments. The majority of the data from risk assessments is from traditional toxicity tests that rely predominantly on mortality and sub-lethal end points, such as oral, dermal and ocular toxicity; immunotoxicity; genotoxicity; reproductive and developmental toxicity; teratogenotoxicity; carcinogenicity, growth, foraging, behavioural changes and among others. These toxicity tests are quite costly and time consuming. Unfortunately, most of these tests do not necessarily capture the delayed toxicity and these do not give an opportunity for reliable prognosis about the ultimate effect on organisms. There is a suggestion that in order to understand the long-term impact that some of the nanoparticles used in agrochemicals may have on human health and environment, more studies should begin to incorporate the genomics and proteomics techniques. These techniques though they involve the state of the art of instrumentation can prove to be faster and cost effective in the long term, particularly in the face of thousands of nanofertilizers and nanochemicals that are anticipated to be generated in the coming decades.

Some uncertainty arises from lack of validated techniques and methods that can detect the presence of nanoparticles in the food matrix. The detection and ultimate characterization of different types of nanofertilizers in agricultural food is necessary in understanding the benefits as well as the potential risks. Although some methods for detection and characterization of, such nanoparticles are currently available, these methods need to be validated in addition to the need for the development of standard materials required in, such methods. Given that there are a number of nanopaticles that are being developed for use in agriculture, need exists for research and development of more and validated methods required in the detection of nanoparticles in agricultural products, especially food crops.

The application of nanostructures or nanoparticles as agrochemicals (fertilizers or pesticides) is systematically being explored, before nanofertilizers could be used in agriculture or farming for a general farm practice. The properties of many nanoparticles are considered to be of potential risk to human health, *viz.*, size, shape, solubility, crystal phase, type of material, exposure, dosage and concentrations. However, expert opinions indicate that food products containing nanoparticles available in the market are probably safe to eat, but this is an area that needs to be more actively investigated. To address the safety concern detail studies are required to know the impact of nanoparticles within the human body once exposed through nanofood. Researchers have to assess and develop proper assessment strategies to assess the impact of nanoparticles and nanofertilizers on biotic and abiotic components of ecosystem. Among the various issues, the accumulation of nanomaterials in environment, edible part of plants might be the important issues before use in agriculture.

It was noted that rats breathing in nano-particles generally have those particles settle in the brain and lungs, which led to a significant increase in biomarkers for inflammation and stress response, and that nano-particles include induce skin ageing through oxidative stress in hairless mice. A two-year study at UCLA's School of Public Health found lab mice consuming nano-titanium dioxide showed DNA and chromosome damage to a degree "linked to all the big killers of man, namely cancer, heart disease, neurological disease and ageing".

The Royal Society report identified a risk of nano-particles or nano-tubes being released during disposal, destruction and recycling, and recommended that "manufacturers of products" that fall under extended producer responsibility. Reflecting the challenged for ensuring responsible life cycle regulation, the Institute for Food and Agricultural Standards has proposed that standards for nanoparticle research and development be integrate across consumer, worker and environmental standards. They also propose that the NGOs and other citizen groups play a meaningful role in the development of these standards. Due to these risks, the use of nanofertilizer and pesticide formulations has to be addressed very cautiously, and this warrants mandatory need to critically analyse and examine the risks involved with the nano-formulations. If these limitations are countered, nanofertilizer technology will become a revolutionizing technology of global farming in future generations.

The initial studies performed for nano-materials have caused serious health hazards and also showed toxic effects when entered into human body. It may cause tissue damage while reaching all the vital organs. For example, silver nanofertilizers are frequently using to plants because of their antimicrobial properties, but studies have considered that it poses serious threat to ecosystem causing membrane damage, reducing the annual growth of grass, depletion photosynthesis in alga (*chlamydomonas reinhardtii*). Moreover, silver nano-particles are usually difficult to recover; some plant species tends to use this nano-particle maximum and accumulates in its tissue exceeding the limit. For example, soybean, an important cash crop in most of the country was produced using manufactured nano-materials with fossil fuel equipment that will allow nanofertilizer material to locally deposit on the crop.

15.2: Health Concern of Nanofertilizers

To address the safety concern detail studies are required to know the impact of nanoparticles within the human body once exposed through nanofood. Researchers have to assess and develop proper assessment strategies to assess the impact of nanoparticles and nanofertilizers on biotic and abiotic components of ecosystem. Among the various issues, the accumulation of nanomaterials in environment, edible part of plants might be the important issues before use in agriculture. The nanosize properties of nanofertilizer, such as surface structure and reactivity are responsible for processes, such as dissolution, redox reactions and generation of reactive oxygen species (ROS). Nanofertilizers particles also may induce ecotoxicological effect on an organism upon prolong contact, the ingestion and inhalation will depend to a larger extent on size of the material. Most nanofertilizer particles are likely to be of human or environmental health concern only when they have diameters of 30 nm or less.

To assess the adverse effect, a number of toxicity test have been conducted using various organisms. Researchers have tried to understand the best dose metrics that would define the toxic limit of a given nanofertilizer. The other factors they are trying as its particle charge, concentration including number of particles, total surface area beside size of the particles. There should be health hazards limit value for each matrics so that risk assessors can compare and combine exposure and hazard information and conclude on the likelihood of health risks of each nanofertilizer type.

Nano-ecotoxicity and risk management of various types of nanoparticles, such as carbon-, inorganic-, and organic- based nanoparticles have been investigated. Due to their great potential to enter into the human systematic circulation system and interact with vital organs, carefully designed and comprehensive toxicity tests involving *in vivo* and *in vitro* have been carried out to assure safety to human and environmental health. Carbon-based nanoparticles have comprehensively being studied on their toxicological impacts.

Generally these nanoparticles have been observed to show some low or no toxicity both for *in vitro* (on cell cultures) and *in vivo* (on organisms). However, in some cases results have shown adverse effects. For example, a number of carbon based nanoparticles shown cytotoxic to human alveolar epithelial (A549) cells, hepatocytes (hep G2 cells), human embryonic kidney cells (HCT 116), and intestinal (p 407 cells) cells. A number of studies have shown that these nanoparticles can have negative effects, such as inhibitory effects against plant growth and against some beneficial microflora. Therefore, to ensure that these nanofertilizer particles are safely used in a manner that human and environmental health is ascertained.

Inorganic nanoparticles are the most diverse group of nanoparticles that include metals, metal oxide and quantum dots (QDs). Nanoparticles, such as TiO_2, ZnO, MgO and combination of other inorganic nanoparticles have been utilized as effective nanopesticides. The nanofertilizers are known to contain nano-zinc, nano-K, nano-N, nano-silica, Fe and TiO_2, zinc-cadmium selenide, manganese/zinc selenide etc. specially designed for control release. The inorganic based nanoparticles have been shown to cause both acute and chronic toxicity. The toxic effects have been observed in plants, animals, microflora, macroflora including cell lines when the critical limit exceeds. Although the ecotoxicological data are available for aquatic and terrestrial organisms, it is predominantly from species deemed highly sensitive. For acute toxicity, nanoparticles, such as Cu, CuO, Se, Zn, ZnO and TiO_2 have been implicated in numerous studies. For instance, silver and copper nanoparticles were absorved to cause adverse effects to both zebra fish and *Daophnia pulex,* while Cu nanoparticles were observed to cause oxidative stress to earthworms, *Eisenia fetida*. Commonly encountered metal oxide nanoparticles, such as CuO, ZnO, TiO_2, Fe_2O_3, SnO_2, CeO_2 etc. have also been implicated in causing diverse effects to organisms. Quantum dots are particularly toxic as they are usually made from already known toxic materials. For instance, cadmium selenide, cadmium telluride, gallium have been shown to cause immunotoxicity, oxidative stress and DNA damage. In most cases, the inorganic based nanoparticles are coated or encapsulated immediately after synthesis to prevent any aggregation and preserve their properties. Usually this surface functionalization can result into behavioural modifications, which in turn have a direct impact on the surface charge, size and reactivity that in turn reduced toxicity.

The organic based nanofertilizer particles are being widely used in agriculture. These nanoparticles can be synthesised as nanowires or nanofibres and may be designed as hydrophilic or hydrophobic depending on the anticipated application. The most important examples of organic nanoparticles are nanocapsules and polymeric nanoparticles. These nanoparticles also used in the formulations of smart delivery nanosystems. For instance encapsulation of fertilizer and pesticides in the organic nanoparticels ensure that there is slow and controlled release of the active ingredient and therefore delivering more effective control over plant nutrition and certain pests at lower doses rates and over prolonged period of time. These reduce the overdosing and hence prevent pollution. Moreover, as smart systems, the nanoparticles incresed the dispersion and solubility, and therby reducing the accidental toxic effects.

There are various mechanisms by which different nanofertilizer particles induce their toxic effects. They are: cell proliferation, necrosis, apoptosis, DNA damage and oxidative stress among others. Thus it is absolutely necessary to investigate the minimum concentration of nanoparticles that can cause adverse effects, particularly in the actual environment that contain other toxicants. The risk assessment should involve aspects, such as additivity, synergistic, potentiation and antagonistic effects. This type of information is very beneficial in terms of coming up with the regulatory framework and policies aimed at protecting human health with regard to nanofertilizer in agriculture. This is because human being exposed to nanoparticles for growing crops through ingestion, inhalation and dermal contact.

15.3: Development of Risk Management Programme

Implementation of a risk management programme in workplaces where exposure to nanomaterials exists can help to minimize the potential of exposure to nano-aerosols. Elements in, such a programme should include:

- Evaluating the hazard posed by the nanofertilizer based on available physical and chemical property data and toxicology or health effects data.
- Assessing potential worker exposure to determine the degree of risk.
- Education and training of worker in the proper handling of nanofertilizers (*e.g.*, good work practices).

- Establishment of criteria and procedures for installing and evaluating engineering controls (e,g, exhaust, ventilation) at locations where exposure to nanofertilizers might occurs.
- Development of procedures for determining the need and selection of personal protective equipment (*e.g.,* clothing, gloves, respirators).
- Systematic evaluation of exposures to ensure that control measures are working properly and the workers are being provided the appropriate personal protective equipments.

15.4: Effects on Nanoparticles Exposure

An individual is exposed to nanofertilizer particles while handling these substances at either in the laboratory during research or during the large scale manufacturing processes consequently the level of hazards exposure will also vary depending on the quantity of the nanofertilizer particles and the nature of the nanoparticles used.

15.4.1: Respiratory exposure

Ultrafine nanoparticles are proven to be 10 to 50 times more damaging to lung tissue than other particles. Nanoparticles exposure in excess amount may cause lung inflammation, lung fibrosis, asthama, breathing problems and possibility of Alzheimer disease. The level of damage depends on their size and how deeply they penetrate the lungs. Studies show that smaller particles cause a stronger reaction in lung tissue, as surface reactivity can harm the surrounding tissue through chemical activity. Even large particles that normally do not cause damage to lung tissue, can cause damage when they are crushed to nanoscale.

Most inhaled particles are exhaled, but those particles that penetrate the lungs more deeply, enter into pulmonary alveoli. Cells called phagocytes are responsible for absorbing foreign substances in the lungs and eliminating them. But if phagocytes absorb nanoparticles, a large amount of nanoparticles could be deposited in the lung tissue without being exhaled again. As the paulmonary alveoli exchange oxygen with blood, the nanoparticles could enter the bloodstream and be absorbed by blood cells. Studies have also shown that when inhaled, some nanoparticles can directly enter into the brain through the nasal mucous membrane, thus by-passing the blood-brain barrier.

15.4.2: Exposure through ingestion

Nanofertilizer particles may be ingested through drinking water, food additives, atmospheric dust on food, toothpaste and dental fillings and implants. Ingested nanoparticles can then be absorbed through Peyer's patches or small nodules in intestinal tissue that are part of the immune defense system. If nanoparticles enter the digestive system and proceed into the blood stream, they could move throughout the body and cause damage. Nanoparticles may also accumulate in certain organs, disrupt and impair biological, structural and metabolic processes and weaken the immune system.

15.4.3: Dermal exposure

Harmful effect arising from skin exposure may either occur locally within the skin or alternatively the substance may be absorbed through the skin and dessiminated via bloodstream, possibly causing systematic effects. Dermal absorbtion of nanoparticles has not been well investigated and suggested that the nanoparticles may penetrate into hair follicles where constituents of the particles could dissolve in the aqueous conditions and enter the skin. Direct penetration of the skin has been reported for particles with a diameter of 1000 nm, much larger than nanoparticles. So, it can be assumed that nanoparticles are more likely to penetrate.

15.5: Social Safeguard Concern

Good work practices can help to minimize worker exposures to nanofertilizer materials. The managerial and safety staff has a big role to play in putting up viable and dynamic safety programme in the field and working place with a high degree of commitments. An integral step in establishing good work practices is having knowledge of the potential hazards in the work place and developing formal procedures that describe action to be taken to ensure the protection of researchers and workers. Educating workers on the safe handling of the fertilizer nanoparticles or nano-object containing materials to minimize the likelihood of inhalation exposure and skin contact. Providing appropriate information as needed on the hazardous properties of the precursor materials and those of resulting nanofertilizer material products with instructions on the measure to prevent exposure is very important to ensure that the adequate safety precautions are in place to handle each product. Transportation of the nanofertilizer material outside the work area without appropriate sealed and labeled containers, and through untrained personnel should be prohibited. Additional control measures, such as use of buffer area, decontamination facilities for workers (if warranted by the hazards) and facilities for showering and changing clothes should be in place to prevent the inadvertent contamination of other areas (including take home) caused by transfer of nanomaterials on the clothing and skin.

In addition to an establishment of nanomaterial safety system, safe handling of nanomaterial by the workers and personnel requires a high degree of awareness, commitment and discipline from the workers. Avoiding handling nanomaterials in the open air in a free particle state and storing dispersable nanomaterials, whether suspedded in liquids or in dry particle form in closed (tightly sealed) containers. Whenever possible, ensures the localization of the nanomaterial at a specific place that reduces its inadvertent exposure. A culture of cleaning work areas at the end of each work shift in the industry as well as in the field, at a minimum, using either HEPA-filtered vaccum cleaner or wet wipping methods at the end of each work shift (in the industry or the laborateris), needs to be developed. Dry sweeping or air hoses should not be used to clean work areas. Clean up should be conducted in a manner that prevents worker contact with wastes. Avoiding storing and consuming food or beverages in workplaces where nanofertilizer materials are handled would prevent unintentional exposure and ingestion of the particles.

15.5.1: Personal protective equipment

No guidelines are currently available on the selection of clothing or other apparel (*e.g.*, gloves, hat) for prevention of dermal exposer to nanoaerosols. As stated earlier, nitrile gloves and mask are generally practised. However, some clothing standards incorporate testing with nanoscale particles and therefore provide some indications of the effectiveness of protective clothing with regard to nanofertilizer particles. Use of respirators is often required when engineering and administrative controls do not adequately keep worker exposures to an airborne contaminant below a regulatory limit or an internal control target. Currently there was no specific exposure limits for air borne exposures to engineered nanoparticles, although occupational exposure limits exist for larger particles of similar chemical composition.

The decision to use respiratory protection should be based on professional judgement that takes into account toxicity information, exposure measurement data, and the frequency, and likelihood of the workers exposure. Preliminary evidence shows that for respirator filtration media, there is no deviation from the classical single fibre theory for particulates as small as 2.5 nm in diameter. While this evidence needs confirmation, it is likely that NIOSH certified respirators will be useful for protecting workers from nanoparticles inhalation when properly selected and fit tested as part of a complete respiratory protective programme,

As, such, safety concerns with nanofertilizer particles must not be ignored because it is one of the ways to make the technology very safe for anyone using them. Since nanofertilizer technology is still a new science, long term safety consequences of the exposure of nanofertilizer particles in the human population and the environment, and also the efficacy of the existing available safety measures and equipments are still not clearly known. A lot of research work in this area is required to be carried out to actually come up with a foolproof safety protocol for handling these materials both in the laboratory and field level as well as under the commercial production and application of these nanomaterial based products.

15.6: Nanofertilizer Quality Control

A tentative guide has been prepared for quality control of nanofertilizers which is under approval of Government of India. The proposed regulatory framework are as follows:

15.6.1: Nano Nutrient Products (Nomenclature / Definition / Scope / Application)

- At the scale of (1–100 nm); what is the most appropriate size for higher efficacy? Accordingly, the definition of Nano Nutrient Products (agri input) may be defined in FCO.
- How size will affect the efficiency of nutrients in plant system as well as determine its application rate and requirement in crops?.
- What are the nutrients which are suitable for application at nano-scale? If all nutrients can be applied or are required to be applied then it may need a separate FCO for addressing it.
- Nutrient uptake in crops varies and depends on several factors, such as soil, moisture, crop demand, fertility status etc. What would be the long term effect of application on inherent soil fertility?
- Nano Nutrient Products (agri input) also to be seen in light of overall input use efficiency, such as that of in terms of 1) Water

use/Application efficiency; 2) Expression of seeds/ plants etc. (Corresponding studies/details is needed) .

- Though, there is broad scope of application of nano-nutrients, its best application mode/method in terms of safety, efficacy and affordability to be promoted, such as a) Soil application; b) Foliar application; Fertigation etc.
- Patents/Ease in Technology availability (It is a critical issue to address anomalies that may arise due to application of Nano Nutrient Products (agri input).

15.6.2: Research & Development (Scope, Application and Limitations)

15.6.2.1: Method of synthesis of the material

- Physical
- Chemical
- Biological
- Aerosol

15.6.2.2: Release mechanism

- Regulated
- Spontaneous

15.6.3: Agriculture specific

- As seed germination
- As Nano fertilisers

15.6.3.1.: Justifying 4 R principle (Right Source, Right Rate, Right Time , Right Place)

- Slow and Controlled release
- Instantaneous action

15.6.3.2: Pesticides, herbicides etc.

15.6.3.3: For water retention (Hydrogel)

15.6.4: Production Technology (Scope and Limitations)

15.6.4.1: Production Technology (Economical / Costly)

15.6.4.2: Process of Manufacture (Nano Technology) - 1) Green; 2) Physical ; 3) Chemical; 4) Biological; 5) Aerosol

15.6.4.3: Production status

- Lab scale
- Pilot scale
- Commercial scale

15.6.5: Experimental trials (To ascertain – Scope and Limitations)

15.6.5.1: Comparative advantage of nano-nutrients vis-a-vis conventional fertilizers / Water soluble fertilizers / Chelates / Slow release fertilizers etc.

15.6.5.2: Crop specificity

- Cereals / Pulses / Oilseeds/ Horticultural and Medicinal crops, Plantation crops, Medicinal and aromatic crops etc.

15.6.5.3: Area specificity

- Irrigated / Rainfed areas

15.6.5.4: Yield advantage over a period of time

15.6.6: Policy

15.6.6.1: Specifications to maintain uniformity and duplicity

- Categorisation - Type , Size, Shape, Source
- Source - Biological, Organic, Chemical, Physical, Aerosol
- Form / State – Solid, Liquid

15.6.6.2: Specifications / Quality Standards

SN	Parameter	Range	Deviation/Tolerance limit
1	Particle size	1–100nm	± 10 % , tolerance limit should be mentioned
2	Particle Purity	66 % (min.)	Only 5% to the minimum level, there is no maximum limit
3	Particle Concentration (Liquid)	200 mg/L(Min.) 5000 mg/L (Max.)	± 10%, tolerance limit should include

SN	Parameter	Range	Deviation/Tolerance limit
4	Moisture % by Wt. (in case of solid)	1.0 (max.)	—
5	Solubility (in case of solid)	99 % (Min.)	—
6	Polydispersity index (PDI value)	< 1.0	—
7	Stability (Shelf life)	6 months – 1 year	± 5%
8	pH	Should mention	—

15.6.7: Grades / Nanoformulations

- Single nutrient / Multi nutrient / Combinations etc.

15.6.7.1: Nano-phosphorus; Nano-Zinc; Nano-fortified N / NP / NPK grades etc.

15.6.7.2: Nano-composites

15.6.7.3: Foliar Grade ; Soil Grade ;Fertigation Grade / Hydroponic Grade; Seed Coating grades

15.6.7.4: Encapsulation – Microbial, Organic, Chemical

15.6.8: Testing Procedure/Designated Labs for validation

15.6.8.1: Sampling procedure

- Screening Test
  - a. Individual Nutrient grades
  - b. Combination grades
- Confirmatory Test
  - a. Individual Nutrient grades
  - b. Combination grades

15.6.8.2: Marketing

i. Packaging size

ii. Sampling procedures

iii. Label claim (said to contain)

15.6.8.3: Safety standards / Precautions

i. During Storage / Warehousing

ii. Application

iii. Disposal of product

15.6.8.4: Compliance

i. Monitoring and Evaluation Committee

ii. Standard procedure guidelines for handling, storage, packaging and transportation

iii. Qualified and certified personnel, trained applicators at field level

15.6.8.5: Exemptions

i. For naturally present nano in existing products

ii. For Microbial, Organic, Green Synthesis products

iii. Biodegradable and Nano

15.7: Public Awareness and Perceptions

Although nanofertilizer is viewed as the most important technology with major potential benefits for future agriculture in the world. It has to be embraced with a precautionary measure that not much is known about its unintended effects on account of being new. A large proportion of the people in the world have very limited knowledge about this technology. Even with little knowledge, a majority of the people believe that benefits of nanofertilizer outweigh the risks involved in the technology, as well as it is extremely essential to replace the present chemical fertilizers to protect soil and environmental health. As expected the level of knowledge of nanofertilizer is much higher among the highly educated than those with least education.

When people know little about the technology, their perception and acceptability will to a large extent depend on social and ethical issues concerning the technology that has to be properly handled by the industry and the researchers. As the knowledge is missing, people can use heuristics, such as trust to assess the risks and benefits of the new technology. Thus people are more likely to accept assurances about the safety of nanofertilizer technologies in agriculture when they have higher levels of trust in the institutions, researchers and industry. Another aspect that affects the public perception about nanofertilizer technology is the way the media reports issues on the technology. In less developed countries, the level of coverage of nanofertilizer technology is very low and this is coupled with low levels of research in the technology. For developed countries, with high levels of knowledge and application of nanofertilizer technology, the reporting or coverage of nanofertilizer technology may be modest probably due to the specialized nature of the field and hence requiring specialist journalists who may be fewer.

Based on the factors that influences the perception and awareness of the nanofertilizer technology in agriculture, there are quite a number of ethical issues that arise. So the details study is needed on the effect of human health and environment to encourage the industry continue using these nanosystems despite the few uncertainties. The public/government institutions should provide more funding for the development of the products along with to remove the ethical issues with logical conclusions despite the current levels of uncertainties. The public should aware about much information in relation to the nanofertilizer in the entire agricultural value chain. The industry who produces and marketed nanofertilizer should release its potential doses as well as made public to aware its negative impact on their health if the proper doses are not followed. There should be regulations set in regulating nanofertilizer in agriculture to increase public perception and acceptability. If these issues are not fully addressed, the negative perception and reluctance of acceptance of this technology may be experienced.

There is no doubt that fertilizers have contributed greatly to the growth and increase in agricultural production. Despite huge benefits in terms of increased agricultural productivity, the excessive use of chemical fertilizers has resulted into serious pollution to aquatic and terrestrial systems. The pollution has also resulted into increased disease burden, particularly to humans and detoriating soil health. The elevated levels of nitrates from chemical fertilizers have been found in both surface and ground water resources in various places across the countries and continents. One of the main reasons for chemical fertilizer pollution is due to yearly progressive increase in their application and only a small fraction of applied chemical fertilizer get utilized by plants. Therefore, the advent of smart nanosystems, such as nanofertilizers, nanobiosensors and nanopesticides are designed to increase solubility, nutrient use efficiency, etc. Enhances bioavailability and promotes targated delivery and controlled release over a long period of time that immensely benefits the entire agricultural value chain. Thus nanofertilizer in agriculture will improve the efficient doses of fertilizers, improve vector and pest management, reduce chemical pollution as well as improve soil and environmental health with disciplined application.

Of course currently, nanofertilizer, nanopesticides and nanobiosensors that are in use have made a huge impact in revolutioning agriculture. However, the use of nanofertilizer in agriculrure has a number of risks unless proper knowledge has been provided to the end users. Therefore, before advocating using throughout the world the following questions remain to be solved:

- Is current toxicity testing protocols sufficient to provide necessary information on delayed toxicity of nanofertilizer particles?
- Which doses are the best described the toxicology of nanofertilizer particles through those that gain entry into human through ingestion?
- Are there currently some validated techniques and methods that can detect the presence of nanofertilizer particles in the food matrix?
- Is there sufficient regulatory framework to ensure safety of nanofertilizer particles related to their use in agricultural fields?
- Is there nanofertilizer particles toxicity data from cell lines sufficient to inform regulatory framework?
- Are there some guidelines on the generation of nanofertilizer risk assessment data to ensure comparability of, such data?
- Are risk assessment protocols used for both aquatic and terrestrial organisms sufficient to provide credible information for the exposure of humans through ingestion?
- What impacts will these nanofertilizers on beneficial soil organisms?
- And finally, to what extent do these nanofertilizer particles accumulate and biotransform upon entry into plants?

All these questions demand new approaches and perspectives in the design of risk assessment methods to ensure that humans and the environment are safeguarded from nanofertilizer particles potential harm, if any, as a result of their application in agriculture.

15.7.1: Limited knowledge

The important issues of nanofertilizer technology are their low or limited knowledge of the general public as well as their low levels of publicity. Despite numerous benefit of the technology, when that technology is unknown, people will only resort to using heuristics, such as trust to inform their perception about risks and benefits. If the industry and public regulators, for instance, FDA, have built a good relationship with the general public in terms of providing adequate oversight through trustworthy information, the public/farmer is inclined to believe when assured that a given product is safe. Furthermore, some ethical issues arise as to how much information the general public is given on the potential risks of the nanofertilizer in agriculture. The role of the media is critical in shaping public opinion and perception about the technology. Unfortunately, only few journalists are well schooled to report appropriately and effectively on the issues of nanofertilizer in agriculture. In order to gain public acceptance and avoid incidences of negative connotation, there should be adequate follow of information. The labeling of agricultural crops containing nanofertilizer should be encouraged to promote the free choice of use of, such products.

15.8: Nano-ecotoxicity and Risks

The nano-properties of fertilizers change their surface structure and reactivities that changes their properties of dissolution, redox reactions and the generation of reactive oxygen species (ROS). These are the properties that can lead to biological/toxicological effects that would not be produced by bulk particles of the same chemical composition. The size is an important factor to induce ecotoxicological effects on an organism upon contact, ingestion or inhalation. In assessing the potential adverse effects, several toxicity tests have been conducted in various media and using a variety of organisms. But sometimes conflicting results are obtained due to the lack of assessment guidelines. In order to reduce, such dichotomy results, researchers have tried to understand the best dose matrics that would define the toxicology of nanofertilizers. Researchers have investigated whether a given nanofertilizer particles induces its toxicity through its particle charge, number concentration, mass concentration, total surface area or simply by size. Some researchers stated that knowledge of dose metrics responsible for toxic effects, it can have a number of advantages that include easy of adaption of the risk assessmsnt data into the regulatory framework that ensures the safe use of, such nanofertilizer particles. It will also provide easy to comparison of study results and hence enable regulators to formulate health-based limit values for each metric. And finally, this can help risk assessors to compare and combine exposure and hazard information and conclude on the livelihood of health risks of each nanofertilizer type.

Although the application of nanofertilizer in agriculture is still largely in the developing stage, there is a great potential to cover the entire food chain from production to processing, preservation, safety packaging, transportation, storage and delivery. For instance, a variety of products exists, such as nanosised fertilizers, nanopesticides, nanopromoters for plant growth, nanosensors etc, which when applied cannot only come into contact with humans, but can actually be consumed along with the agricultural products. Due to their great potential to enter into the human systematic circulation system and interact with vital organs, carefully designed and comprehensive toxicity tests involving *in vivo* and *in vitro* have been carried out to assure safety to the human and environmental health. Generally, nanofertilizers have been observed to shown some low or no toxicity in majority of cases, but some other cases, the toxicity results have shown adverse effects. Therefore, to ensure that the particular nanofertilizer can safely used in a manner that human and environmental health is ascertained, extensive and comprehensive risk assessments that include techniques which can capture delayed toxicity are required.

15.8.1: Assessment methods of nanofertilizer toxicity

There has been an extensive evaluation of the toxic effects of nanofertilizer particles and currently these evaluations are still ongoing. There are numerous mechanisms by which different nanofertilizer particles induce their toxic effects and these include cell proliferation, necrosis, apoptosis, DNA damage and oxidative stress among others. Interestingly, these same mechanisms have also been shown to be caused by environmental toxicants, such as metal ions, pesticides and other industrial chemicals. Thus in investigating the minimum concentration of nanofertilizer particles that can cause adverse effects, particularly in the actual environment that contain other toxicants, the risk assessment should involve aspects, such as additivity, synergistic, potentiation and antagonistic effects. This kind of information is hugely beneficial in terms of coming up with regulatory framework and policies aimed at protecting human health with regard to nanofertilizer in agriculture. There is no doubt that the current assessment of the risks posted by nanofertilizer particles has a number of inherent limitations and uncertainties. The degree of uncertainty to a large extent is dependent on the application to which nanofertilizer particles will be subjected. As observed, the genetical risk in the application of nanofertilizer in various fields has been reasonably assessed. But the current risk assessment data in agriculture do not seem to be sufficient for both industry and cosumers to make informed choices about the use of this technology.

This insufficiency of data leads to some significant uncertainties that relate to consumer and environmental safety, which is critical in the regulatory framework, and a necessary ingredient in giving public confidence in the products. In addressing the current state of uncertainties, there are number of critical questions that need to be answered. Moreover, the testing protocol needs to be examined thoroughly including the delayed toxicity of the material. The question of whether the current toxicity testing protocols are sufficient to provide necessary information on delayed toxicity of nanofertilizer is one that speaks the adequacy/inadequacy of the design of the risk assessment. The majority of the data from risk assessment is from traditional toxicity tests that rely predominantly on mortality and sub-lethal endpoints, such as oral, dermal and ocular toxicity; immunotoxicity, genotoxicity, reproductive and developmental toxicity, teratogenic toxicity, carcinogenicity, growth, foraging, behavioural changes and among others. These toxicity tests are quite costly and time consuming. Unfortunately, most of these tests do not necessarily capture the delayed toxicity and these do not give an opportunity for reliable prognosis about the ultimate effect on organisms. In order to understand the long term impact some of the nanofertilizers used in agrochemicals should begin more studies using genomics and proteomics techniques. This technique though they involve the state of the art of instrumentation can prove to be faster and cost effective in the long term, particularly in the face of thousands of nanochemicals that are anticipated to be generated in the coming decades. The aspect of the dose matric that best describe the toxicology of nanofertilizer particles, particularly, has been the subjective of debate among nano-toxicologists for quite some time. Traditionally, mass has been used a dose metrics for most risk assessments for most nanofertilizers. However, other dose metrics, such as surface area, number of particles, volume and size have also been investigated on their influence on toxicity, irrespective of chemical compositions. While in some cases, a particular dose metric could be responsible for the observed toxic effect, in other cases, another dose metric may be responsible. This create some uncertainty, and thus, risks assessments for nanofertilizers need to ensure that all factors of a given nanoparticles type that lead to some toxic effects are clearly understood. This is particularly important because different nanofertilizer particles have different risks, uncertainties and ethics.

Some nanofertilizers are soluble, while others are insoluble and some may be biopersistent. Some uncertainties arise from lack of validated techniques and methods that can detect the presence of nanoparticles in the food matrix. The detection and ultimate characterization of different types of nanoparticles in agricultural food is necessary in understanding the benefits as well as the potential risks. Although few methods for detection and characterization of, such nanoparticles are currently available, these methods need to be validated in addition to the need for the development of standard materials required in, such methods. There are more validated methods required in the detection of nanoparticles in agricultural products, especially food crops. Although many countries are now setting definitions and regulatory frameworks for nanofertilizer technology, the very nature of nanoparticles in many ways makes it quite challenging in coming up with separate legislation that deals with these miniature materials away from their bulk counterparts. For example, both the United States Food and Drug Administration (US FDA) and the United States Environmental Protection Agency (US EPA) have not recognized nanomaterials as the new chemicals and that nanomaterials do not require any new oversight. Ironically, these bodies (especially US FDA) require manufacturers of food products to demonstrate that the food ingredients and food products are not harmful to health. There is no regulation that specifically covers nanoparticles, which could harmful only in nanosised applications. In the similar fashion, it is interesting to note that the European Unions main regulation covering nanotechnology application is the REACH (EU Regulation on Registration, Evaluation, Authorization and Restriction of Chemicals). Generally, as the technology is new, there are currently no clear regulations governing the production, use, labeling and disposal of nanomaterials. With the predicted increase in the production and commercialization nanomaterial for use in agriculture, there is a need for clear cut legislation and policies to protect and foster public health and confidence. Now, the majority of the risk assessment data is from *in vitro* animal tests. In as much as these tests can yield some useful information necessary to inform the regulatory framework, they are costly and time consuming. Additionally, the traditional tests normally involve one type of nanoparticles at a time. But humans will be exposed to these nanoparticles used in agriculture together with other chemical contaminants. The data from these tests therefore may have an inherent degree of uncertainty. Recently, there has been some suggestion using genomics, proteomics, transcriptomics and metabolomics (the omics technique) as throughout techniques utilizing cell lines to cope with the rate of the production of the nanomaterials. Here again, there would be quite a number of uncertainties. For example, how reliable is the data from, such techniques in terms of extra-polatability and pre-divtability to human biology and physiology? Particularly, what is the degree of uncertainty for these data obtained from isolated cell lines kept in culture medium without the benefit of cross talk and interaction from other organs, as would be the case in the *in vivo* tests, have? The protective regulatory framework should always take into account the uncertainty to assure public confidence and trust.

15.8.2: New visions in plant science

There are quite a number of reasons why application of nanofertilizer technology in the agriculture is still at an infancy stage in comparison with other fields. The major ones include potential consumer health risks and a lack of unifying regulations and guidelines on risk assessment of nanofertilizer. The use of nanofertilizer technology in agriculture more than any application can lead to the introduction of nanoparticles in the human biology and physiology. Therefore, when risk assessment is not guided by unifying guidelines and regulations, then the risk quotient may be high and this can make the technology non-attractive to industries and consumers alike. To harmonize the nanofertilizer risk assessment data and ensure comparability, there is need for some guidelines of these data. Currently, one of the challenges relating to the usability of nanofertilizer risk assessment data in regulatory framework is the somewhat conflicting nature of the toxicity results of different researchers. When there are specific guidelines to follow, during the processes of conducting risk assessment of nanomaterials, the degree of uncertainty is minimized and regulatory framework can easily be formulated, particularly for agriculture.

Risk assessment protocols used for both aquatic and terrestrial organisms have contributed greatly in understanding the effect of nanoparticles to organisms and to large extent have provided credible information required for the development of safety guidelines on nanofertilizer technology in general. However, with regard to nanofertilizer in agriculture, a sensitive field, new protocols and research designs of evaluating safety of nanoparticles are required. In addition to safeguarding human health as benefits of nanofertilizer in agriculture increase, the safety of beneficial soil microorganisms which enable nutrient cycling and hence help to maintain basic soil fertility, need to be protected. Thus there is need to carry out comprehensive nanofertilizer risk assessment for all the nanoparticles type envisaged to be used in agriculture. And finally, more works need to be done in investigating whether or not nanofertilizer particles can bioaccumulate and biotransform in plant materials.

15.8.3: Perception of the people

A majority of US and the European public as well as in India have equally very limited knowledge about nanofertilizer. These researchers concluded that although public have limited knowledge, a majority believed that benefits of nanofertilizer technology outweigh the risk. It should be quite obvious that if public knowledge of nanofertilizer technology in, such highly advanced societies, such as US and Europe, where literacy level is relatively high is limited; then the situation is worse in other parts of the world that are also grapping with high literacy levels. As expected, the level of knowledge of nanofertilizer technology is much higher among the highly educated than with the least education. When people know little about the technology, their perception and acceptability will be a large extent depends on how social and ethical issues concerning the technology are handled by the industry and the researchers. Because, when knowledge is missing, people can use heuristics, such as trust, to assess the risk and benefits of the new technology. Thus, people are more likely to accept assurances about the safety of nanofertilizer technology in agriculture projected by the research workers and the industry people. Another aspect that affects public perception about the nanofertilizer technology is the way the media reports the issues on the technology. In less developed countries the level of coverage of nanofertilizer is very low and this is coupled with low levels of research in the technology. For developed countries with high levels of application of nanofertilizer technology, the reporting or coverage of the technology may be modest probably due to the specialized nature of the field and hence requiring specialist journalists who may be fewer. Moreover there are quite a number of ethical issues regarding the human health and environment may arise. They are:

- Should the industry continue using these nanofertilizers despite the uncertainty?
- Should there be a mandatory requirement for labeling of nanoenabled agricultural products, particularly food stuffs?
- Is it ethical that public/government institutions should continue funding for the development of the nanofertilizer products for the use in agriculture despite the current levels of uncertainty?
- How much information should the public be made aware in relation to the technology in the entire agricultural value chain?
- Is it ethical for an industry to release nanofertilizer for use in agriculture to the public who have no idea about the potential negative impact on their health?
- Should there be regulations set in regulating nanofertilizer technology in agriculture to increase public perception and acceptability?

If these issues/questions are not fully addressed in the present development stage of nanofertilizer, it may create negative perception and hence reluctance of acceptance of this technology.

One of the main reasons for chemical fertilizer pollution is due to yearly progressive increase in their application. In some cases, only a small fraction of applied fertilizer gets utilized by plants. Therefore, by introduction of nanofertilizer the entire agricultural value chain may be benefited due to more solubility, enhances bioavailability, promotes targeted delivery and controlled release over a long period of time. This technology have the potential to efficient enough in very small dosage, improve the pest management, reduce chemical pollution and ultimately decreases the contact with agricultural operators. But one has to know the exact nanoparticles needed for a particular agro-ecosystem along with the required doses of application precisely. It is also essential to know the answers of all the questions raised to ensure that humans and the environment are safeguarded from nanofertilizers potential harm as a result of their application in agriculture.

Further studies

Aufan, M., Rose, J., Bottero, J., Lowry, G.V., Jolivet, J., Mark, R. and Wiesner, M. R. (2009). Towards a definition of inorganic nanoparticles from an environmental, health and safety perspective. *Nature Nanotech.* **4:** 634–664.

Siegrist, M,. Stampfli, N., Kastenholz, H. and Keller, C. (2008). Perceived risks and benefits of different nanotechnology tools and nanotechnology food packaging. *Appetite.* **51:** 283–290.

Grover, M., Singh, S. R. and Venkateswarlu, B. (2012). Nanotechnology: Scope and limitations in agriculture. *Int. J. Nanotech. Appl.* **2:** 10–38.

Agrawal, S. and Rathore, P. (2014). Nanotechnology pros and cons to agriculture: A review. *Int. J. Curr. Microbiol. Appl. Sci.* **3(3):** 43–55.

Handford, C. E., Dean, M., Spence, M., Henchion, M., Elliott, C. T. and Campbell, K.(2015). Awareness and attitudes towards the emerging use of nanotechnology in the agri-food sector. *Food Contro*l. **57:** 24–34.